Wild Medicinal Plants

Wild Medicinal Plants

Dr. M.P. Singh, *Ph.D., FMA*
University Professor and Chairman
Department of Forest Sciences
Birsa Agricultural University,
Ranchi – 834 006
(Jharkhand)

2011
DAYA PUBLISHING HOUSE®
Delhi - 110 002

ISBN 9789351242321

Published by : **Daya Publishing House®**
A Division of
Astral International Pvt. Ltd.
– ISO 9001:2008 Certified Company –
4760-61/23, Ansari Road, Darya Ganj,
New Delhi - 110 002
Phone: 23245578, 23244987
Fax: (011) 23260116
e-mail : dayabooks@vsnl.com
website : www.dayabooks.com

Laser Typesetting : **Classic Computer Services**
Delhi - 110 035

Printed at : **Chawla Offset Printers**
Delhi - 110 052

PRINTED IN INDIA

Preface

As old as civilization and origin of earth is the history of plants and medicines. In the form of hymns (mamtron) in the Rigveda, Many references of plants and thus formulations are found. In the world which is considered as one of the oldest epics of Hindu. Till today the knowledge of plants and medicines has undergone various stages. Although, various systems of medicine are prevalent in India yet Indian system of medicine *i.e.*, Ayurveda stands for science of life, has attained the world wide importance. Till to day it is the best system of medicine and developed countries with the passage of times are focusing their attention in selecting few of the herbs and carrying research and development work. These countries are using the extract of these medicinal plants in allopathic medicine in the form of syrup and capsules. Stash life long security against with no side effects. But, many of these valuable and useful plants are going extinct because of over exploitation and extensive use of fresh plants. An endeavour is to be made to make people aware of potential of the medicinal plants from all angles so that these life saving plants might be protected from their total destruction. Besides, there is an urgent need to identify the natural plant wealth available for civilization and production of these plants and it should be encouraged by government, particularly in forest area and orchards. To grow these medicinal plants after making them fertile, the waste land should be used.

To provide the information about the common wild medicinal plants of India with ethnobotanical information on their uses, in this book, an attempt has been made. According to Bentham and Hooker system of classification, the plants are arranged with some modification suiting to present environment. With description general distribution, parts used and medicinal uses with folk medicinal uses and main Ayurvedic and Unani preparation, each plant species is described. With appendices, the work has been provided on list of medicines of common bazaar available under common names and ancient systems of plant classifications besides a glossary on medical terms. After Bennet (1987), the nomenclature of the plant has been updated and their botanical family names are provided according to International Code of Botanical Nomenclature as assigned in the war of Airy Shaw (1973).

A special mention, however, I would like to make that many of the medicinal plants in raw forms are poisonous and self medication with old plants is not at all available on various plant drugs, the information furnished here in ethnomedicinal uses are yet to be proved clinically in many of the cases and these should be administered under the strict prescription of a qualified doctor. Hope that this monograph will be useful for research workers, plant taxonomists, besides medical students and to common man.

The works of various authors have been consulted, for which author is grateful to them. As far as personal assistance is concerned, the author is thankful to Dr. N. Sanjppa, Director, Botanical Survey of India, Kolkata, Professor V.L. Chopra, Member Planning Commission, Govt. of India, New Delhi for their valuable suggestions. The author is also thankful to Sri Ramjee Prasad, Chief Librarian, Central Library, Birsa Agricultural University, Ranchi for their help in various ways.

The author is indebted to his brother in laws Sri Deep Narayan Sinha and late sister Mrs. Urmila Sinha for their blessings and moral encouragement. The author is also thankful to his son Mr. Saurav and daughter Miss Ankita, to whom he could not pay much attention, which he deserved while writing the book. In the last but not least, my sincere thanks are due to my wife Mrs. Meena Singh, who has always been a source of inspiration to me to complete this book.

M.P. Singh

Contents

Chapter 1
General Introduction

For its sustenance and survival since his existence on the earth the man has been dependent on nature, particularly on the plants. In ancient time, he knew how to relieve his sufferings by using the plants growing around him. The civilizations records show that a number of drugs used today were already in use during ancient times. Its credit goes to Indian *Rishis* and Physicians, who were acquainted with a large number of medicinal plants compared to the other countries in the world. As compared to Indian material medica. The material medica of the Greeks, Romans, Egyptians, Babylonians, Persians, Chinese and Arabians did not possess extensive uses and knowledge of medicinal plans and drugs.

To the remote past in Vedic Period. The history of Indian medicine can be traced back *Rig Veda*, the oldest repository of human knowledge written about 4500-1600 B.c., described about 99 medicinal plants, *Yajur Veda* listed 82 plants and *Sam Veda*, described various medicinal plants already mentioned in *Rig Veda*, particularly, the Soma plant. *Atharva Veda*, important amongst all four Vedas, dealts with 288 plants almost all with medicinal ingredients which were used to cure deadly diseases. Later derivations from *Vedas*, the *Brahmans* dealt with 129 plants and *Kalpa Sutras* described some about 519 plants. Those plants which were not dealth with in the *Rig Veda*, *Atharva Veda* and the *Brahmans* in later Sanskrit texts like Kalpoa sutras were described.

Account on many drugs and their uses *Ayurveda*, the science of life which is considered as an *Upveda* (about 2500 B.C.) give a more detailed. In fact *Ayurveda*, is the foundation of the ancient medical science of life and art of healing. The eight divisions of *Ayurveda* were followed by the comprehensive works of *Charak* (800 B.C.) and of *Sushruta* (800 B.C.), which gave a detailed description of the material medica as it was known to ancient Vaidyas. *Charak Samhita*, the first recorded treatise on ayurveda was an edited version of an old scientific treatise by *Agnivesha*, who wrote on ayurvedic medicine based on teaching of *Ayurveda* from his preceptor, *Atreya*, the great sage. This work primarily dealts with medicine and its seventh chapter is devoted entirely to the consideration of purgatives and emetics. Vidyalankar (1991), had mentioned that there were six pupil of *Acharya Punarvasu*. They were *Agnivesha, Jatukarana. Parashara, Kshreepani, Bhel* and *Harit* around 800 B.C. Each of them described his own thoughts in the form of Samhita. Bhel presented his own thoughts in *'Bhel Samhita'* which now-a-days, is scattered form. The other pupil of *Punarvasu, Harit* also wrote *'Harit Samhita'* that is available today. But Sharma (1981) is of opinion that the present *'Harit samhita'* is not the original one because the pattern of its description does not ressemble with other samhitas of that time. Although, the other learned researchers considered this work as much older as the other Samhitas. The other *Samhitas; Agnivesha Samhita'* of *Agnivesha, 'Jatukaran Samhita'* of *Jatukaranm, 'Parashar Samhita'* of *Parashar and 'Ksheerpani Samhita'* of *Ksheerpani* are not available today. *'Kashyap Samhita'* or *'Vrahad Jivakiya Tantra'* was also written nearly of Charaka's time but it is also available only in scattered form (Vidyalankar, 1991). *Acharya Charka* has divided the drugs into fifty Vargas (groups) which are mentioned as follows:

I. Jeevaneeyam, II. Brimhaneeyam, III. Lekhaniyam, IV. Bhedaneeyam, V. Sandhaaneeyam, VI. Deepaneeyam, VII. Balyam, VIII. Varnyam, IX. Kanthyam, X. Hridyam, XI. Triptibhanam, XII. Arsoghnam, XIII. Kustaghnam, XIV. Kandughnam, XV. Krimighnam, XVI. Vishaghnam, XVII. Sthanyajananam, XVIII. Sthanyasodhanam, XIX. Sukrajananam, XX. Sukrasodhanam, XXI. Smehopagam, XXII. Sedopagam, XXIII. Vamanopagam, XXIV. Virechanopagam, XXV. Aasthaapanopagam, XXVI. Anuvaasanopagam, XXVII. Sirovirechaniyam, XXVIII. Chardinigrahanam, XXIX. Trishnaanigrahaanam, XXX. Hiccaanigrahanam, XXXI.Purisha Sangrahaneeyam, XXXII. Purisha Virajaneeyam, XXXIII. Mootra Sangrahaneeyam, XXXIV. Mootra Virjaneeyam, XXXV. Mootra Virechneeyam, XXXVI. Kaasahara, XXXVII. Swaasahara, XXXVIII. Swayathuhara, XXXIX. Jwaraharam, XXXX. Sramaharam, XXXXI. Daahaprasamanam, XXXXII. Seethaprasamanam, XXXXIII. Udardaprasamanam, XXXXIV. Angamardaprasamanam, XXXXV. Sulaprasamanam, XXXXVI. Sonitaasthaapanam, XXXXVII. Vedanaasthaapana, XXXXVIII. Samgnaasthaapana, XXXXIX. Prajaasthaapana, XL. Vayasthaapana.

With the advanced state of knowledge while *Sushurta Samhita*, described like surgery but there is a detailed chapter on therapeutics. It contains information on about 385 *vanaspati* or plants (Vidyalankar, 1991). *Sushruta* the drugs divided into thirty eight Ganas (groups) by sushruta which are mentioned as follows:

I. Vidaarigandhaadigana, II. Aaragwadhaadigana, III. Varunaadigana, IV. Veeratarwaadigana, V. Saalasaaraadigana, VI. Rodhraadigana, VII. Arkaadigana, VIII. Surasaadigana, IX. Mushkakaadigana, X. Pippalyaadigana, XI. Elaadigana, XII. Vachaadigana, XIII. Haridaraadigana, XIV. Syaamaadigana, XV. Brihatyaadigana, XVI. Patolaadiganam, XVII. Kaakolaadigana, XVIII. Ooshakaadigana, XIX. Saaribaadigana, XX. Anjanaadigana, XXI. Parooshakaadigana, XXII. Priyangwaadigana, XXIII. Ambashtaadigana, XXIV. Nyagrodhaadigana, XXV. Gudoochyaadigana, XXVI. Utpalaadigana, XVII.

Mustraadigana, XXVIII. Triphala, XXIX. Trikatukam, XXX. Aamalakyaadigana, XXXI. Trapwaadigana, XXXII. Laakshaadigana, XXXIII. Kaneeyahpanchamoola, XXXIV. Mahaapachmoola, XXXV. Dasamoola, XXXVI. Valleepanchamoola, XXXVII. Kantakapanchamoola, XXXVIII. Trinapanchamool.

In 7th century *Vagbhatta* wrote an important treatise *Astanga Hridaya* and *Astanga Samgrah*. HENCE, there is a controversy areses, whether the author of *Astang Hridaya* and *Astanga Samgrah* were the same or different because *Vagbhatta* himself has stated in his book his grandfather's name as *Vagbhatta*. But Saraswati (1995), is of opinion that these seem to be the same as both the books follow the same pattern. In brief the only difference is that the earlier one *i.e., Astanga Hridaya* is whereas the later is the detailed versioin. *Vagbhatta* has divided drugs into thirty two groups (Ganas):

I. Vamanaoushadhagna, II. Virechanaoushdhagana, III. Niroohana Dravyagana, IV. Seershavirechaneeyagan V. Vaataharaganas, VI. Pittaharaganas, VII. Kapharaganas, VIII. Jeevaneeyaadigana, IX. Vidaayaadigana, X. Saaribadigana, XI. Padmaakadigana XII. Parooshakadigana, XIII. Anjanaadigana, XIV. Patolaaadiagana, XV. Guduochyaadigana, XVI. Aaragwadhaadigana, XVII. Asanaadigana, XVIII. Varanaadigana, XIX. Ooshakaadigana, XX. Veerataraadigana, XXI. Rodhraadigana, XXII. Arkaadiagana, XXIII. Surasaadigana, XXIV. Mushkaakaadigana, XXV. Vatsakaadigana, XXVI. Vachaadigana, XXVII. Privangawaadigana, XXVIII. Ambashtaadigana, XXIX. Mustaadigana, XXX. Nyagrodhaadigana, XXXI. Elaadigana, XXXII. Syaamaadigana.

The above mentioned three works are known as '*Vriddha trai*', of *Ayurveda* and the period of their creativity *i.e.*, the period from 800-1000 B.C. is considered as the golden ear in the Indian system of medicine. Another, triad known as '*Laghu trai*' of *Ayurveda* are the compilations by *Madhavakar* (12th century), who wrote '*Madhavanidana*', concerned mainly with diagnosis; *Sharngadhara* (14th century), who wrote '*Sharngadhara Samhita*', a systematic ayurvedic material medica; and *Acharya Bhava* Mishra (15th century), a native of Benaras who wrote '*Bhavprakasha*' which contains more than 600 drugs including some foreign drugs too. Besides, a large number of *Nighantus* or pharmacy lexicons on medicinal herbs were also written. The important ones include '*Baangsen Samhita*' of *Acharya Bangsen* which contains information on various ayurvedic formulations (Sharma and Sharma, 1981) and '*Shivkosh*' of *Shivdutt Mishra* which was written in 1677 and had described about 4860 synonyms out of which 2860 are *Ayurveda* related and includes information on medicinal plants. The other work of *Sodhal* from Gujarat is '*Godh Nighrah*' which contains information on more than five hundred eighty five ayurvedic formuations.

By *Deodas Kashiraj*, the *Raja* of Benaras, the oldest '*Nighantu*' seems to be written who is also believed to be the incarnation of *Dhabvabtgru*. He is said to have taught his '*Dhanvantri Samhita*' to his pupils among whom *Sushruta* was most renowned. He is also said to have written another book, known as '*Raj Nighantu*' on drugs, but some writers believe it to have been written by another *Vaidya, Dhanvantri* by name, who lived during the reign of *Vikramaditya*. About 400 herbs in this book are described and many later authors had drawn upon his source. Later, various *nighantus* were compiled, of which '*Raj Nighantu*' or '*Churamani*' written by *Narhari Pandit*, a native of Singhpur in Kashmir, *'Madanpala Nighantu'* of *Madanpala;* and In 1896 *'Saligram Nighantu'* of *Lala Salaigram,* a native of Moradabad written are important.

To mention here it is interesting that during ayurvedic period, the chemistry of natural products isolated both from flora and fauna was well understood at least for practical purposes. Nagarjuna who is considered as a learned person is Hindoo chemistry, was the inventor of *Kajli* (a compound of sulphur and mercury) and art of calcinatioins (*Bhasma*). He was not only a renowned *Vaidya* but was an authority on astronomy, chemistry and magic as well. He was born in a poor Brahmin family and wrote *'Rasa Ratnakara, Arogyomanjari* and *Kakshaputa. Bhoja Prabandha,* a treatise written about 980 A.D., contains a reference to inhalation of medicaments before surgical operations and an anesthetic called, *'Sammohini'*, is said to have been used in the time of Buddha. The other *Vagbhatta,* in 13th century has written *'Rasratan Samuchhaya'*, which deals with the chemistry of many ayurvedic formulations and information on medicinal plants.

Hindu medicine flourished well, from this period down to Mohammedan invasion on India,. But after the period of the *Tantras* and *Siddhas*, the glories of the Hindu medicine rapidly declined because during invasion on India by Mohammedan, no original work could be done and Hindu medicine gradually began to decay. The Buddhistic doctrine of *'Ahimsa'* also influenced the work as no work could be done in the field of surgery as well. Consequently the thinking of "study and practice of the healiang art led the pollution" and "to touch the body is sinful" etc. influenced the work. However, Greeks, Scythians and Mohammedans invaded India successively and enriched their material medica by coming in touches of Indian antiquity.

Degeneration set is all round with the decline of Buddhism *i.e.* in knowledge, learning and practice of both medicine and surgery, decline was in full swing at the time of the *Mohammedan* invasion. This decline became more fast because invaders brought their own healing system which was well advanced in its timing, thus, the ayurvedic system of medicine perished. Then, Arabic System or Unani system of medicine came into vogue which was better system of relief. The history of Unani system of medicine in India can be studied under two periods: Pre-Mughal period and Mughal period and British period. During 1296-1316, in *Alauddin Khilji's* period, Hakim Sabruddin Damishqui and Hamid Motraz occupied a high rank. The system also existed during the reign of *Sultan Altamish. Hakim Zia Muhammad* was the Court Physician of Muhammad Bin Tuglaq. The famous books on Unani System of medicine like *Tibb-e-shahabi, Kafaya Mujahidiya* and *Madan-ush-shifa Sikander* were published during the pre-mughal period. In mughal period. *Hakim Ali Geelani, Hakim Ananullah Khan, Mirza Muhammad* were the court physicians of emperors *Akbar, Jahangir* and *Muhammad Sher* respectively. The famous books like *Ummul Ilag, Mizanut Tibb, Qurabadeen-e-quadri, Shah-e-geelani* were published in this period. Unani or Arabian medicines were prevalent during the kingdom of *Pathans* and *Moughuls* dynasties but with the fall of moughuls it too declined. Although the Britishes looked after this system of medicine, the Nawabs of some states patronized this system. More famous were *Nizam* of *Rampur, Nawab* of *Tank* and *Maharaja* of *Patiala*. Besides, *Hakims* of *Delhi, Lucknow* and *Hyderabad* also took major signs to save Unani system of medicine and their decline both utilized the material medica of each other. But during the amalgamation of old Hindu system of medicine and Arabian system of medicine, which lasted for many year. .

On Indian material medica the effect of their material medica can be felt clearly. First time Opium was included in *nighantus* as its references are not met earlier. *Hakims* took advantage of the plants introduced from Arab and Afghanistan and prepared mixed *nighantus* comprising of Indian system of medicine and Arab system of medicine (Bhandari, 1993). Due to this both systems declined but a rich store of combined material medica was left behind. Some important contributions include *'Taleem a Sharif, Abyunsur-Maffakkas* and *Dakhir–A- Khajarmushahi*. Later, Europeans, first the Portuguese then French and lastly English invaded the India and downfall was still further marked. Dr. Garcia d aorta, who came in India in 1534, perhaps was the earliest European Physician to describe some Indian drugs. He was appointed medical adviser to the Poretugese Viceroy during 1554-55 and traveled all over India and studied the qualities of plants growing there. He published a treatise on drugs used in Indian Medicine in Dutch language in 1563 in Goa, which later on was translated into English. The various drugs mentioned theirin where those in use during 16th century. By Chatterjee and Pakrashi (1991) described the drugs which are given as follows:

As anthelmintic *Acacia catechu* used, in eye diseases, stomach/intestine, teeth/gum, throat affections; Aloe used as anthelmintic, in liver/kidney/bladder infections, purgative, wounds/cuts; *Banana* used in fevers; *Bettle* used in stomach/intestine, teeth/gum infections; *Bhang* used as Aphrodisiac, sedative; Bittergourd used in fevers, as diuretic, in liver/kidney/bladder affections, and as sedative; *Box elder* used in childbearing, painsm wounds/cuts; *Camphor* used in burns, as sedative, analeptic stimulant; *Cardamom* used in teeth/gum infections; *Cashew nuts* used as anthelmintic, anti-asthmatic; *China root* used in calculus of kidney/bladder, as aphrodisiac, in venereal diseases, wounds/cuts; *Cinamon* used in nervous problems, stomach/intestine affections; *Clove* used in fevers, pains, teeth/gum affections; *Coconut* used in nervous disorders, arthritis, as purgative; *Convolvulus turpethum* used as purgative; *Curcuma longa* used in eye-diseases, itches; *Curcuma zeodoaria* used in pains, snake-bite/poison; *Ferula assafoetida* used as aphrodisiac; *Manna* used as purgative; *Mango* used as anthelmintic; *Neem* used as anthelmintic, in nervous problems, wounds/cuts, as antiseptic, in skin diseases and as insecticide; *Nutmeg* used in child bearing, nervous problems; *Opium* used as aphrodisiac, in stomach/intestine infections; *Piper cubeba* used as aphrodisiac, in stomach/intestine affections; *Purging cassia* used as mild purgative; Sandal wood used in fevers, liver/kidney/bladder affection; *Styrax benzoni* used as aphrodisiac, in stomach/intestine affections; *Sweet flag* used in childbyearing, nervous problems and in wounds/cuts *Tamarind* used as purgative.

In India, Western system of medicine was introduced with the British rule, which was appreciated and accepted by people due to its advanced technology, especially due to its surgical achievements. Britishes brought their own material medica and there was further amalgamationand use of new medicinal plants, but some works relating to ayurvedic and its related concepts also came in British period. The important ones among include 3 volume treatise entitled *'Hikmat Prakash'* of *Mahadev* written in 1773. This work is in Sanskrit language and had many Unani formulations and it was translated from *Hikmat* in Farsi language. This work consists of three parts; the second part of which dealth with pharmacology. In 1867 *Vishnu Vasudev Godbole* published another important treatise under the title '*Nighantu Ratnakar*' in two parts (Sharma, 1981). In 1896 Shaligram of Moradabad published '*Shaligram Nighantu*' and it is supposed to be the last *nighantu* as after this no *nighantu* published ever. However, the very fact that the Indian System of Ayurvedic Medicine survived all trials and tribulations through centuries bears enough testimony to the efficacy of this inditgenous system of medicine.

The earliest British contributor Jones was (1799), perhaps was who wrote a memoir '*Botanical Observations on Selected Plants*'. This was followed by Flemmings (1810), '*Catalogue of Indian Medicinal Plants and Drugs*'; Ainsile (1813), '*Materia Medica of Hindoostan*'; Roxburgh (1834), '*Flora Indica*'; O' Shaughnessy (1841), '*The Bengal Dispensatory*'; and Drury (1858), '*Useful Plants of India*', which later revised in 1873 and '*Hand Book of the Indian Flora*' (1864-69). Hooker and Thomson (1855) were the first who could describe botanically the plants of India in their work '*Flora Indica*'. Later, Hooker (1875-97) explored the plant wealth of India and published a monumental work '*Flora of British India*' in seven volumes, which is a mile stone in the floristic history of India. Waring (1868), was first to publish '*Pharmacopoeia of India*' supplemented by Mohidden Sheriff in 1869 and later by Dymock, Warden and Hooper (1890-93). It is most careful and useful compilation containing a mass of information regarding the uses of the indigenous material medica in the Eastern and Western medicine. Waring (1897), compiled another work '*Remakrs on the uses of some of the Bazar Medicines and Common Medical Plants of India*' which gives information on the various medicinal uses of plants of Indian subcontinent. The other document '*The Dictionary of Economic Products of India*' (1889–93) was originally projected by J.N. Mukherji, but subsequently completed by Dr. George Watt, the Reporter to Economic Products to the Government of India during 1889-1904. This work is considered most important as it not only incorporated the earlier work on the indigenous plants but also utilization of the new results on clinical trials.

'*Indigenous Drugs of India*' of Dey (1896), *Materia Medica of India and their Therapeutics*' of Khory and Katrak (1903) and '*Indian Medicinal Plants*' of Kirtikar and Basu (1913), were mostly summaries and compilations from the above mentioned works. With the illustrations the latter work is supplemented which greatly help the workers in differentiating them from plants which they were not aware of. In 1935 its revised and almost completely rewritten edition by Blatter *et al.*, was published. Dutt (1922), published '*The Materia Medica of Hindoos*' in which various informations regarding medicinal plants were given. Sanyal (1924) wrote on the medicinal uses of the plants in his work '*Vegetable Drugs of India*'. '*Bharat Bhaisajaya Ratnakar*' compiled by Nagin Das Chagan Lal Shah during 1924-37 is another important compilation of various ayurvedic formulations reffered in different ayurvedic books. Another work '*Vanaspati*' by Majumdar (1927) described the uses of plants and their uses in Indian sub-continent. Chopra (1933) started a project at Calcutta School of Tropical Medicine, Calcutta, which resulted in the publication of an useful and informative volume on 'Indigenous Drugs of India' which was later revised in 1958 by the same author with his colleagues. Meantime, Roberts (1931), also wrote on different medical aspects of plants and their products in his work *'Vegetable Materia Medica of India and Ceylon'* and Bose (1932) in *'Pharmacopoea Indica'* about the latest discoveries of that time on medicinal plants. Swamy (1936), published *'Sandigdh Nirnay Vanoushdhi Shastra'* in seven volumes in which he had described various controversial plants with their medicinal uses. Besides, Nadkarni in 1908 wrote a treatise, on *'Indian Plants and Drugs'* which later in 1927 was modified and published as *'Indian Materia Medica'* and has been revised in 1954, provides information on 2671 medicinal plants. Vaish (1940), published, *'Abhinav Booti Darpan'* which is an illustrated work and is considered a very important treatise in the field of Ayurveda. Later, Bhandari (1936-48), published *'Vanoushdhi Chandrodaya: An Encyclopedia of Indian Botanies and Herbs'* which in addition given information on important preparations of Indian drugs made out from the plants besides their medicinal uses. This work has been revised and its new edition which was published in 1993, had updated information. Singh (1948), prepared a guide *'Vanoushdhi Darshika'* for the students of botany and forestry and later in (1955), he surveyed the forest of Bihar and published his work under the title *'Bihar Ki Vanaspatiyan'*. Singh (1969), published *'Vanoushdhi Nirdeshika* (Ayurvediya Pharmacopoeia) later in 1983 which was revised.

A Recent invention added Sharma's five volume recent works (1976-1981) *'Dravaya Guna Vigyan'*; Jain and De Filipps (1991)'s *'Medicinal Plants of India'*; a voluminous work of Warrier (1993-96) on *'Indian Medicinal Plants'*; Sharma's (1996), *'Classical uses of Medicinal Plants'*; Agarwal (1997)'s *'Drugs Plants of India'* and of Kaushik and Dhiman (2000)'s *'Medicinal Plants and Raw Drugs of India'*. Recently, Dhiman (2003) has discussed various plants of sacred nature having in addition, the medicinal potential too in details in his work *'Sacred Plants and their Medicinal Uses.'* His other work published in 2004, under the name *'Common Drug Plants and Ayurvedic Remedies'*, also describes common medicinal plants of Indian sub-continent along with their medicinal uses. Another work of Dhiman (2004b). *' Medicinal Plants of Uttaranchal State'* discusses important plants of higher altitutes.

After independence Government activities are concerned, Council of Scientific and Industrial Research (CSIR), New Delhi, during 1948-1976 published *'Wealth of India'* which is considered an updated edition of the Dr. Watt's work of 'The Dictionary of Economic Products of India'. It is supplemented by illustrations and plates that make it a superior over previous work of Dr. Watt. Its revision has been launched to update the Wealth of India and its three volumes have already been published. Chopra *et al.* (1956), work on *'Glossary of Indian Medicinal Plants'*; Chopra *et al.* (1969)'s *'Supplement to Glossary of Indian Medicinal Plants'*; Asolkar *et al.* (1992), *'Second Supplement to Glossary of Indian Medicinal Plants with Active Principles'* and Chatterjee and Pakrashi (1991-2001)'s *'The Treatise on Indian Medicinal Plants'* have been brought out by the CSIR, New Delhi.

Ayurveda committee has been constituted by Government of India in 1962 with a view to maintain the uniform standards in preparation of drugs and to prescribe working standards for compound formulations including tests for identifying purity and quality of the drugs. In this field Central Institute of Medicinal and Aromatic Plants (CIMAP) at Lucknow is also serving. Pharmacological Laboratory for Indian Medicines, established at Ghaziabad in serving as a centre for standard setting cum-drug testing laboratory for Indian medicine including Ayurveda, Siddha and Unani System of medicine. Central Council of Indian medicine was established for working in Ayurvedic, Siddha and Unani system of medicines. National Institute of Ayurveda was established in 1976 at Jaipur (Rajasthan) in collaboration with Government of Rajashtan is working as a national centre for promoting the Ayurveda. Besides, Central Council for Research in Ayurveda and Siddha was constituted in 1978 to initiate, aid, guide, develop and coordinate scientific research in different aspects of fundamentals of Ayurvedic and Siddha system of medicine. Institute of History of Medicine and Medical Research, Delhi and Central Council of Research in Unani Medicine (CCRUM) established in 1979 are working for the coordination and scientific research in Unani medicine which in fact indirectly incorporates Ayurveda. The National Botanical Research Institute, Lucknow, has also published Kapoor and Mitra work on '*Herbal Drugs in Indian Pharmaceutical Industry*' in 1979. Besides, the Indian Council of Medical Research, New Delhi, is also working towards the development and research in the field of medicines.

Chapter 2

Fundamentals Medicine of Ayurvedic and Unani Systems

Medicine of Ayurvedic System

Britannica Encyclopaedia has defined ayurvedic medicine as "an example of well–organized system of traditional health care, both preventive and curative. It is still a form of health care in large parts of the Eastern world especially in India, where a large percentage of population uses this system exclusively or combined with modern medicines". It is the science which imparts knowledge of Ayus (life), provides longevity, contains relevant information and discussed all allied topics (Sharma, 1998). For alleviation of diseases the objectives of this system of medicine are that the persons should remain full of health endowed with strength and vigour and well protected from disorders and the sick should be treated.

Ayurveda Parts

In eight parts, the medical knowledge of Ayurveda is divided which are known as *Astanga Hridya* of the ayurveda. These include:

1. Kaya chikitsa (medicine)
2. Kaumarabhrtya (pediatrics including obsterics and gynaecology)
3. Agadantantra (toxicology)
4. Salyatantra (surgery)
5. Salaykayatantra (medicine and surgery of supraclavicular diseases).
6. Bhutavidya (dealt with bhutas *i.e.,* invisible agents)
7. Vajikarana (aphrodisiac) and
8. Rasayana (dealt with promotion of health and life.)

With the advancement of knowledge gradually now it has been doubled and now has sixteen parts. The additional parts includes are mention as follow:

1. Sarira (anatomy-physiology),
2. Drvayaguna (pharmacology),
3. Bhesjakalapana (pharmacy),
4. Rasasastra (dealing with mercurial),
5. Nidana or Rogavijana (pathology),
6. Svasthavrtta (preventive and social medicine),
7. Manasa roaga (psychiatary) and
8. Parasutitantra and Striroga (obstetrics and gynaecology).

Almost bhutavidya practically has become obsolete, while with the advancement of knowledge

Based on Basic Principles

Upon certain principles of physical, chemical and biological sciences. Ayurveda is based in addition, it gives special consideration to spiritual aspects of life in the understanding and treatment of human diseases. There are four concepts in ayurveda which together guide the preventive, promotive and curative aspects of the practice of ayurveda. These concepts are, *Panchmahabutas, Tridosha, Dhatus and Malas*.

Tridoshas Principle

The entire biological process of the living organisms is controlled by three essential factors which include *Vata* (vayu), *Pitta* and *Slesman* (Kapha). These are collectively referred as *Tridosha*.

Form the verb-root *'va'* the word *'vata'* is derived meaning to move, to inform and to impel which are the natural actions of vata. It explains all the biological phenomena which are controlled by the functions of the central and autonomic nervous systems.

From the verb-root *'tapa'* the word *'pitta'* is derived meaning to heat. It is the manifestation of energy in the living organisms that helps digestion, assimilation, tissue building, heat production, blood-pigmentation and activities of the endocrinal glands etc.

From the verb-root 'slisa' the word 'slesman' is derived meaning to embrace. It implies the functions of heat regulation and also formation of various preservative fluids as for example mucus and synovia etc. The main function of the *slesman* is binding, to provide firmness, heaviness, virility and strength to the body.

The Dhatus

It refers to different vital organs or parts of the body. There are seven *Dhatus* in the body which include *Rasa* (body liquids), *Rakta* (blood), *Mamsa* (muscular tissues), *Meda* (adipose tissues), *Asthi* (bone tissues), *Majja* (nerve and bone tissues) and *Shukra* (generative tissues including sperm and ovum).

The Malas

The *malas* deals with the waste products of the body. The food consumed by the body brings it into existence and builds further the above seven *dhatus*. During this metabolic process, each organ produces a specific wast or *mala* as hair, nails, stool, sweat and urine etc.

The Drug action

Rasa, Vipaka, Veerya, Prabhava and *Guna*, are the basis of drug action in ayurveda system. Chatterjee and Pakrashi (1991), have described their different types as under.

The Rasa is based on the taste which induces different physiological activities when herbal drugs are given depending upon their constituents in the body. The different types of tastes include: *Madhura rasa* (sweet), *Katu rasa* (pungent), *Amla rasa* (sour), *Tikta rasa* (bitter), *Lavan rasa* (salty) and *Kasaya rasa* (astringent). The Madhura rasa is used as a pleasant, brain and heart tonic and as galacatagoue. *Katu rasa* is used as sialagogue, appetizer, and in dyspepsia. *Amla rasa* is used as sialagogue, appetizer and digestive. *Tikta rasa* is used as appetizer and to produce dryness in the mouth. *Lavan rasa* is appetizer, digestive, diuretic, expectorant and sialagogue. *While Kkasaya rasa* is used as diuretic, in soreness of throat and dryness of mouth and to heal the wounds.

The Vipaka is the metabolism of the products arising out of the biochemical changes of food and drugs during the gastro-intestinal digestion.

Potency of the drug used. Is Veerya further, it may be *sheetaveerya* and *ushnaveerya*. Sheetveerya drug stabilizes excretory function, stops haemorrhage and increases vigour and vitality and aggravates *vata* and *kapha* and subdues *pittta*. While ushnaveerya drug leads the storing up of energy, easy digestion, diaphoresis, emesis, fatigue and thirst and also osubdues *vata* and *kapha* but excites *pitta*.

The Prabhava is the specific and characteristic influence of a drug.

The Gunas are the physical properties of the drugs. These may be further subdivided as follows:

- ☆ The *"Guru"* (Heavyness) can induce the feeling of heaviness, dullness and fatigue and increase the quantities of waste products and also impairs digestive function.
- ☆ The *"Laghu"* (lightness) leads the lightness of the body and helps in digfestion.
- ☆ The *"Sheeta"* (cold) produces cooling effect, promotes *vata* and *kapha* but act against *pitta* and impedes blood flow and reduces syncope.
- ☆ The *"Ushna"* (hot) increase body temperature, induces thirst, promotes *pitta* but it acts against *vata* and *kapha* and is used to improve blood circulations, to increase amount of urine, excreta and sweat, and to stimulate appetite and digestive functions.
- ☆ The *"Snigdh"* (unctuous) produces soothing effect on the body and promotes kapha and eliminates waste products and acts against *vata*.

- The *"Ruksha"* (dryness) induces uncomfortableness in the body and acts against *kapha* but promotes *vata* and reduces vigour and virility and libido.
- The *"Manda"* (dullness) acts slowly and pacific and deranaged dosha.
- The *"Tikshna"* (sharpness) is sharp, highly potent and it removes morbidity, promotes *pitta* and also acts against *vata* and *kapha*.
- The *"Shtira"* (immobility) renders the physiological functions normal and regularly. Sara (mobility) stimulates excretory system.
- The *"Mridu"* (softness) stimulates *Kapha* and depresses *vata* and *pitta*. *Kathina* (hardness) causes stiffness and firmness of the body and *excites vata*.
- The *"Vishda"* (clarity) removes slimness and promotes *vata* and heals ulcers. *Picchila* (sliminess) makes the object slimy and increases body weight, promotes fracture and wound healing and stimulates excretory system.
- The *"Shalkshana"* (smoothness) promotes tissue synthesis and increases *kapha* and *pitta* and stimulates formation of excretory products. *Khara* (roughness) leads to emaciation and retards formation of waste products and aggravates *vata*.
- The *"Sukshma"* (minuteness) penetrates through all parts of the body and stimulates *vata*.
- The *"Sthoola"* (bulkiness) obstructs vessels, tubes and the channels of the body. *Sandra* (solidity) nourishes the body and;
- The *"Drava"* (fluidity) pervades the entire body.

The Unani System of Medicine

In Greece this system of medicine was originated in Greece during 460-377 BC and was further enriched and developed by the Arabians and Persians. In India as mentioned earlier was introduced in India about a thousand years ago and is still being in vague today.

On the humoural theory the Unani system of medicine is based which pre-supposes the presence of four humours namely include,

1. *Dam* (blood),
2. *Balgham* (phlegm),
3. *Safra* (yellow bile) and;
4. *Sanda* (black bile).

Accordingly the temperament of persons is expressed by the *sanguine, phelmatic, choleric* and *melancholic* according to the preponderance of them in the body. A unique humoural constitution every person has go which represents the healthy state of humoural balance of the body. Unani system plays a vital role when the individual experiences humoural imbalance. That the proper balanced diet and digestion can bring back the humoural balance, it believes.

For the prevention of diseases there are six essential pre-requisites as per Unani system which are collectively knows as *Asab-e-sitta zarooriya*. These are air, drinks, foods, bodily movement and repose, psychic movement and repose, sleep and wakefulness and excretion and the retention.

Chapter 3
Geography, Climate and Vegetation: India

Geography of India

Between the latitudes 8°4′ and 37°6′ North and longitudes 68°7′ and 97°25′ East, India lies on the western fringe it is situated of the Indian Ocean and lies midway between South-East Asia and the South-West Asia. On the north by Himalayas, it is bounded in the East by Myanmar and Bangladesh, in the South by Indian Ocean and in the West by Pakistan. Its three sides are guarded by Oceans *i.e.*, Bay of Bengal in the South-East, Indian Ocean in the South and Arabion Sea in the South-West. Almost through the centre of the country the Tropic of Cancer runs and divides it in two-halves- Tropical zone and Sub-tropical Zone. Indian Peninsula southern part divides the northern part of Indian Ocean into two-the Arabian Sea and the Bay of Bengal.

The Indian Union, the total area is 32, 87, 263 sq. km. It measures about 3,214 km from north to south and about 2,933 km from East to West. It has a land frontier of 15,200 km and a coastline about 7,516.6 km, including the coastline of Lakshadweep Islands and Andaman and Nicobar Islands. In the world areawise, it is the seventh largest country.

Physical Major Division

Physical Major divisions are:

1. Northern Mountains
2. Great Northern Plains or the Indo-Gangetic Plain
3. Plateau of Penisular India
4. Western and Eastern Ghats

5. Western and Eastern Coastal Plains
6. Thar Desert (Great Indian Desert)

1. The Northern Mountains

With intervening valleys the mountains of the north are young fold mountains comprising three main almost parallel ranges. These Northern Mountains include the Himalayas, the Trans-Himalayan Ranges and the Eastern Hills or Purvanchal.

a. *Kknown as the Great Himalaya the northern most range of the Himalayas is the Himadri Range or the Inner Himalayas* and has an average height of 6000 metres.

   An are-like form for 2,500 km from it extends the Indus Valley in the north-west to the Brahmaputra in the north-east. The Nanga Parbat Peak (8,126m) overlooking the River Indus and the Namcha Brawa Peak (7,756m) overlooking the River Brahmaputra are the traditional limits of the Himalays. The width of these mountains increases from 150 km in the east to 400 km in the west.

   Several peaks of the Himadri are there which exceed 8,000 m, all snowcapped and have glaciers originating from them. Mt. Everest 8,848 m, the world's highest peak, is located on the northern borders of Nepal. Kanchenjunga (8,598) in Sikkim is the highest Himalayan peak in India. Other peaks are Dhaulagiri (8,172m), Annapurna (8,078m) and Nanda Devi (7,817m).

   To the south of the Himadri Range *the Lesser Himalayas, Middle Himalayas or Himachal Range* lies and has peaks of over 5,000 m like Pir Panjal Range in Kashmir and the Dhaula Dhar Range which stretches from Jammu and Kashmir across Himachal Pradesh and into Uttaranchal. The Kashmir Valley, the Kangra Valley and Kulu Valley lying between the Himachal and Himadri Ranges, and Shimla located, are easily accessible in Himachal Pradesh.

   The southern most range forming the Himalayan foothills the *Outer Himalayas or Siwalik Range* represents which have an altitude of less than 1500m and width between 15 to 50 km. Most likely these are formed by the accumulation of materials eroded from higher areas. The Siwalik are known as the store house of timber. Between the Siwalik Range and Himachal Range to the north are located narrow Valleys called *Duns*. A typical example is The Dehra Dun Valley.

   Into three regions the Himalayas are usually divided from west to east. The Western Himalayas include the Himalayas in Jammu and Kashmir and Himachal Pradesh. The Himalayan region in Uttaranchal and the Eastern Himalayas include the mountainous regions in Sikkim, West Bengal, and Arunachal Pradesh, the Central Himalayas cover.

   For many months though snow-covered of the year, Zojila in Kashmir, Shipkila in Himachal Pradesh, Nathula in Sikkim and Bomdila in

Arunachal Pradesh are very important high altitude passes.

b. Beyond the Great Himalayas lies *The Trans-Himalayas*. The Karakoram Range is the most prominent range and extends from the Pamir Knot eastwards into Tibet where it is known as the Kailash Range (K2). Mt. Godwin Austin the world's second highest peak (8,611 m), is located in the Karakoram Range. There are many large snowfields and glaciers in the Karakoram Range. In the region the Siachen Glacier is the largest.

   Two parallel range of lower altitude south of the Karakoram are called the Ladakh and Zaskar Ranges. Between them the River Indus flows in a narrow valley.

c. Along the eastern borders of India *Purvanchal of Eastern Hills* lies in a north-south direction. Their average elevation is less than 3,000 metres. They are known by different local names such as Patkai, the Naga Hills, the Manipur Hills, Lushai Hills, the Mizo Hills, the Tripura Hills and Garo Hills. The Khasi-Jaintia Hills represents the steep edge of the Shillong Plateau which is structurally similar to Penisular Plateau.

   From the Pamir Knot the range radiating westwards is the Hindu Kush. The other mountain ranges the extend south-westwards almost down to the sea are the Sulaiman and Kirthar Ranges in Pakistan. In between these ranges the Khyber Pass, the Gomal Pass and the Bolan Pass lie.

2. The Indo-Gangetic Plain or Great Northern Plains

Between the mountains of the North and the Penisular Plateau of the South the Great plain lies. Extending over a length of 2,400 km and a width 240 km in Bihar to 500 km in Punjab it forms a great curve from the Arabian Sea to the Bay of Bengal. In the remote past the plain was a huge depression in which the rivers of Northern India have deposited layers of alluvial sediments. The plain is divided into two parts by the Aravalli Hills which form the water divide between the east flowing rivers and the west flowing rivers. Indus, Gang and Brahmaputra are the three major rivers.

a. *River System* of *the Indus:* The river system of Indus 2,700 km long, has its source in the Kailash Range which is about 100 km north of Mansarovar Lake.

   In a north westerly direction it first flows through Tibet, then turns south through the Ladakh range, cuts a deep gorge and enters Pakistan through Kashmir. It received the water of the '*Panchandi*', five rivers of Punjab, namely the Jhelum, Chenab, Ravi, Beas and Sutlej, which except for Sutlej rise in the Himalayas. The River Sutlej originates in the lake Mansarovar region in Tibet. It crosses the Himalayas through a deep, narrow gorge and enters the plains. Across the desert plain of Sind into the Arabian Sea, the Indus then flows.

b. *River System* of *the Ganga*: The river system of Ganga 2,480 km long rises in the Gangotri Glacier in the Himalayas at a height of 6000m cutting through the Siwalik Range and at Haridwar, it enters the great plain of India.

Ganga Tributaries: In Yamunotri the Yamuna rises and after flowing 1300 km, it joins the Ganga on its right bank at Allahabad (Prayag). The Chambal, Betwa, Ken join the Yamuna on its right bank. In the Deccan Plateau of Penisular India these rives rise.

The Gomti, Ghaghara, Gandak and Kosi join the Ganga on its left bank while the Son which drains the plateau region in the south joins the Ganga on its right bank. The Ganga plain slopes very gently from Harayana into the Bay of Bengal and is covered by thick alluvial sediments and traversed by meandering rivers. Floods are quite common due to shifting water channels. In the delta region the river divides into a number of distributaries, one of the largest being the River Hooghly. The seaward face of the delta has tidal estuaries, sandbanks and islands, knowsn as the Sundarbans. The River Damodar flows into the River Hooghly.

c. *River System of the Brahamputra*: The Brahmapautra which rises in the glacier south-east of the Mansarovar Lake, flows through Tibet as the Tsangpo. After flowing through deep gorges it enters the lowlands of Assam. The valley slopes from east to west and the lowland is frequently flooded during the monsoon. The southern part of Ganga-Brahmaputra delta is marshy and is known as the Mangrove Swamps or Sundarbans. It joins the lower course of the Ganga in Bangladesh and flows into the Bay of Bengal.

3. Penisular India Plateau

In shap it is triangular with the broad base in the north and apex towards Kanyakumari. It is an ancient landmass which has been denuded over millions of years. The plateau is mainly made up of old igneous and metamorphic rocks. It begins from the Aravalli Range which is a remnant of one of the oldest fold mountains in the world. The Aravallis extend for 800 km in a north-east to south-west direction from Delhi to Palanpur (Gujarat). Mt. Gurushikar (1722 m) is the highest peak.

This can be divided into three parts.

a. Between the Aravallis and Vaindhyas The *Malwa Plateau*. The Vindhyas rise steeply from the Narmada Valley. The Satpura Range is considered a horst lying between the Narmada and Tapi river Valleys which are narrow, rift valleys.

b. The *Chota Nagpur Plateau* in Jharkhand which is to the north-east of the Deccan Plateau is very rich in minerals.

c. The *Deccan Plateau* which lies south of the Narmada river has its northern boundary formed by the Vindhyas. It is triangular in shape, has an average height of 600 m and slopes from west to east. It is a dissected plateau cut by river valleys. The interior of the Deccan with Bombay the Thal Ghat and Bhor Ghat gaps connect.

The intense volcanic activities of past lead the eruption of lava along fissures in the Deccan Plateau. The north-western part of this plateau in Maharashtra, adjoing Madhya Pradesh and Gujarat, is known as the *Deccan Trap Region*. It is formed by thick, nearly horizontal lava sheets making a series of steps and is composed of old hard crystalline rocks. The Deccan Plateau has many ranges, such as the Vindhya, Satpura, Rajmahal, Mahadeo Maikal, Ajanta and Satmala in the north and the Nilgiris or the Blue Mountains and the Cardamom, Annamalai and Palni Hills in the South.

The *Anaimudi* in the Annamalai Hills is the highest peak in penisular India, being 2695 m high. In the *Nilgiris, Dadabetta* (2637 m) is the highest peak.

4. Ghats: Western and Eastern

Abruptly the Western Ghats rise from the Malabar and Konkan Coast and stretch from the mouth of the River Tapi to Kanakumari, 1600 km parallel to the coast. The northern half, a step sided edge, is known as *Sahayadri*. The average heights 1200 m but parts rise to 240 m. These are several passes and gaps in the Western Ghats, the most important being Thal Ghat and Bhor Ghat, east of Mumbai and Palaghat, south of the Nilgiri Hills. In the northern part of the Narmada and Tapi Rivers, the Ghats are broken.

The Eastern Ghats which form the eastern boundary of the Deccan Plateau stretch from the Mahanadi Valley uptothe Nilgiri Hills are lower than the Western Ghats, with an average height of 450 m, rarely exceeding 1200m. The Eastern Ghats are not acontinuous chain of mountains as they are dissected by the Rivers Mahanandi, Godavari, Krishna, Kaveri, Penner which descend gently towards the plains and flow into the Bay of Bengal. The Eastern Ghats disappear between the Godavari and Krishna Rivers.

5. Coastal Plains: Western and Eastern

The western plain, stretching form Kachchh in the north to Kanyakumari, are narrower than the eastern coastal plains. In the north nowhere are they more than 65 km wide. The costal strip is shared by the states of Gujarat, Jaharashtra, Goa (Konkan Coast), Karnataka (Kanara Coast) and Kerala (Malabar Coast). The coastal plains continue uninterruptedly in Kerala State with a maximum wide of about 100 km. The sandbars along the coast have formed lagoons parallel to the coast. These are called *backwaters* and are useful for irrigation.

The Rann of Kachchh was once a shallow arm of the seas which has now silted up and is a tidal flat covered with a few metres of sea water. The Kathiawar Peninsula (Saurashtra) lies between the Gulf of Kachchh and the Gulf of Khambhat. The Narmada, Sabarmati, Tapi and Mahi flow into the Gulf of Khambhat. The important ports are Kandla, Mumbai, Goa, Mangalore and Kochi (Cochin).

The eastern plain stretches from the mouth of the River Ganga to Kanyakumari between the Eastern Ghats and the Bay of Bengal. The Eastern Coastal strip is broader than the Western Coastal strip and is continous. Most of the plain is covered by the deltas of the Mahanadi, Godavari, Krishna and Kaveri rivers. The palin is very fertile but the deltas are unsuitable for harbours as their mouths are always silted. The northern half of the coast is called ' The Northern Circars' while the southern half, from Nellor to Kanyakumari is known as the 'Coromandel Coast'. The important harbours are Vishakahapatnam, Paradeep, Chennai and Tuticorin. The important lakes are Chilka Lake in Orissa, Kolleru in Andhra Pradesh and Pulicat Lake at north of Chennai.

A Peninsular India Major Rivers: The Western Ghats from the main watershed in the peninsula. The Mahanadi, Godavari, Krishna and Cauvery are the main east-flowing rivers which enter the Bay of Bengal. These rivers have many atributaries. Some of the east-flowing rivers like the Penner, Palar and Vaigai are non-perennial.

There are many short, swift flowing rivers which enter the Arabian Sea. Of these, the Sharavati in Karnataka and Periyar in Kerala are well known. The Narmada and Tapi rivers flow in narrow rift valleys with the Satpura forming a block mountain in between, and hence do not have long tributaries.

- ✰ On the Maikala Range from spring, *the Narmada* (1280 km) rises
- ✰ In the Satpura Range *the Tapi* (704 km) rises.
- ✰ In the Maikala Hills *the Mahanadi* (880 km) has it source.

The Islands: In the Bay of Bengal, Arabian Sea and the Gulf of Mannar there are a number of small islands. The Andaman and Nicobar Islands extending for a length of more then 500 km are the summits of the mountain ranges rising from the ocean floor. Some of these islands, like Barren Islands, are of volcanic origin. The Ten Degree Channel separates the Andaman group fo islands from the Nicobar group. Indira Point in the Nicobar Islands is the southern most point of the Indian Union.

In the Arabian Sea the Lakshadweep Islands are tiny islands of coral origin located about 350 km off the Kerala Coast, and built by small marine organisms called coral polyps which live in shallow tropical ocean waters. Some of these islands which are ring-shaped are called atolls.

Between India and Sri Lanka, the Rameshwaram Island lies In South of Rameshwaram there are other small coral islands lying parallel to the coast.

6. Thar Desert (Great-Indian Desert)

Between the Aravallis in the east and the Indus plains in the west, thar Desert lies mainly in Rajasthan where this arid region merges with the Sind Desert of Pakistan. It is a fairly flat low- lying region and lies in north western Rajasthan. It is sandy waster, interrupted by sandy ridges and shifting sand dunes. There are many basins of inland drainage containing salt lakes. The Sambhar Lake is the largest inland salt lake in India. The Luni is the only large river but carries little water. Here, the rainfall is less than 25 cm.

Climate of India

The physical features of India have a great influence on its climate. The heavy rainfall occurs on the windward side of the Western ghats, the hills of Assam, and the Himalayas. The rainfall is moderate in the plateau of the Peninsula and the Ganga plains. While the Southern Punjab and Western Rajasthan are the driest regions.

There are two transitional periods between the two principal monsoon; the hot weather before the beginning of the south-west monsoons and the retreating southeast

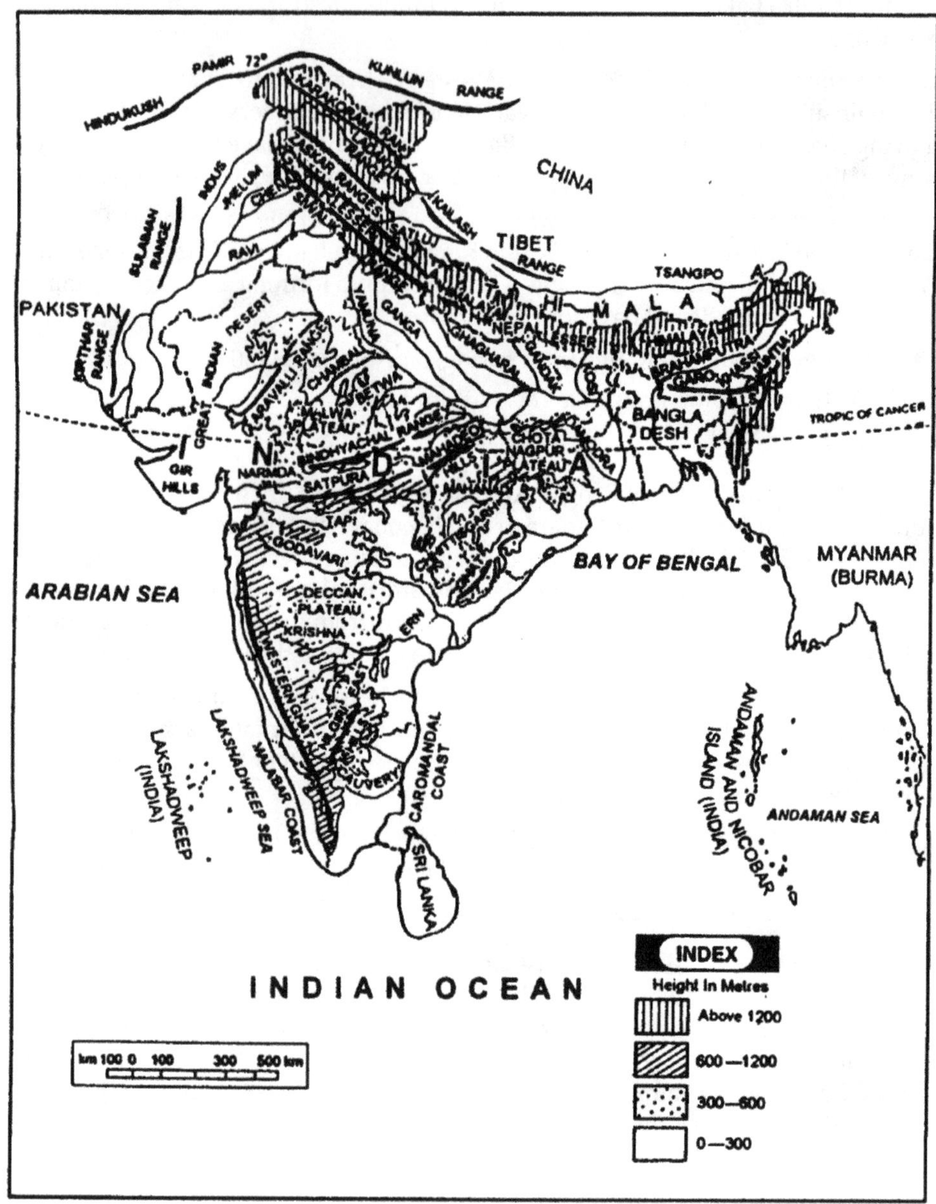

Map 1: Physical India.

monsoon seasons. There are four distinct principal seasons in India. Pichamuthu (1967), has described their characteristic features as follows .

1. Hot Weather

It starts in march and ends in May. There is a continuous and rapid rise of temperature and fall in air pressure in North India and a decrease in temperature in the Southern Indian ocean during these months. In March, there can be seen a highest temperature of 38°C in Deccan plateau and about 38 °C-43 °C in Gujarat and Madhya Pradesh. I May, the highest temperatures occur in North India, especially in the desert regions of the Northwest, where the maximum temperature is over 48 °C and the dust storms are of commonly occurrence. The lowest air pressure between Northwest India and Chota Nagpur causes Southern winds to blow across the West Bengal coast and North westerly winds across the Bombay coast, which often results in violent winds, torrential rain, and hail. In West Bengal, these, 'Northwester' often attain the intensity of tornadoes and cause destructions.

2. Southwest Monsoon Weather

The period of this weather starts from June ends with September. At the ends of May, a fairly deep low pressure of winds extends from West Rajasthan to West Bengal. The Southeast trade winds from the South of the equator blow northwards into the Bay of Bengal and the Arabian Sea. These are influenced by the air circulation over India and deflected inland as Southwesterly winds which give rise the cool and humid Southwest monsoon. The Southwest monsoon bursts on the Kerala coast at the beginning of June and monsoon gradually extends northwards and spreads over most of the parts of India by the end of June. The rainfall during June and July occur extensively. The agriculture of the country depends mainly on the amount and distribution of rain during these months. A part of the monsoon wind advancing Northwards from the Bay of Bengal towards Myanmar is deflected by the Arakan hills westwards ups the Ganga plain. After crossing the deltaic coast of Bengal, these winds are moved to Assam and Chittagong hills and as a result of this, a very heavy rainfall occur in this region. A part of the monsoon current is then turned Westwars by the Himalayan ranges and consequently all along their lower slopes from Sikkim to Kashmir where there is almost continuous rainfall during this season. The winds of Southwest monsoon season in the Arabian sea are obstructed by the Western ghats and leads to a heavy rainfall in the coastal region west of the ghats. After crossing the ghats, the monsoon advances over the Deccan plateau and Madhya Pradesh, and meets the current from the Bay of Bengal. The other parts of the Arabian Sea branch of the monsoon crosses the coast of Saurashtra and Kutch and reaches the Aravalli hills after passing the Arid Zone of Rajasthan. In Eastern Punjab, these winds join the current deflected Westwards from the Bay of Bengal and produces moderate to heavy rains in the Western Himalayas, Eastern Punjab and Eastern Rajasthan. The monsoon retreats back from Northern India by the second week of September.

3. Retreating Southwest Monsoon Weather

It begins from October and ends in November. During this season, there is a dry weather in Northern India but there is general rainfall on the Coastal parts of Southern State and over the Eastern half of the Peninsula. Cyclonic storms often form in the Bay of Bengal and advances towards the East coast of the Peninsula.

4. Cold Weather

It starts from December and ends in February when temperature in Asia is lowest. The Northeastern monsoon prevails over the Indian continent and sea areas. This is characterized by the presence of clear sky, find weather, light northerly winds, low temperature and humidity. There may be shallow cyclonic depressions from West to East across Northern India giving rise to mild to heavy rains in the Punjab plains and considerable snow-fall in Kashmir valley. The rainfall and temperature are highest in Northwest of India and lowest in the South and East during this season.

Vegetation of India

The natural vegetation of an area depends primary on climate, mainly the rainfall, the temperature, soil, elevation and other local factors. India possesses a wide range of temperature and there is also a great variation in altitudes from sea-level to the loftiest mountains of the world. Therefore, India has practically all the climatic zones from the torrid to the arctic between the coastal plains and the mountains of India. The humidity and rainfall range from the lowest in the desert of Rajasthan to the highest in the Cheerapunji hills of Assam with 1080 cm annual rainfall has the reputation of being the rainest part of the world. Broadly speaking the natural vegetation of India may be classified broadly into the following types.

1. The Tropical Evergreen Forests or Tropical Evergreen Rain Forests

The Tropical Evergreen Forests or Tropical Evergreen Rain Forests usually occur in areas receiving more than 200 cm of annual rainfall and having a temperature of 25 to 27° C. The trees are evergreen as there is no period of drought. Though heavy rain is experienced only during the rainy season, the soil retains enough moisture for the rest of the year.

Trees are tall, above 60 metres high and yield valuable hard wood. The crowns of trees may form two or three layers, as in the equatorial forests. Heavy rain and high temperatures encourages luxuriant growth of trees, climbers, epiphytes, bamboos and ferns. Leaves are broad to give out excess moisture by evapo-transpiration, to increase surface area, to receive more sunlight and manufacture more food. There is dense undergrowth.

These tropical evergreen forests occur on the western slopes of the Western Ghats *i.e.* the Western parts of Maharashtra, Kerala, Karnataka and Tamil Nadu. They also occur in West Bengal and on the lower slopes of the Eastern Himalayas and on the Jaiantia, Garo, Khasi and aother hills and the Andaman and Nicobar Islands where they are not economically exploited on a large number of species occur in "mixed stands' and the forests are dense and not easily accessible.

The economically useful species are rosewood, mahoagany, ebony, sisham, toon and bamboos. Rosewood is valuable, find grained and hard and is used in making expensive furniture, especially in Kerala, Karnataka and Tamil Nadu where it is widely grown. Ebony is a hardwood with a coal black colour. It is very heavy due to its high specific gravity and is also very strong and durable.

2. Tropical Monsoon Deciduous Forests

Tropical Monsoon Deciduous Forests are found in areas receiving a rainfall of 100 to 200 cms a year; with a distinct dry season, a distinct rainy season and a small

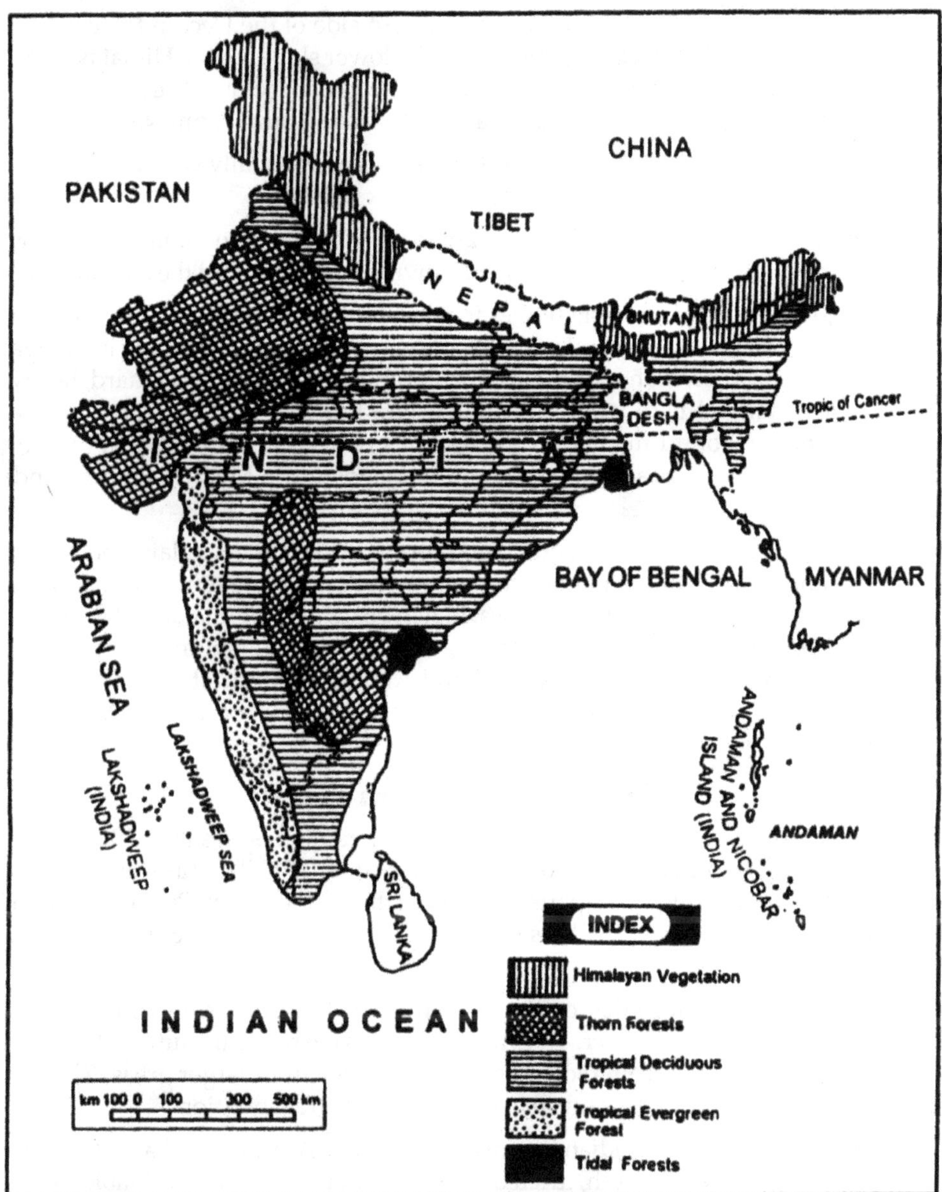

Map 2: Natural Vegetation: India.

range of temperature. In other words there is no distinct bitterly cold winter. These trees shed their leaves for six to eight weeks during the period of drought usually from March to May to prevent loose of moisture through evapo-transpiration as also to reduce surface area. In these area the rainfall is not heavy and temperatures are high.

These forests are found on the wetter western side of the Deccan Plateau, the north-eastern part of the Deccan plateau and the lower slopes of the Himalayas, on the Siwalik Hills from Jammu in the west to West Bengal in the east. They cover parts of Chattisgarh, Orissa, Bihar, Andhra Pradesh, Karnataka and Tamil Nadu.

These tropical monsoon deciduous forests are commercially very valuable for building purposes and furniture making.

The principal trees of these species are teak, sal, sandalwood, mahua, khair, mango, jackfruit, wattle and bamboo, semal, myrobalan, arjun and the banyan tree.

Teak provides hard, durable timber for ship building, house construction and furniture. It resists white ants and nails do not rust in it. It is grown mainly in Mahya Pradesh, Orissa, Bihar, Maharashtra, Karnataka and Tamil Nadu. *Sal* is hard, heavy wood which is immune to white ants. It is useful for railway sleepers and house construction. It is found mainly on the foothills of Himalayas in Uttar Pradesh, Uttaranchal and Bihar, as also in Tripura, Assam, Orissa and Madhya Pradesh and Chattisgarh.

Sandalwood trees provides sandalwood for handicraft and sandalwood oil for perfumes and soap. They are mainly grown in Karnataka and Tamil Nadu. *Semal* is soft, white, timber used for packing cases, match boxes and toy-making. It is grown widely in Assam and Bihar. *Myrobalan* grows in abundance in Tamil Nadu. Maharashtra, Bihar, Orissa, West Bengal and Andhra Pradesh. Its timber is very strong. The fruits are used for tanning leather, dyeing cotton, wool and silk. A number of dyes are obtained from it.

3. The Scrub and Thorn Forests

The Scrub and Thorn Forests occur where the rainfall is less than 100 cm but more than 25cm. Temperatures, too, must be more than 20°C. They are found in the drier areas of the north-western parts of the India that is, Rajasthan, Punjab, Western Uttar Pradesh, Gujarat and the hills of Penisular India and have scrub and thorn forests.

The chief characteristics of these plants are long tapering roots to tap underground resources of water. Trees are stunded as the rain is unsufficient for tree growth and are widely scattered so as to draw water from larger areas. Thorny branches, small fleshy leaves and sharp spines reduce transpiration.

Typical species are the Indian Khajuri (Date palms), Acacia and Khair, Dates are edible. They may be eaten fresh, dried or stuffed with nuts. Khair is valuable for its timber and products such as dye for tanning.

4. Semi-deseri and Desert Vegetation

Semi-deseri and Desert Vegetation is found where the rainfall is below 25 cm and the temperature is 25°C to 27°C. Consequently xeophytic types of vegetation is found in West Rajasthan, Sind, Kachchh and Saurashtra in Gujarat, the drier parts of Southern Punjab and the dry, rain shadow areas of the Decan Plateau, the Ladakh region and the Middle and Lower Basin of the River Indus are also characterised by sparse vegetation.

Since this xerophytic vegetation is adapted to the climatic conditions, it is characterized by thorny bushes growing apart from each other. The plants have deep roots, to tap water from great depths, thick fleshy stems to store water so as to survive during the long drought, absence of leaves to reduce transpiration from the surface area and waxy leaves to close the pores.

5. The Mountain Forests

The Mountain Forests are mixed forests of deciduous and coniferous type, depending on elevation and rainfall. They grow usually at an altitude of 1500 m on the hills of Southern India and at an altitude of 1000 to 3000m in the Himalayas, stressing from Kashmir to Assam.

On the slopes of the Eastern Himalayas we find broad-leaved evergreen trees growing *e.g.* oak. At the higher altitude of the Himalayas, bushes, shrubs and thickets of rhobodendrons grow. Beyond 6000 m only snow is found.

Where temperatures are cooler and rainfall low, coniferous forests flourish. Trees are conical in shape so that the snow can slide. They are evergreen because of less rain and lower temperature. Needle shaped leaves protect transpiration, low seasonal rain supports tree growth, while they have more wood than leaves, as the wood than leaves, as the wood stores the food. There is little under-growth due to less rain, low temperature and there is no simultaneous falling of leaves.

The main plant species are spruce, cedar, silver fir, pine, deodar, magnolia and laurel. Populars, walnuts, birch, juniper and elm also occur. Cedar, pine and eucalyptus grow in most hill stations, especially the Nilgiris. Chir (pine) is used for extraction of *resin* and *turpentine* and is used in varnishes and medicines. *Soft wood* is useful for making paper, pulp, matches, packing cases and planks.

Deodar is useful for railway sleepers. The wood is soft and used for making paper. The trees are not so dense and grow straight and tall. Resin is derived from the pines of the Himalayas and hills of Assam. It is used for soap manufacture and in paper mills as well. Eucalyptus is also used for medicinal purposes.

6. The Manager or Tidal Forests

The Manager or Tidal Forests are found along the edges of the deltas of the Rivers Ganga, Mahandi, Godavari and Krishna. Mangrove. Tidal or Littoral Forests flourish due to the flooding of these areas by the sea during high tide. These lowlands have thick deposits of mud and silt of saline water during high tide. Some of these forests are dense and impenetrable. Fewer varieties are found in this area. The Sunderbans are the largest existing natural mangrove in the world spread over 6000 sq. kms. It is a dynamic ecological system.

Mangrove forests are well-developed in the tidal zone of the Ganga delta in West Bengal where they are knows as the Sunderbans from the Sundri trees which grow in abundance. In this area the casuarinas tree is also found. The mangrove tree has stilt-like supporting roots which are exposed at low tide. Gorjan and Lintal are other species found in these swamps. The wood is hard and durable and is used for boat building and fuel.

Main Types of Natural Vegetation

I. Tropical Evergreen or Evergreen Rain Forests

Amount of Rainfall and Temperature	*Area of Occurrence*	*Characteristics and Adaptation*	*Plant Species Found*	*Uses*
+ 200 cm Temperature 25ºC to 27ºC	a) Western Slopes of the Western Ghats *i.e.* Western parts of Maharashtra, Kerala, Tamil Nadu b) On the lower slopes of the Eastern Himalayas c) North East India, Lushai, Jaintia Garo, Khasi Hills and in Bengal d) Wetter parts of Andaman and Nicobar Islands	a) The trees are evergreen as there is no period of drought. Though heavy rain is experienced only during the rainy season, the soil retains moisture enough for the rest of the year b) Trees are tall, above 60 metres high and yield hard wood. c) Heavy rain and high temperatures encourage luxuriant growth of trees; climbers, epiphytes, bamboos and ferns in an efforts to each the sun. d) Leaves are broad to give out excess moisture by evapotranspiration, to increase surface area, to receive more sunlight and to manufacture more food. e) Dense under growth.	Rosewood, Ebony, Sisam Toon and Bamboo	The wood is hare and valuable. Rosewood is valuabe, finer grained and hard and is used for making expensive furniture especially in Kerala, Tamil Nadu and Karnataka, where it is widely grown.

II. Tropical Monsoon Deciduous Forests

Amount of Rainfall and Temperature	*Area of Occurrence*	*Characteristics and Adaptation*	*Plant Species Found*	*Uses*
150-200 cm a year Temperature 20ºC	a) Wetter Western side of the Deccan Plateau b) North Eastern part of the Deccan Plateau c) Lower slopes of the Himalayas	Decidusous trees shed their leaves for 6 to 8 weeks during the period of drought. a) Deciduous trees shed their leaves because of a lack of rainfall from March to May, to prevent loss of moisture through evapotranspiration and also to reduce the surface area. In these areas rainfall is not heavy and the temperatures are high. b) Very useful trees as they provide valuable wood for making furniture and building purposes.	Teak, Sal, Sandal Wood, Mahua Khair, Mango, Jackfruit, Bamboo, Semul, Myrobalan and Arjun.	*Teak*: Hard durable timber for ship building, house construction and furniture. Resists white ants, nails do not rust in it. Grown in M.P., Mahara-shtra and Tamil Nadu. *Sal*: Hard heavy wood immune to white ants. Useful for railway sleeprs and house construction. Found in Bihar, U.P., Orissa, M.P., Tripura and Assam. *Sandalwood* provides sandalwood or handicrafts and sandalwood oil or perfumes and soap: mainly grown in Karnataka. *Semul*: Soft white timber used for packing cases, matchboxes and toy making. Found in Assam and Bihar. *Myrobalan*: Fruit used for tanning leather, dyeing cottonwool and silk, Grows in M.P.,Orissa and Andhra Pradesh.

III. The Scrub and Thorn Forests

Amount of Rainfall and Temperature	*Area of Occurrence*	*Characteristics and Adaptation*	*Plant Species Found*	*Uses*
Less than 100 cm a year more than 25 cm Temperature 20ºC	a) In drier parts of the north western Indian subcontinent *i.e.* Rajashthan, Punjab, Western U.P. b) Gujarat c) On the drier hill slopes of peninsular India	a) The trees are stunted as rain in insufficient for tree growth. b) Wiely scattered so as to draw water from larger areas. Trees often have sharp spines to reduce transpiration.	Indian date, Khajuri, Acacia, Khair	a) Dates are edible. b) Khair is valuable for its timber and products such as dye or ranning.

IV. Semi-Desert and Desert Vegetation

Amount of Rainfall and Temperature	*Area of Occurrence*	*Characteristics and Adaptation*	*Species and Uses*
Less than 25cm a year Temperature 25ºC to 27ºC	West Rajasthan, Kutch and Sauashtra in Gujarat, drier parts of southern Punjab and in the dry parts of the Deccan Plateau.	Vegetation is Xerophytic. Mostly thorny bushes plants grow far apart from each other. They have deep roots to tap water from depths, thick fleshy stems to store water so as to survive the long drought, absence of leaves to reduce transpiration from the surface area, waxy leaves to close the pores.	Thorny bushes, wild Berries, Cactii, Kikar and Babul (acacias). Latter yield gum and the bark is used for tanning hides and skins.

V. Mountain Forests

Area of Occurrence	*Characteristics and Adaptation*	*Plant Species Found*	*Uses*
Mixed forests of deciduous and coniferous type, depending on elevation and rainfall. Grow usually at altitude of 1500 m, on the hills of Southern India of 1000 m. in the Himalayas, stretching from Kashmir to Assam. At the higher altitude of the Himalayas, bushes, shrubs and thickets of Rhododendrons. Beyond 6000 m. only snow is found	Where temperature are cooler and rainfall low, coniferous forests flourish. a) Trees are conical in shape so that the snow can slide. b) They are evergreen because of less rain and lower temperature. Needle shaped leaves reduce transpiration. Low seasonal rain supports tree growth. More wood than leaves, as the wood stores the food. Little undergrowth due to less rain, low temperature and no falling of leaves. The wood is soft and is used for making paper. The trees are not so dense and grow straight and tall.	Spruce, Cedar, Silver Fir, Pine, Deodar, Magnolia and Laurel. Popular, Walnuts, Birch and Elm also occur. Pine and Eucalyptus grow in most hill stations, especially the Nilgiris.	*Chir pine* is used for extraction of resin and terpentine, *soft wood* is used for paper, pulp, matches, packingcases and planks. Deodar is useful for railway sleepers.

VI. The Mangrove Tidal Forests

Area of Occurrence	*Characteristics and Adaptation*	*Plant Species Found*	*Uses*
Along the edges of the deltas of the Ganga, Mahanadi, Godavari and Krishan.	Fewer varieties in an area. These forests are found in the submerged coastal plains. Grow best in salty water. Some of these forests are dense and impenetrable. In Bengal, these are known as Sunderbans from the Sundri tree.	Sundri and Casuarina	The wood is hard and durable and used for boat building and fuel.

Chapter 4
Description and Discussion of Plant Species

Family: ADIANTACEAE

Genus: *Adiantum* Linn.

From Greek word, the generic name has been derived Adiantos meaning the maiden hair fern and A= unwetted and Dinots = capable of being wet, referring the nature of the plant.

Terrestrial and tufted ferns with long-creeping or short rhizomes and ascending scales, the genus comprises The scales. The scales are usually brown to black and narrow. Stipes are inarticulate, scaly at base only, dark and polished. Fronds are bright grey. Lamina is broad, simple to 1-pinnate or pinately decompound with the final pinnules, dimidiate or flabellate, usually herbaceous, rarely membranous or coriaceous and glabrous or commonly hairy and rarely glaucous. Sori are marginal, globose to linear and separate and confluent. On the inner face or reflexed marginal outgrowths the sporanagia are borne along the veins.

About 200 species according to Airy Shaw (1973) are found which are cosmopolitan in distribution, especially in tropical America. From India of them 8 have been recorded.

Species: *Adiantum capillus-veneris* Linn.

Description of Plant

A delicate graceful fern. Stipes are blackish, 10-25 cm long, suberect, naked and polished. Fronds are bipinnate with a short terminal pinna and numerous erect-patent lateral ones on each side, the lowest being slightly branched; segments are cuneate and 1.5-2.5cm broad. The sori are roundish or obreniform which are placed in the roundish sinuses of the crenations.

Throughout the greater part of the year, sori are formed

Distribution

In South India, Maharashtra, Punjab and Bihar it is common. It also occurs mainly in the Western Himalayas ascending upto 2400 m and extending to Manipur.

Plant Parts Used

Whole plant is useful.

Uses

The plant is demulcent, expectorant, diuretic, emmenagogue and febrifuge and is used an hair tonic. Its fronds are pounded with honey and are beneficially used in catarrhal affections. Alcoholic extract of the fern is given internally in the treatment of hard tumour in spleen. Besides, an infusion of its leaves with tea is given in menstrual cycle complaints.

Folk Uses

The juice of entire plants is dropped in blisters of the mouth throughout the different parts of the country.

Preparation

Hanspadi Panak.

Family: MARSILEACEAE

Genus: *Marsilea* Linn.

The generic name has been assigned in the honour of Count Marsigli, the founder of Academy of Science at Bologna.

These are small herbs with a slender creeping rhizome terminating in a 3-sided special cell which give rise to 2 dorsal rows of leaves and a vertical rows of roots. Mature leaves are cruciform, consist of two continuous pairs of opposite leaflets. Sori are numerous on a gelatinous receptacle attached to the wall of the bilateral sporocarp by its ends and extruded in the form of ring. Each sorus is with few microsporangia containing normally one functional megaspore and many megasporangia with numerous microspores.

About 60 species are found in tropical and temperate parts of the world. Only 9 species have been reported from India.

Species: *Marsilea minuta* Linn.

Description of Plant

It is an aquatic fern with acreeping rhizome. Leaves are quadrifoliate, very variable in size and erose or entire. Petioles are 2.5-5.0 cm long, obscurely connate at the base and adnate to the whole base of the sporocarp. Sporocarp is bean-shaped and sori 10-12 or more.

Sori Formation

During November-January.

Distribution

It is found throughout India, usually at the edges of the canals and ponds and grows as a weed in marshy places.

Plant Parts Used

Whole plant, leaves, fruits (sporocarps) and spores are used.

Uses

The whole plant and its leaves are used a s sedative in insomnia. These are also used in the treatment of epilepsy and behavioural disorders. The fruits and spores of the plant are used in homeopathic medicines.

Preparation

Sunishankachangeri ghrit and Changeri ghrit.

Family: EPHEDERACEAE

Genus: Epherdra Linn.

The generic name *Ephedra*, is the Classical name of the plant.

The genus comprises of erect or sub-scandent rigid shrubs; branches opposite or fascicled, terete, striate with opposite scales at the nodes which are rarely produced into linera leaves. Cones are formed in axills of decussately opposite bracts. Female cones are with single naked, orthotropus ovule. Seeds are usually oblong, plano-convex with dry testa.

There are 40 species known to occur in the warmer temperate North and South America and Eurasia including India.

Species: *Ephedra gerardiana* Wall. Ex Stapf (Syn. *E. vulgaris* Hook, A. Rich.)

Description of Plant

It is a low growing, rigid and tufted shrub usually with a guarled stem and erect and green branches attaining a height of 15.0-120cm. Cones are 0.6-0.8cm long, subsessile and often whorled; mature with fleshy red succulent bracts and 1-2 seeded. The seeds are biconvex or plano-convex.

The cones are formed and matured during May-August.

Distribution

It is found in drier parts of temperate and alpine Himalayas, from Kashmir to Sikkim at an altitude between 2133-4876m.

Plant Parts Used

Whole dried plant in the form of panchang is used.

Uses

The plant is alterative, diuretic, stomachic and tonic in action and its decoction is used as a remedy in rheumatism and syphilis. The juice of its fruits is given in affections of respiratory tracts besides, liquid extract for controlling asthmatic paroxysm. A chemical is obtained from this plant which ressembles with Ephedrine of Chinese plant, 'Ma-huang' which is termed as pseudo-ephedrine. This is used for the relief in attacks of asthma and causes no side-effect as ephedrine.

Folk Uses

A powder of its stem is used as a snuff in headache in different parts of the country.

Preparation

Shavasari churna.

Species: *Ephedra major* Host (Syn. *E. nebrodensis* Tineo)

Description of Plant

This is a low, rigid and dense tufted shrub with woody stems and green, erect and ascending branches. Internodes are smooth, slender, pedunculate and about 1.2-2.5cm long. Male cones are few-fid, solitary or 2-3 togethr and females are solitary and short pedunculate. The fruits are globose and 0.6cm in diameter.

The cones are formed and matured during May-August.

Distribution

It is known to occur wild in Himachal Pradesh.

Plant Parts Used

Whole plant is used medicinally.

Uses

A decoction of stem and roots of othe plants is used as a remedy in rheumatism and syphilis and their liquid extract for asthmatic paroxysm. Besides, juice of its fruits is used in respiratory tracts infections.

It is also used as a source for obtaining ephedrine.

Family: CUPRESSACEAE

Genus: *Juniperus* Linn.

The generic name *Juniperus* is the Latin name of the juniper tree.

The genus comprises of evergreen, aromatic and monecious shrubs or trees. Leaves are more often subulate or linera and 3-nately whorled. The cones consist of 1-4 whorls or scales, one of which as a rule is fertile. In ripening, the whole becomes a fleshy mass enclosing the seeds and forming a good imitation of a true berry. Seeds are 1-3 with a thick hard testa and often connate in to a hard mass.

About 60 species are known to occur in Northern hemisphere. 5 of them have been reported from India.

Species: *Juniperus communis* Linn.

Description of Plant

It is a bushy plant attaining a height of 70-80 cm; prostrate at high elevations with broader leaves. Leaves are in whorls of 3, spreading, 0.6-1.2cm long, pungent and whitish above, convex or obtusely keeled beneath with a more or less prominent cushion on the branchlets. Fruits are black, 0.6-0.8cm in diameter and glaucous with scarious empty scales at the base.

Cone formation starts in August-September and ripen in the second year.

Distribution

This is known to occur wild in Western Himalayas from Kumaon westwards between an altitude of 3810-4267m.

Plant Parts Used

Fruits and the oil obtained from its fruits are used medicinally.

Uses

The fruits and oil, commercially knows as 'Juniper oil' both are considered carminative, diuretic and stimulant and are used in dropsy, gleet, gonorrhoea, leucorrhoea and in some skin diseases and disorders of urino-genital tracts. Its oil should not be given in acute nephritis.

Juniper berries are roasted, ground and are used as a substitute of Coffee.

Folk Uses

In Western Himalayas, native people rub the powder of its fruits on rheumatic and painful swellings and apply the ashes of its bark in some skin diseases.

Preparation

Hapushadi churna.

Species: *Juniperus macropoda* Boiss.

Description of Plant

It is a small tree with sudistichous slender branches. Leaves are dimorphic subulate of lower branches and of upper and branchlets are pungent, scale-like, imbricate closely appressed ovate with a dorsal large gland. Male cones are at the ends of branchlets, 0.3-0.4cm long and closely set with imbricate scales. Fruits are subglobose, black, glabrous, 0.6cm across and 2-5-seeded.

Cones are formed and matured during July-August.

Distribution

It is found in Himalayas westwards upto 4267m.

Plant Parts Used

Fruits and oil are used medicinally.

Uses

Its fruits as carminative, diuretic and stimulant and they are used in dropsy, gleet, gonorrhoea, leucorrhoea and in skin diseases.

The oil of fruits is used as a substitute and adulterant to Juniper oil, obtained from *Juniper communis*.

Folk Uses

Its twigs are burnt as incense and its fumes are inhaled which are supposed to relieve delirium conditions in fever.

Family: PINACEAE

Genus: *Abies* Mill

The generic name is the Classical name of the plant.

These are tall trees with more or less bifarious linear, I-nerved leaves. The cones are erect and cylindric, scales are thin and deciduous leaving a woody axis, supporting scale free under ovaliferous scale and sometimes produced beyond it.

There are 50 species known to occur in North temperate regions of the world; 6 of which have been recorded from India.

Species: *Abies spectabilis* (D. Don) Spach. (Syn. *A. webbiana* Lindl.)

Description of Plant

This is a tall evergreen tree upto a height of 60m and 3-10m in girth. The bark is silvery white or grayish brown; scaly when young becomes dark or brownish grey and fissured when old. Leaves vary in size, densely set in 2-4 ranks on stout branchlets, dark green, lustrous, and coriaceous persisting upto 10 years. Male catkins are 1.3-1.8cm long and ellipsoid. The female cones are 14-20 x 4-6cm, cylindrical and violetpurple when young and become brown when old with fan shaped scales and slightly exerted or concealed bracts. The seeds are 0.8cm long, oblong-obovate and angular with broad notched wings.

The cones formation starts during April-May and they mature in autumn of the same year.

Distribution

It is known to occur in the Himalayas from Kashmir to Assam between 1600-4500m; also grown ornamentally.

Plant Parts Used

Leaves are usable.

Uses

Its leaves are considered carminative, expectorant, stomachic and astringent. The fresh juice of its leaves is given infantile fever during dentition and in bronchitis in the doses of 5-10 drops in water or mother's milk. A powder of its leaves in a dose of 1.8gm mixed with the juice of *Justicia adhatoda* (Bansa) and honey is used in cough, asthma and haemoptysis.

Folk Uses

In Bengal, the juice of its leaves is administered as a tonic in cases of parturition. Besides, an infusion of the leaves is also used in hoarseness.

Preparation

Talisadya churna, Talisadya vati, Sudarshan churna, Lavanbhaskar churna and Kanakasav.

Genus: *Cedrus* Trew

The generic name of the plant is based on its Greek name Cedron which is a small river in Judea, the original home of *Cedrus species*.

These are handsome evergreen trees. Leaves are triquetrous in dense clusters and acicular and are jointed near the base. The cones are with deciduous scales leaving a columnar axis.

According to Airy Shaw (1973), r species are found in the world; of which only 1 has been recorded from India.

Species: *Cedrus deodara* (Roxb. Ex. Lamb.) G. Don (Syn. *C. livani* Barrel var. *deodara* Hook, f.)

Description of Plant

This is a tall evergreen tree upto 75m height with thick and fissured bark. Leaves are dark green with wavy margins. The wood is oily, aromatic, fine textured, fairly evergreen and straight; sap wood white and heart wood light yellowish-brown or brown. The cones are erect, 8.0-12.5cm long and 7.5-10.0 cm broad, obtuse and scales closely imbricate, broadly cuneate; upper edge thin and rounded and broader than long, deciduous leaving the axis of the cone standing erect on the branches. The seeds are 0.8-1.6 cm with triangular wings.

The cone formation starting and maturation is during September-December.

Distribution

It is found in North-western Himalayas from Kashmir to Garhwal between 1219-3048 m.

Plant Parts Used

Bark, wood, resin and oil of the plant are used.

Uses

The bark is known as astringent and is used in fever, diarrhea and dysentery. The decoction of its wood is used successfully in fever, diarrhea, dysentery and urinary diseases in the doses of 0.7-2.8 gm per day. Oleoresin and a dark coloured oil, turpentine, obtained from its wood are used as a diaphoretic agents. The oil known as 'Kelon ka tel' in the doses of 1.8 gm is considered emetic and is used in ulcers, leprosy, flatulence and rheumatism.

Preparation

Devdardi kwath, Devdaradi vati, Devdardi churna, Rasnadi kwath, Rasna panchak kwath, Satyadi kwath, Sudarshan churna, Sarvajwar lauha, Punarnava mandur, Karanjadl yog, Laghu vis garva taila, Khadirarista, Chandraprabha vati and Devdas vista.

Genus: *Pinus* Linn.

The genetic name has been derived from Greek word Pinos, used by Theophrastus for a Pine tree.

This gymnospermic genus comprises of evergreen morpecious trees. Leaves are dimorphic. Male flowers are in spikes sterinal column is ovoid, oblong or cylindric; anthers are in many series and 2 cened. Female cones are globase or avoid bracts are spirally imbricate, ovuliferous scales are omuch larger tyhan the bracts and ovules are 2. Ripe cones are ovoid or oblong, scales persistent and formed of the enlarged thickened usually woody ovuliferous scale, the tips of which often squarish and with a boss. The seeds are 2, reversed and usually winged with 2 or more cotylendons.

About 70-100 species have been recorded which are confined to Northern temperate region and the mountains of North tropics; of these, 5 species are found in India.

Species: *Pinus roxburghii* Sarg.
(*P. longifolia* Rox.)

Description of Plant

It is a large tree with symmetrical branches high up on the trunk and forming a round head of light foliage and reaching upto 30 m in height. Normally, it is evergreen but becomes completely or partially deciduous in dry season. The bark is drak grey and deeply fissured. Leaves are 20-30 cm long, arranged in clusters of three, finely toothed and light green. Male flowers are arranged in form of cones which are 1.5cm long. Female cones are 10.0-20.0 x 7.5-12.0 cm, ovoid and are solitary or 2-5 together and woody. The seeds without wings are 0.7-1.3 x 0.5-0.6 cm and winged; wings are membranous and long.

The cone formation is started in January, next year during June-July its cones become old and in next summer, burst and seeds are shed out. About 26 month are spared to complete the whole process. The dry cones remain long on the tree.

Distribution

It is found in the Himalayas from Kashmir to Sikkim and in the Shiwalik hills at the altitude of 450-2400 m. Sometimes, raised in gardens as an ornamental plant.

Uses

The wood is stimulant, diaphoretic and is useful in the burning of the body, cough, fainting and ulcerations. The oleoresin (gandhbiroza), is considered to be anodyne and diuretic and is beneficial in coryza and oedema. This is advised to be given in the dosess of 1.8-5.4 gm four times daily in gonorrhoea. It is expecetorant and useful in chronic bronchitis and is especially recommended in gangrene of lungs, also given as carminative in flatulent colic and to arrest minor haemorrhages in tooth sockets and nose. Externally, it used as a rubefacient in rheumatic infection and for deep seated inflammation, especially of abdomen (Ambasta, 1986) and is applied as a plaster to buboes and abscesses for suppuration.

The oil obtained from the wood mixed with mustard oil is rubbed in rheumatic pain. Its vapours relieve the spasm of whooping cough and asthma and internally, it is given in indigestion, loss of appetite and intestinal worms.

Folk Uses

Its seeds are roasted and eaten by the people.

Preparation

Chandandi vati. Also used in the preparation of ointments and plasters.

Family: TAXACEAE

Genus: *Taxus* Linn

The generic name *Taxus* is the Latin name of a Yew tree.

The genus comprises of evergreen trees or shrubs with tought wood. Leaves are linear, distichious and flattened. Male and female cones are solitary and axillary.

About 10 species occur in North temperate region, South to Himalayas, Philippines, Celebes and Mexico. Only one of which is known to found in India.

Species: *Taxus baccata* Linn.

Description of Plant

It is a small evergreen or medium sized tree with fluted stem and reddish grey bark. Leaves are 2.5-3.5cm long, linera, flattened, distchious, acute and narrowed into a short petiole. Male cones are subglobose and solitary in leaf axis and females are also solitary, axillary with a single erect ovule. The fruits are red coloured and are with woody testa.

Cone formation is started during March–May and they mature in September-November.

Distribution

It is found in temperate Himalayas between 1828-3352 m and in Khasia hills.

Plant Parts Used

Leaves and fruits are used medicinally.

Uses

The leaves are considered antispasmodic, emmenagogue and sedative and these are used in asthma, bronchitis and hiccough. Besides, a tincture is made from the young shoots of the plant which is used to cure headache with giddiness, feeble faltering pulse, coldness of the extremities, diarrhea, general prostration and severe biliousness.

The bark of the plant is used as a substitute for tea.

Folk Uses

In Garhwal Himalayas, Gangwal people use its bark in cold and cough. In Khasi hills (Meghalaya), its leaves with ginger are made into a paste which is applied on tumours by the Garo and Jaintia tribal people.

Preparation

Talisadi churna and Talisadi vati

Family: RANUNCULACEAE

Genus: *Aconitum* Morgan

The generic name of the plant is based on Greek word Akon means means an arrow; as in ancient times, the juice of the roots of *Aconitum* plant was used as an arrow poison. In Greek mythology, the name *Aconitum* was derived from the Mount *Aconitum* where Hercules had a fight with Cerverus, a mad dog, progeny of typhoon and the serpent woman Echidna. It was believed that this plant originated from the deadly saliva of Cerberus.

The genus comprises of erect, rarely of twining herbs. Leaves are palmatipartite or entire. Flowers are irregular, blue white or yellow in racemes. The follicles are sessile with numerous rugose or wrinkled seeds.

There are 300 species occur in North temperate region of the world. 30 of which are known to be found in India.

Species: *Aconitum balfourii* Stapf.

Description of Plant

It is a biennial herb with erect, straight, robust, simple and terete stem reaching a considerable height. Leaves are scattered, 6-10; lower decayed at the time of flowering, intermediate and upper leaves rather distant and pubescent when young. Flowers are blue in straight, 30 cm long, many-flowered racemes. Sepals are blue, pubescent, upper most helmet-shaped; carpels 5, oblong, yellowish tomentose and combining in the flower, then slightly divergent. The follicles are oblong, slightly divergent above, otherwise contiguous, loosely hairy or glabrate and 0.1x0.4 cm. Seeds are dark brown, trigonous and broadly winged.

Flowering and Fruiting

Flowering is during July–August and fruiting in August-September.

Distribution

This is found in sub-alpine and alpine Himalayas from Garhwal to Nepal between 2133–4266 m.

Plant Parts Used

Its tuberous roots are used medicinally.

Uses

The tuberous roots of the plant are used as antiperiodic, diaphoretic and diuretic (Uniyal, 1989).

Species: *Aconitum bisma* (Ham) Rapaics (*A. Palmatum* D Don; Caltha bisma)

Description of Plant

A perennial herb with tuberous roots and erect and simple stem. Leaves are scattered, rather distant, 10 or rarely more; lowest usually withered at the time of flowering , glabrous or upper most finely pubescent. Flowers are bluish or variegated white and blue in a very loose 10-20 cm long leafly panicle or raceme; carpels 5 and quite glabrous. The follicles are subcontiguous or somewhat diverging in the upper part, oblong, obliquely truncate and 2.5-3.0 x 0.5-0.6 cm. Seeds are blackish, obovoid and about 0.3cm long.

Flowering and fruiting is during July-August.

Distribution

It is known to be found in alpine Himalayas of Sikkim and the adjoining part between 3048-3487 m.

Plant Parts Used

The roots of the plant are usable.

Uses

Its roots as non-poisonous and to possess antiperiodic and tonic properties. In combination with pepper, its roots are used internally as a remedy for pains in the bowels, diarrhea and vomiting besides, anthelmintic against intestinal worms.

It is also used to adulterate other species of *Aconitum* usually.

Species: *Aconitum chasmanthum* Stapf. Ex Holmes

Description of Plant

A perennial herb with tuberous and paired roots and erect and simple stem reaching a height upto 0.6-1.2m. Leaves are numerous, usually more distant in the lower part and crowded in the upper part; lower ones on about 7.5cm long petioles and uppers shortly petioled; cauline leaves are orbicuilar-reniform. The flowers are blue or whitish in long, narrow, stiff and dense or loose racemes; carpels 5, glabrous, rarely sparingly hairy on back and conniving. The follicles are oblong, 0.1cm long and glabrous with brown, obovid-obpyramidal, 0.3 cm long and 3-winged seeds.

Flowering and Fruiting

Flowering is during July-August and fruiting in September-October.

Distribution

It is found in sub-alpine and alpine Himalayas between 2133-3657 m.

Plant Parts Used

Its tuberous roots are used.

Uses

Its roots are considered anodyne, antiphlogistic, antipyretic and sedative. These are given internally in small doses in fever to lower temperature and for relieving pain in place of *A. napellus*.

The roots of the plant are used as a substitute of *Aconitum ferox* and *A. napellus*.

Species: *Aconitum deinorrhizum* stapf.

Description of Plant

It is a biennial herb with paired tuberous roots and a considerable height. Stem is erect, straight, simple, terete, sparingly and finely crispo-pubescent in upper part, otherwise glabrous. Leaves are 10-12, scattered, lower usually decayed at the time of flowering and upper ones rather distant, sparingly hairy when young. The flowers are blue in straight racemose inflorescence; carpels usually 3, oblong and combining in the flower. The fruits are obconic with 0.3cm long seeds.

Flowering and Fruiting

Flowering is during August-September and fruiting in September-October.

Distribution

It is known to occur in Himachal Pradesh, Uttar Pradesh and Nepal above 2133 m.

Plant Parts Used

The tuberous roots of the herb are used medicinally.

Uses

Its roots are considered poisonous but as cited in Asolkar *et al.* (1992), these along with its leaves are used in acute headache, rheumatism and rheumatic fever.

Species: *Aconitum falconeri* Stapf.

Description of Plant

It is an erect, biannual herb with tuberous roots and simple and erect stem reching a height upto 1.0 m. Leaves are scattered, 10 or more, upper ones crowded if many, intermediate usually distant and lowest decayed at the time of flowering. The flowers are blue in erect, stiff and dense raceme which is 15-20 cm long usually; carpels glabrous or nearly so. The follicles are erect and contiguous or slightly diverging upwards, oblong, rounded at the top and 0.1 x 0.4 cm. Seeds are obconic, brown, 0.3-0.4cm long and winged.

Flowering and Fruiting

Flowering is during July–August and fruiting in August-September.

Distribution

It is found in sub-alpine and slpine zones of the Himalayas in Garhwal between 3000-3900 m.

Plant Parts Used

The roots are used medicinally.

Uses

The roots are known to possess antiperiodic, antidiabetic, antiphologistic, antipyretic, anodyne, diaphoretic and diuretic properties. These are soaked in cow's urine for a week, then washed, dried and powdered, about 0.5 gm of which is mixed with *Piper nigrum* and is administered internally in acute non-malarial fever and diarrhea.

Folk Uses

Bhotiyas in Garhwal Himalayas, use its roots oand flowers in rheumatic joints and skin-diseases (Uniyal, 1989).

Species: *Aconitum ferox* Wall. Ex. Ser.

Description of Plant

It is perennial herbs with tugerous. Leaves are scattered, orbicular-cordate to reniform; palmately 5- lobed with pubescent blade and petioled. The peducles are straight bearing flowers on both sides. The flowers are dirty blue and are borne in a dense terminal raceme, about 10-25cm long and helmet-shaped with a short sharp beak. The follicles are oblong with long and obpyramidal to obovoid seeds.

Flowering and Fruiting

Flowering is during in July-August and fruiting during October-December.

Distribution

It is found to grow wild in alpine Himalayas and in Kashmir upto 3600 m.

Plant Parts Used

Underground stems and the roots of the plant are used.

Uses

A liniment of its underground stems and of roots in small doses of 0.023-0.018 gm is used in nasal catarrh, uvula hypertrophy, sore throat, gibbous, paralysis and in chronic fever, while in larger doses in acts as a narcotic powerful poison and sedative. Internally, the tincture of its roots (1 in 8 of alcohol in doses of 2-5 drops) is used in combination with other drugs for the treatment of fever and rheumatism and also as a remedy for cough and asthma.

Its roots are cardiac stimulant, hypoglycaemic, diaphoretic and antiphologistic. A powder or liniment prepared from the root is spreaded over the skin in case of arthritis and scabies.

Folk Uses

It is much used by the native people for body pain after proper curing and the leaves are used to improve the flavour of country liquor.

Preparation

Mritunjaya ras, Hinguleshwar ras, Anand bhairav ras, Jwarmurari ras, Panchabaktra ras, Soubhgaya vati, Rambana ras, Kaphketu ras, Swalpa kasturi-bhairav ras, Agnitundi vati, Sanjivani vati and Laghu vis garbha taila.

Species: *Aconitum heterophyllum* Wall. Ex. Royle

Description of Plant

It is a perennial erect and showy herb with about 30-90cm height. The stems are simple or branched from base, glabrous and puberulous above, broad, ovate or orbicular or somewhat 5-lobed and toothed, above 3-fid or entire. The flowers are 2.5cm long, blue or greenish blue with purple veins and helmet shaped.

Flowering and Fruiting

Flowering is during in July-August and fruiting during October-December.

Distribution

It is very common in the alpine and sub-alpine belts of Himalayan north-western regions at altitude between 1800-4500 m.

Plant Parts Used

Dried tuberous roots are used.

Uses

The roots are known as aphrodisiac, digestive, febrifuge and bitter tonic (Chatterjee and Pakrashi, 1991) and are used in throat infections, abdominal pain and gastralgia in the doses of 0.7-2.0 gm. plain powder of its roots mixed with honey is given to children suffering from cough, coryza, fever and vomiting.

Folk Uses

In Garhwal Himalayas, the people use its roots for fever and also given to children for stomach pains and worms.

Preparation

Ativishadi chruna and Balchaturbhadra.

Species: *Aconitum napellus* Linn.

Description of Plant

It is a perennial herb. The stems are often decumbent in small stage, glabrous or slightly pubescent. Leaves are variable in size, palmatipartite; segments linear. The flowers are dull greenish-blue and are borne in simple or sparingly compound racemes. The follicles are 3-5 with numerous seeds.

There are four varieties of this species *viz.*; *napellus* proper, *rigidum, multifidum* and *rotundifolium*.

Distribution

It is known to found in alpine and sub-alpine belts of the Himalays in India.

Plant Parts Used

Leaves and roots are used.

Uses

The roots are generally used after processing in combination with other medicines in small doses as tonic and as a remedy for fever, nervous debility, rheumatism and cardiac diseases. But in larger doses, it becomes toxic, milk is prescribed as antidote for its poisoning. Aconitine, one of main alkaloids of its roots, is used externally in various forms of neuralgia, tetanus, acute and chronic rheumatism, gout, erysipelas and in affections of the heart characterized by increased action. It is also given internally in a very small doses in fever to lower the temperature, for relieving pains, cystis, diarrhea and cough irritation but its repeated internal dose of 1.0-2.0 mg may cause toxic symptoms.

Genus: *Coptios* Salisb

The generic name is taken from Greek word, Kopto meaning to cut; referring to the divided leaves of the plant.

The genus comprises of small stemless herbs with perennial rootstock. Leaves are ternatisect. Flowers are regular, small and white coloured on slender leafless scapes. The follicles are many seeded. Seeds are with black crustaceous testa.

Species: *Coptis teeta* Wall.

Description of Plant

it is a small stemless herb with golden yellow and horizontal, perennial rootstocks. Leaves are glabrous and ternatisect; leaflets 5.0-7.5 cm long, ovate-lanceolate and pinnatifid, lobes incised. Flowers are small, white and 1-3 pedicelled and are borne on slender leafless scapes. The follicles are many seeded.

Flowering and Fruiting

Flowering is during in September–October and Fruiting: Later.

Distribution

It is known to be found in temperate regions of Mishmi hills at the northern frontier of Assam.

Plant Parts Used

The roots of the plant are used medicinally.

Uses

Its roots as bitter, stomachic and tonic which are used efficaciously in debility, atonic dyspepsia and in mild forms of intermittent fevers. A paste of its roots with '*rasavanti*' is used as a collyrium or eye salve in catarrhal and rheumatic conjunctivitis. Besides, a paste of its roots is also applied on sore.

Preparation

Surmas and various collyriums.

Genus: *Delphinium* Linn.

The generic name is the ancient Greek name Delphinion, for Larkspur and is derived from Greek words Delphis, a dolphin, referring to the nectary of the plants to the imaginary figure of Dolphin.

These are annual or perennial erect herbs. Leaves are palmately lobed, The flowers are white, blue or purplish in racemes or panicled. Sepals are 5, free or collaring at the base and dorsally spurred behind and petals are 2-4 and small. Stamens are many. The follicles are 1-7 with numerous wrinkled or plaited seeds.

About 250 species are found in North temperature regions of world; of which 15 have been reported form India.

Species: *Delphinium denudatum* Wall

Description of Plant

This is an annual herb reaching a height of about 60-90 cm with many branches. Radical leaves are orbicular, 5.0-15.0 cm across, 5-9 partite; segments cuneate-obovate, bipinnatifid, lobes linera-oblong or oblong; cauline leaves few, uppermost 3-partite; segments linera-entire. The flowers are pale-blue, few in number and are borne in lax much branches racemes. The follicles are 3, inflated and sparsely hairy or glabrous with 1-7 seeds in each follicle.

Flowering and Fruiting

Flowering is during in May-July and Fruiting: Later.

Distribution

It is commonly found in the western Himalayas, from Kumaon to Kashmir between 2440-3600 m.

Plant Parts Uses

Seeds and roots are usable.

Uses

Its roots are bitter, stimulant and alterative and are used in toothache, rheumatism, impotency and syphilis, A decoction of its roots in the doses of 3.5-7.0 gm per day is used as a tonic after fever. Besides, combined with other herbal medicines, it is also used a s a remedy in cardiac and cerebral diseases.

Its roots are used as an adulterant to Aconite.

Folk Uses

The roots of the herb are chewed to relieve in toothache. Dards tribal people in Kashmir Himalayas, put its dried root piece in tooth cavities for pain and a poultice of the same is used for rheumatic pains (Singh, 1996). Besides, its dried roots are chewed as stimulant and also in toothache by the local people in Har-ki-Doon.

Genus: *Thalictrum* Linn.

The generic name is derived from Greek words, Thamos = a shrub and Kalamos = a reed; referring to the shrubby nature of the plant.

The genus comprises of perennial glabrous herbs with erect, rarely of partially decumbent and branched stems. Leaves are alternate, pinnate or bipinnate; leaflets stalked, orbicular or ovate, more or less distinctly 3- lobed and crenate or bluntly toothed. Flowers are small, regular and often polygamous in panicles. The fruits consist of small head of ribbed and more or less flattened and rarely terete achenes.

About 150 species are found distributed in North temperate zone, tropical south America and tropical South Africa; 20 of them are reported from India.

Species: *Thalictum foliolosum* DC.

Description of Plant

This is a tall perennial rigid herb with glabrous stems reaching a height upto 1.2-2.5m. Leaves are exstipulate, pinnately decompound; leaflets 0.4-0.6 cm, rarely

2.5cm long and orbicular. Flowers are white or pale green or tingly purple and polygamous in much branched panicles. The achenes are usually 2-5, small, oblong, acute at both ends and sharply ribbed.

Flowering and Fruiting

Flowering is during in October-November and Fruiting in November-December.

Distribution

It is known to occur throughout Himalayas between 1540-2438 m and in Khasi hills between 1219-1828 m.

Plant Parts Used

Its roots are used medicinally.

Uses

The roots of the plant are considered aperient, diuretic, febrifuge, purgative and tonic and these are used as a good remedy for atonic dyspepsia and in convalescence after acute diseases. In Unani system, its roots are used as a good application in piles, nail trobles and the discolouration of the skin. Besides, a cold infusion of roots is used as a lotion for opthlamia.

The roots of the plant are used as a substitute and adulterant to *Coptis teeta* .

Family: MENISPERMACEAE

Genus: *Anamirta* Colebr.

The generic name is derived from its Tamil name 'Anai-Amyrtavalli' meaning elephant tinospora, as this genus is closely allied to Tinospora but is bigger and larger.

This monotypic genus is known to occur in Indomalayan region of the world.

Species: *Anamirta cocculus* Wt. and Arn. (*A. paniculata* Colebr.)

Description of Plant

This is a woody climber with vertically furrowed corky grey bark. The stem is about 2.5 cm across. Leaves are broadly ovate or suborbicular, glaucous, long petioled and 7.5-20.5 cm long. The flowers are green or yellow about 1.5cm in diameter and are horne onbranched drooping panicles. The drupes are reniform, transversely or obliquely ovoid, wrinkled and blackish or reddish brown with globose yellowish grey seeds.

Flowering and Fruiting

Flowering is during in summer and Fruiting: In the rainy season.

Distribution

It is found in Khasia hills, Orissa, East Bengal and in Peninsular India.

Plant Parts Used

Seeds, fruits and leaves are usuable parts of the plant.

Uses

The oil derived from its seeds is used eternally as an application in ringworm and other skin diseases. A ointment, prepared from its fruits is applied locally in the treatment of scabies. The seeds in the doses of 0.35-0.70 gm twice daily are used in night sweat of phthisis. Besides, snuff powder prepared from its dried leaves is considered as a good remedy in malarial fever.

Besides, its fruits and seeds are also used as a substitute for Hops (*Humulus lupulus*) in the manufacture of Beer.

Folk Uses

The herb is massaged over skin in various skin diseases in different parts of the country.

Genus: *Cissampelos* Linn.

The generic name has been taken from Greek words Kissus = ivy and Ampelos = a vine, referring to the habit of the plant.

The genus comprises of twining or suberect shrubs. Leaves are often peltate. Male flower is cymose; sepals are 4 or 5-6, erose, petals are connate, forming a 4-lobed cup and anthers are 4 and connate. The female flowers are racemed and crowded in the axils of leafly bracts; sepals are 2 and 2-nerved and adnate to bracts, staminodes absent and ovary is one. The drupes are ovoid with horseshoe-shaped endocarp, compressed and dorsally tubercled. The seeds are curved.

About 30 species occur in tropics. One species is found in tropical and subtropical India.

Species: *Cissampelos pareira* Linn.

Description of Plant

It is a slender, twining and climbing perennial herb. Leaves are peltate, cordate, softly pubescent on both the surfaces with 7-11 veins and 3.0-8.0 cm across. The flowers are small, unisexual, greenish and are crowded in the axils of the leafly bracts. Male flowers are greenish-yellow in axillary, fasicled, pilose cymes or panicles. Female flowers are in pendulous racemes of 6.0-10.0cm long. The fruits are drupes, ovoid, endocarp reniform and compressed with hollowed tubercled sides on the back. Seeds are 0.05-0.06 cm long and curved.

Flowering and Fruiting

Flowering is during in rainy and autumn season and Fruiting: In the winter.

Distribution

It is found distributed throughout the tropical and subtropical India.

Parts Used

Leaves and roots are used.

Uses

Its roots are antipyretic, diuretic and laxative and are used in leucorrhoea, gonorrhoea and chronic inflammation of the bladder. A decoction of the roots is given in enlargement of spleen and pounded roots are used internally in indigestion and headache. The roots are also used to lower the blood pressure. The leaf juice is used in eye problems. Externally, the plaster of its crushed leaves is applied on various skin ailments, burns, wounds and also on forehead in cold and fever. Besides, various plant parts are used in fever, epilepsy, mad convulsions, delirium febris, haematuria, gravel, bronchitis and cholera.

The extract of its roots and stem is used as central nervous systems depressant. The pills prepared from the root of plant, pepper, asafetida and ginger in 4:5:3:6 ratio mixed with honey are useful in indigestion and colic in the doses of 0.2-0.3 gm per day.

Folk Uses

Its pounded roots are applied on leucoderma and paste of leaves, flowers and roots is used for itch and small pox . The decoction of entire plant is given in urinary calculi and roots rubbed with 'ghee' are given internally in poison. The powder of the roots along with two grains of *Hordeum vulgare* is given thrice a day to cure malarial, typhoidal and general fever. The juice of the leaves mixed with powder of black pepper is given in diphtheria and the decoction of the whole plant to women for prolapus of uterus.

Preparation

Shaddharan Yog, ambashtadi churna, Gangadhar churna, Kutajashtak kwath, Kutaj-ghanvati and Majun jograj guggul.

Genus: *Tinospora* Miers

The genetic name has been taken from Greek words, Tino=to avenge, Mis, prefix meaning wrong and Cium = a column, but derivation is unexplained.

The genus comprises of climbing shrubs. Leaves are alternate. The flowers are in axillary or terminal racemes or panicles. Sepals are 6 and 2-seriate; inner larger and membranous. Petals are 6 and smaller. Male flowers: stamens 6, filaments free, tips thick. Female flowers: staminods 6 and clavate, ovaries 3 and stigmas forked. The drupes are 1-3. The seeds are grooved ventrally or curved round the intruder and with 2-lobed endocarp.

There are about 40 species found to be distributed in tropical Africa, South-East Asia, Indo-Malayan region and Australia; of which 3 species have been recorded form India.

Species: *Tinospora cordifolia* (Willd.) Miers ex Hook.f. and Thoms

Description of Plant

It is a large, glabrous and climbing succulent shrub with rocky bark. Leaves are petioled, membranous, cordate with broad sinus and 5.0-10.0 cm long blade. The flowers are small, yellow and unisexual; male flowers are fascicled and females are usually solitary. The fruits are ovoid drupes, succulent, lustrous and red when ripe. Seeds are curved.

Flowering and Fruiting

Flowering is during in summer and Fruiting: In winter.

Distribution

It is found throughout the tropical India ascending upto an altitude of 300m.

Plant Parts Used

The leaves, stem, roots and fruits are used.

Uses

Its leaves are prescribed in fevers. Its stem is known as bitter, stomachic and antipyretic. A powder of its stems is made into an infusion, which is used as alterative and aphrodisiac. Stem juice is given in fever and mixed with honey in jaundice and the decoction is used for rheumatic fever and vomiting due to excessive bile secretion. The decoction mixed with the fruits of Piper longum and honey is used in fever associated with cough in the doses of 0.35-1.05 gm per day and in combination with the stem of *Piper longum* and honey is used to control heart palpitation due to flatulency and with combination of sweet oil, is administered in elephantiasis. A kind of starch, known as 'Giloe ka sat' is prepared from the aqueous extract of its dry stems which is used as tonic and also as antidiarrheal, antidysenteric and antacid substance in the doses of 0.7-2.0 gm per day.

The roots are considered as a powerful emetic and are used for visceral obstruction. The watery extract of the same is used in leprosy. The pulverized fruits are considered as tonic and are used in jaundice and rheumatism. The fresh plant is considered more efficacious than the dried ones. Its watery extract known as 'Indian quinine' is used as febrifuge. Besides, various parts of the plant are used in general debility, dyspepsia, fever and urinary diseases. The alcoholic extract of the whole plant possesses antihepatitic and antifibrotic activities and can be used in liver disorders.

Folk Uses

A necklace of small pieces of its stem known as 'kamlanimala' is worn in some parts of India as a cure for jaundice.

That stem of the plant and roots of *Solanum surattense* are made into a paste. The pills are prepared from this paste which are used in the treatment of fever for three days by the tribals of Varanasi. Assamese of North Lakhimpur district use 2 teaspoonful of the decoction of its stem in gastric problems three times a day after meals and decoction with honey, 2 teaspoonful in night after dinner for leucorrhoea.

Preparation

Gudachyadi Kwath, Guduchi louha, Guduchayadi churna, Amritarista, Kirattiktodi kwath, Sudarshan churna, Sanjivani vati, Amritastak kwath, Phalatrikadi kwath, Guduchi taila, Pippalimuladi kwath, Panchbhadra kwath, Rasnasaptak kwath, Yograj guggulu, Kaisur guggulu, Rasnapanchak kwath, Sansmani vati, Arkharabhara, Sat giloe and Arq maul laham mako kashiwala.

Family: BERBERIDACEAE

Genus: *Berberis* Linn.

The generic name is based on its Arabian name, Berberys.

The genus comprises of spiny shrubs with yellow wood. Leaves are imparipinnate or in fascicles of unfoliolate leaflets in the axils of 3-5-partite spines; often themselves reduced to spines. Flowers are yellow and bisexual usually in racemes or fascicles, rarely solitary. The fruits are berries with few seeds.

About 450 species are found distributed in North and South America, Eurasia and North Africa. Out of which 70 species are recorded from India.

Species: *Berberis aristata* DC

Description of Plant

It is an erect and glabrous spinescent shrub attaining a height of 3-6m. Leaves are acute to obtuse, entire or toothed and ellipticobovate. Flowers are yellow and are borne in corymbose racemes. The berries are bright red and obovid or oblong-obovoid.

Flowering and Fruiting

Flowering is during in April-June and Fruiting: During August-October.

Distribution

It is found in the Himalayas from 1829 m to 3048 m and in Nilgiri hills.

Plant Parts Used

Root bark, stem and fruits are used.

Uses

The root bark is alterative and deobstruent and is used in skin disease, menorrhagia, diarrhea, jaundice and affections of the eyes. A paste of its root bark is applied externally on ulcers for healing and an extract prepared from root bark is used as a local application in affected parts of the eyelids and opthalmia. The decoction of its root bark is given in malarial fever and in enlargement of liver and spleen. The root of the shrub combined with opium, rocksalt and alum is considered an useful anti-inflammatory agent.

A decoction of its stem mixed with that of haldi (*Curcuma longa*) is recommended in gonorrhoea. Its fruits are edible and are given to children as a mild laxative.

Commercially, it is sold in the market under three name *viz.*, Daruhaldi, Rasaut and Zirisk. Daruhaldi, is the pieces of 2.5-5.0cm diameter covered with a soft corky light brown bark beneath this there is a hard layer of stony cells, forming complete coating to the stem and it is marked by longitudinal furrows corresponding to the medullary rays which are prominent and close grained and contain many stony cells. All parts of the wood are impregnated with yellow colouring matter which is freely soluble in water. Wood extract known as 'rasaut' is a dark brown extract of the consistency of opium, having a bitter and astringent taste and is readily soluble in water. Zirisk is a moist mass of small black and almost abortive fruits but some of them may contain one or two seeds with a thin roughish testa beneath which is a membraneous covering exists.

Preparation

Darbyadi kwath, Darbadi taila, Darbadi leha, Sudarshan churna, Rasanjandai churna, Punarnava mandur, Laghu vis garva taila, Chandandi vati, Khadirarista and Phalatrikadi kwath.

Family: PAPAVARACEAE

Genus: *Argemone* Linn

The generic name has been taken from Greek word Argema meaning a white spot in the eye refers, the juice of the plant which was used as a remedy in diseases of the eyes.

The genus comprises of erect glaucous herbs with yellow juice. Leaves are with spiny teeth. Flowers are large and showy. Sepals are 2 or 3 and Petals are 4-6. Stamens are many and ovary is 1-celled, style is short, stigma 3-6 lobed and ovules many. Fruits are prickly capsules which show dehiscence at the top by the short valves. Seeds are numerous.

About 10 species occur in Western and Eastern United States, Mexico and West Indies. Of these, 2 species are reported from India.

Species: *Argemone maxicana* Linn

Description of Plant

It is a robust and prickly annual herb about 45-120 cm high with yellow sap. Leaves are bristly with slightly incised undulated margins and white veins. Flowers are bright yellow and about 2.5-5.0 cm in diameter. Petals are 4-6. Fruits are 1.9–3.8 cm long capsules which dehiscent with apical pores. Seeds are numerous, globose, netted and black.

Flowering and Fruiting

Flowering is during in February–March and Fruiting: In April-May.

Distribution

It is considered a native of America, now naturalized throughout India and found upto an elevation of 1500 m; often growing along road sides.

Plant Parts Used

The seed oil, leaf-juice and roots are used.

Uses

The seeds of the herb are pounded in 'Mahua oil' and are applied to eczema and itching. Ground seeds mixed with mustard oil are also used in treatment of itching. The plant contains milky juice, which is used to cure blisters, excoriations and indolent ulcers. This is also useful as alterative in syphilis, gonorrhoea, leprosy and other cutaneous diseases. Externally, the leaf juice relieves blisters and heals excoriations, herpetic conditions and indolent ulcers. The root extract is said to posses antifungal properties. Its roots are burnt to provide heat in treatment of piles and are also given in Pan (Piper betel) for severe stomachache.

Mustard seeds are often adulterated with argemone seeds and the mustard oil obtained in such a way is probably responsible for out break of epidemic dropsy because of the presence of toxic principle in argemone oil. Internal application of argemone oil is not safe as it causes severe diarrhea and induces toxicity and it should never be used in opthalmia.

The plant is also used as a substitute for Ipecacuanha.

Folk Uses

It roots are used as antidote for snake-bite by the inhabitants of Pauri Garhwal. The Gonds and Kolams of Andhra Pradesh, apply the ashes of its burnt seeds filtered with a find cloth in ophthalmic infection. In Bhavnagar (Gujarat), yellow juice of the plant is applied externally on boils, syphilis and scorpion stings.

Family: BRASSICACEAE

Genus: *Sissymbrium* Linn.

Sisymbrium is the old Greek name of a sacred herb. In Greek mythology, Ovid advises that Venus should be worshipped with garland of Sisymbrium, myrtle and rose.

The genus comprises of glabrous or more or less hairy or hispid annual or perennial herbs. Leaves are entire or toothed. The flowers are white, yellow or purplish and are borne in lax sometimes, bracteate racemes. The pods are elongated, cylindric or compressed with unmargined and 1-seriate seeds.

About 90 species are known to occur in Eurasia, Mediterranean parts, south Africa, North America and Andes. Only 3 of them have been reported from India.

Species: *Sissymbriun irio* Linn.

Description of Plant

It is an anjual or perennial herb attaining a height upto 90cm. The stems are glabrous or slightly pubescent below. Leavs are stalked, runicate or pinnatified, lobes are auricled, remote, spreading and toothed; terminal is large, sometimes hastate. The flowers are yellow and minute with slender pedicels. The pods are about 3.5–5.0 cm long with many seeds.

Flowering and Fruiting

During November-February.

Distribution

It is found distributed in Northern India, Rajasthan and Punjab.

Plant Parts Used

Seeds are usable.

Uses

Its sees are considered stimulant, restorative and expectorant and are used in asthma. Externally, a poultice of the seeds is applied as an stimulating agent.

Folk Uses

An infusion of its leaves is given in affections of throat and chest by the native people.

Preparation

Khubkalan avaleha and Banafshadi kwath. Besides, it is used in combination with other drugs for the preparation of various types of 'majuns' in Unani yogas.

Family: CLEOMACEAE

Genus: *Cleome* Linn.

The generic name has been derived from Greek word Kleio=to enclose, referring to the floral segments.

The genus comprises of herbs. Leaves are simple or digitately 3-9-foliate. The flowers are in racemes. Sepals are 4 and spreading. Petals are 4. Stamens are 4-many and sessile on the disk. Ovary is sessile or sub-sessile. Style is short or absent. Ovules are many. The fruit is a oblong or linera capsule. The seeds are reniform.

About 150 species occur in tropics and sub-tropics; 14 species are found in India, mostly in drier region.

Species: *Cleome gynandra* Linn.
(*Gynandropsis pentaphylla* (Linn.) DC.; *G. Gynandra Briq.*)

Description of Plant

It is an erect glandular, pubescent and showy herb about 30-90 cm in height. Leaves are digitately 3-5 foliate, leaflets are subsessile, unequal, elliptic-oblong or obovate. The flowers are yellow or white and are borne in corymbose racemes. The capsules are cylindric, striated and 5.0-10.0 cm long. Seeds are reniform, rugose, 1.2-1.3 cm across and black or brown.

Flowering and Fruiting

During August-November.

Distribution

It is a common weed found abundantly throughout the warmer parts of India.

Plant Parts Used

The leaves, seeds and roots are of medicinal use.

Uses

The bruised leaves act as rubefacient and vesicant and are applied in boils and scorpion-stings. The juice of leaves is beneficial in otalgia (earache) and convulsions. The leaves are also used as a remedy for muscular pain, headache and intestinal wounds. It seeds are anthelmintic, sudorific, carminative and antispasmodic and as poultice are applied to sores having maggots and as infusion are given in cough and in powder form are used in piles. The decoction of roots is also useful in fever.

Its seeds are used in Indian medicine in the same way as the mustard seeds.

Folk Uses

In Andhra Pradesh, the fresh leaves are eaten as vegetable to remove intestinal worms.

Species: *Cleome icosandra* Linn. (*B.* viscose Linn.)

Description of Plant

It is viscid pubescent annual herb wth soft branches. Leaves are digitately compound; leaflets are 3-5, unequal, nearly sessile, ovate and entire, The flowers are white and about 1.2cm long. Petals are imbricate. Stamens arise from a disk. Ovary is sessile with short style. Fruit is a slender capsule which is about 2.5-7.5 cm long. The seeds are small, granular and black or dark brown.

Flowering and Fruiting

During July-December.

Distribution

The plant is found distributed throughout the greater part of India upto 1524 m.

Plant Parts Used

The seeds and leaves are generally used.

Uses

Its seeds as carminative, anthelmintic, rubefacient and vesicant. As anthelmintic, about 0.25-1.25 gm powder of its seeds is given twice a day for two days followed by administration of one dose of castor oil as purgative to the children and these are also said to be useful in splenn and liver diseases and in syphilis. The seeds are ground into a paste and are applied as a poultice in chronic painful joints and their extract is employed to kill maggots in unhealthy sores.

The leaves are known as rubefacient, sudorific and vesicant. The fresh leaf-juice diluted with water is given in small quantity as sudorific in fever and mixed with honey, sesamum oil and rock salt is dropped in ear in inflammation, pain and pus formation in the middle ear. The fresh leaves after cooking are given as vegetable to patients suffering from oedema.

Its seeds are regarded as an efficient substitute for mustard seeds.

Family: VIOLACEAEAE

Genus: *Hybanthus Jacq.* Nom. Cons.

The generic name is derived from Greek words Hybes=hump-backed and Anthos=flower; referring to the nature of its flowers.

The genus comprises of herbs and undershurbs. Leaves are alternate, rarely opposite. Flowers are orange or purple coloured and axillary. Sepals are 5, subequal;

petals 5, of which lower larger. Anthers are connate or free, 2 or 4 of them gibbous or spurred. The capsules are 3-valved. Seeds are globose with crustaceous testa.

There are 150 species found in tropical and subtropical parts of the world. Only 2 species have been recorded from warmer parts of India.

Species: *Hybanthus enneaspermus* (Linn.) f. Muell (*Ionidium suffriticosum* (Linn.)R.& S.; *I. enneaspermum* (Linn.) Vent.; *Viola enneasperma* Linn.; *V. suffriticosa Linn.*)

Description of Plant

This is a small diffuse annual or perennial herb with a woody base. Leaves are sub-sessile, linera-lanceolate and 3.5-5.0cm long. The flowers are red and solitary. The seeds are longitudinarry striated and yellowish-white.

Flowering and Fruiting

During October-December.

Distribution

This is found throughout the warmer parts of India from Uttar Pradesh southward to the Deccan Peninsula.

Plant Parts Used

Whole plant is used.

Uses

The plant is used as a tonic and diuretic and its leaves and tender stalks as demulcent. Besides, its roots are given in bowel complaints.

Folk Uses

The powder of roots is given in bowel complaints in different parts of the country.

Genus: *Viola* Linn.

The generic name Viola is the latin name for Pancy and Violets; according to Greek mythology, Gr. Vion, violet spring to life as food for Io, one of Jupiter's mistresses whom he was obilized to transform into a cow.

The genus comprises of herbs, rarely of shrubs. The flowers are often dimorpohic. Sepals are produced at the base and petals erect or spreading; the largest spurred or saccate at the base. Anthers are connate, connectives of 2 lower often spurred at the base. Styles are clavate or truncate, top straight or oblique and stigma obtuse, lobed or copular. The capsules are 3-valved with numerous ovoid or globose seeds.

There are 500 species of cosmopolitan distribution, chiefly in North temperate regions but many in Andine; 15 of which have been recorded from India.

Species: *Viola odorata* Linn.

Description of Plant

This is a small glabrate or pubescent perennial herb with short tufted root stock. The root is as thick as a crow's quill, crooked, with a number of radicals having a spongy bark. Leaves are all radical, ovate, 1.2-2.5 cm in diameter and toothed. The flowers are violet coloured and aromatic and become brown or brownish-yellow when old.

Flowering and Fruiting

During April-September.

Distribution

It is found distributed in Kashmir Himalayas at an altitude between 1524-1828m.

Plant Parts Used

Panchang (*i.e.*, all five parts *viz.*, roots, stem, leaves, flowers and fruits) and flowers, commonly known as 'Gule banfshah' are used.

Uses

Panchang, commonly known as 'banafshah' is aperient, antipyreticm, cooling, demulcent, diaphoretic, emollient, expectorant, febrifuge and purgative in action. This is used in piles, asthma, constipation, cough, fever, headache and in skin diseases. Its flowers are considered emollient and demulcent and are used in billiouness, lung problems, prolapse of the rectum and uterus and in restraining suppuranation besides, cough, kidney problems and liver affections.

Folk Uses

A syrup is made from its petals which is given as a remedy for coughs and tighness of the chest to children. In Kashmir valley, people take its dried root powder with hot water at bed time as laxative.

Preparation

Banafshadi kwath, Sharbat banafshah, Kamira banafshah, Gulkand banafshad and Rogan banafshah.

Family: POLYGALACEAE

Genus: *Polygala* Linn.

The generic name has been derived from Greek word, Poly meaning much and Gala meaning milk; referring to its supposed reputation as a good cattle fodder.

The genus comprises of herbs, rarely of shrubs. Leaves are alternate. Sepals are usually persistent; inner 2 larger and petaloid. Petals are 3 and united at base into staminal sheath; inferior keel-shaped and usually crested. Stamens are 8; anthers open by pore. Ovary is 2-celled with one pendulous ovule in each cell. The capsules are 2-celled and 2-seeded.

There are 500 species of cosmopolitan distribution and also found Newzealand, Polynesia and arctic zone; 20 of which have been recorded from India.

Species: *Polygala chinensis* Linn. (*P. telephioides* Willd.)

Description of Plant

A glabrous or pubescent annual erect or diffuse, procumbent herb about 7.5-22.5

cm in height. Leaves are linera-oblong or oblong-lanceolate and 0.8-5.0 cm long. The flowers are light yellow and are borne in short, extra-axillary, few-flowered racemes. The capsules are ciliate, oblong and obliquely obcordate at apex. Seeds are silky.

Flowering and Fruiting

Flowering: During rainy season and *Fruiting*: In August-September.

Distribution

It is commonly distributed throughout the country upto 1524 m.

Plant Parts Used

Leaves and roots are used.

Uses

An infusion of its leaves is recommended in asthma, chronic bronchitis and catarrhal affections. Its roots are considered as expectorant and febrifuge and are used in cough, dizziness and dypsnea.

Folk Uses

An infusion of the herb is given in catarrhal affection. Besides, it is also eaten in western parts of India.

Family: FLACOURTIACAEE

Genus: *Gynocardia* R.Br.

The generic name is derived from Greek words, Gyne=female and Kardia=heart; referring to its heart shaped ovary.

This monotypic genus is known to be distributed in Assam and Myanmar (Burma).

Species: *Gynocardia odorata* R. Br.

Description of Plant

This is an evergreen, glabrous and dioecious tree attaining a height of 9.0-12.0 m and a girth of 0.9-1.8 m with a clear bole upto 6.0 m. The branches are slightly drooping and slender. Bark is greenish-grey and wood is hard, tough and pale-yellow to light brown. Leaves are bifarious, ovate-oblong and about 15.0-25.0 cm long. Flowers are pale-yellow, fragrant and are borne in fascicles. The fruits are globose about 7.5-12.5 cm in diameter with hard rind. Seeds are obovoid or oblong and about 2.5cm long and are embedded in gelatinous pulp.

Flowering and Fruiting

During summer-rainy season.

Distribution

This is found in the forests of Sub-Himalayan tracts from Sikkim and Khasi hills in India upto 1219 m.

Plant Parts Used

Seed oil is used.

Uses

An oil obtained from its seeds is used in leprosy and other cutaneous affections and is applied externally on chest as expectorant and internally in scrofula, scabies, chronic rheumatism and in secondary syphilis, Besides, pulp of its fruits is also used as fish poison.

Earlier, the seed-oil obtained from this plant and other plant species *viz.*, *Hydnocarpus kurzii* and *Hydnocarpus laurifolia* was known as 'chaulmogra oil'., Now, it has been practically proved that the composition of seed oil of this plant species is chemically different from that of latter two plant species. But still the seed oil obtained from this species is used as a substitute for real 'chaulmogra oil' obtained from Hydnocarpus laurifolia.

Folk Uses

Chaulmogra oil is applied externally on various type of skin affections in eastern parts of the country.

Genus: *Hydnocarpus* Gaertn.

The generic name is derived from Greek words Kydn=a tuber and Karpso=a fruit, indicating the shape of the fruits.

The genus comprises of trees. Leaves are serrate or entire with deciduous stipules. The flowers are few together or solitary or fasciculate in the axills of fallen leaves on short thick woody branchlets. Sepals are 5, imbricated in bud and petals are 5 with scales at the base. Male flowers: stamens 5-8, filaments free and glabrous. Female flowers: ovary one-celled with many ovules. The berries are globose with numerous seeds embedded in pulp of the fruits.

About 40 species are found in Indomalayan region; of which 7 have been reported from India.

Species: *Hydnocarpus kurzii* (King) Warb. (*H. heterophyllus* Kurz; *Traktogenos kurzii* King)

Description of Plant

This is an evergreen tree upto 15m in height with tall trunk and narrow crown of hanging branches. Leaves are thinly coriaceous, entire, 17.5-20.5 cm and lanceolate or oblong-lanceolate; petiole geniculate at upper end. The flowers are mostly dioecious and pale yellow and are borne in axillary cyme. Sepals are 4 and petals 8, broadly ovate and ciliate; each with a flat and fleshy pubescent gland at the base. The fruits are choclate-brown, globose and 6.5-7.5 cm across with numerous seeds.

Flowering and Fruiting

Flowering: During May-June and Fruiting: In November-December.

Distribution

It is found in the forests of Assam and Chittagong, frequentally forming the gregarious patches in the forests.

Plant Parts Used

Sees and bark of the plant are used.

Uses

An oil obtained from the seeds of the plants which is known as 'chaulmogra oil' is used in leprosy and other skin diseases both externally and internally. The bark of the plant is said to be used as a febrifuge and an infusion made format has the odour of the essential oil of bitter almonds. Besides, its fruits are also used as a fish poison.

Its oil is adulterated with that of *Gynocardia odorata*.

Folk Uses

'Chaulmogra oil', is applied on different types of skin infections in eastern parts of India. Hill tribals in Sikkim use its seeds as a food after boiling these with water.

Species: *Hydnocarpus laurifolia* (Dennst.) Sleumer (*H. wighitana* Blume; *Munnicksia laurifolia* Dennst)

Description of Plant

This is a tall evergreen dioecious tree upto 15 m or more in height with fluted stem and rough brown bark. Leaves are 10.5-25.5 cm long, membraneous or thinly coriaceous, oblong, ovate or elliptic and somewhat serrate. The flowers are small, white about 1.2-2.5 cm in diameter and are borne solitary or in fascicles. The fruits are globlose, about 5.0-10.0 cm in diameter with 15-20 striate, subovoid and obtusely angular seeds.

Flowering and Fruiting

Flowering: In January-May and Fruiting: Almost throughout the year.

Distribution

It is met in tropical forests along the western ghats, the Konkan southwards and below the ghats in Kanara and Malabar in damp conditions, especially near water. Common in Travancore upto 609 m.

Plant Parts Used

Seeds are used.

Uses

Seed oil, known as 'chaulmogra oil' is used in leprosy and other skin diseases and 15 drops gradually increased upto 30 drops of its seed-oil are recommended to be injected intra-muscularly or intravenously in the form of ethyl esters or salts of chaulmogric and hydnocarpic acids in leprosy and seed oil mixed with lime water is applied in sprains in rheumatic joints. A paste of its seeds mixed with sulphur, camphor, lime juice and seed oil of *Jatropha curcus* is applied externally in wounds and ulcers.

Besides, seeds are also used as a fish poison.

Folk Uses

Seed oil is applied externally in various types of skin diseases in eastern parts of the country.

Preparation

Chaulmogra ointment is prepared by mixing one part of the oil with four parts of vaselin which is applied in various skin diseases.

Family: PODOPHYLLACEAE

Genus: *Podophyllum* Linn.

The generic name has been derived from Greek works, Podos=a foot and Phyllon=a leaf, referring to theshape of leaves.

The genus comprises of herbs with poisonous roots and leaves. Leaves are palmately lobed. The flowers are white or rose coloured and large. Sepals are 3-6 and petals 6-9, rarely 4. Stamens are as many or twice of the petals. Ovary is simple with any ovules. The berries are with numerous obovoided seeds.

About 10 species are found distributed in Himalayas and East Asia. Only one species has been reported from India.

Species: *Podophyllum hexandrum* Royle (*P. emodi* Wall. Ex. Hook. F. and Thomas)

Description of Plant

This is an erect glabrous and succulent herb with creeping perennial rhizome and bearing countless roots and attaining a height of 30-60cm. Leavaes are 2 or 3, peltate, palmate and orbicular-reinform with lobed segments. The flowers are solitary, white or pink and cup-shaped. The berries are elliptic or oblong, 2.5-5.0 cm across and orange or red coloured with numerous seeds embedded in the pulp.

Flowering and Fruiting

Flowering: During June-July and Fruiting: In August-September.

Distribution

It is found in the inner range of the Himalayas from Kashmir to Sikkim at altitude between 3000-4200 m.

Plant Parts Used

Whole plant, roots and rhizomes are used.

Medicinal Uses

The resin, obtained from the herb is used as a hepatic stimulant and also in cold and bilary fever and constipation. It is also used as an anticancerous drug. Besides, its roots and rhizomes are used as cholagogue, purgative and hepatic stimulant.

Folk Uses

The fruits of the plants are eaten by the natives in most of the parts of Himalaya. Gangwal tribes in Garhwal Himalayas, take roots for urticaria, dyspepsia and also to exert destructive action on cancerous tissues (Nagi and Pant, 1990).

Family: CLUSIACEAE

Genus: *Calophyllum* Linn.

The generic name is derived from Greek words Kalos= beautiful and Phyllon=leaf, referring to the beautiful leaves of the plants.

The genus comprises of trees with shining and coriaceous leaves. Wood is reddish, Flowers are in axillary or terminal panicles. Sepals are 4; petals absent or 4-8, stamens indefinite with filiform filaments and ovary is 1-celled usually with one ovule but may contains few ovules. The drupes are with thick and fleshy cotyledons.

About 112 species are found distributed in the world; 4 in tropical America and West Indies; 8 in Madagascar and Mauritius and 100 inaIndomalaya region, Indochina, Pacific and tropical Australia. Of which only 14 have been recorded from India.

Species: *Calophyllum inophyllum* Linn.

Description of Plant

This is a medium-sized evergreen tree reaching a height of 6.0-7.5 m with reddish-brown or reddish-white wood. Leaves are elliptic, elliptic-lanceolate or obovate angles, 10.2-20.5 x 7.5-10.9 cm; upper surface deep green and lustrous. The flowers are white and scented and are borne in axillary racemes. The fruits are subdrupaceous, rotund, smooth, yellows and about 1.3-3.5 cm across.

Flowering and Fruiting

Flowering: In July-August and Fruiting: During September-October.

Distribution

It is found on the sea-shores in India, particularly in Maharashtra, Karnataka, Orissa and Andamans; also cultivated in the plains.

Plant Parts Used

Seeds and bark are used.

Uses

The seed-oil is applied externally in rheumatism and in scabies and other skin diseases. The bark of the plant is considered astringent and its decoction is given in internal haemorrhage. Besides, resin obtained from its bark is also used as emetic and purgative.

Its flower buds are used as a substitute and adulterant of the drug Iron wood, Nagkeshra.

Folk Uses

The leaves of the plant are soaked in water for sometime and these are applied to inflamed eyes. Besides, gum mixed with stripes of bark and leaves is steeped in water and the oil which rises to the surface, is applied in sore of eyes.

Genus: *Garcinia* Linn.

The generic name is given is the honour of Lauren Garcin (1683-1751), a Surgeon and Naturalis, who at the suggestion of H. Boerhaave, Professor of Botany at Leyden, undertook three voyges to the East Indies in the services of the Dutch East India company.

The genus comprises of trees, rarely of shrubs with yellow or white resinous juice in bark. Leaves are evergreen rarely stipulate. The flowers are solitary or cymose, sepals 4 or 5 and petals 4 or 5, imbridcate. Main flowers: stamens many and are collected in a ring or in several bundles which surround a rudimentary ovary, rarely free. Female or bisexual flowers: 2-12 celled ovary crowned by large peltate stigma with one ovule in each cell. The fruits are coriaceous with numerous seeds embedded in pulp.

About 400 species are found in tropical regions, especially in Asia and in South Africa. Out of which 22 species have been reported form India.

Species: *Garcinia indica* Choisy

Description of Plant

This is a slender tree with drooping branches. Leaves are red when young, 5.0-10.0 cm long, thickly membraneous and lanceolate or oblanceolate. The flowers are small; males in terminal 3-7- flowered often pedunculate cymes and females are solitary with 5-7-celled ovary. The fruits are globose, purple coloured and 3.7cm across with seeds embedded in red acid pulp.

Flowering and Fruiting

Flowering: In November-February and Fruiting: During April-May.

Distribution

It is found in the forests of Andhra Pradesh, Maharashtra and Chennai.

Plant Parts Used

Bark, fruits and seeds are used.

Uses

The bark is used as astringent and the fruits are antiscorbutic, cooling, cholagogue, emollient and demulcent and are eaten as cardiotonic. Besides, its seeds yield an edible fat called 'Kokam butter' which is also considered astringent, demulcent and emollient and is used in various cutaneous affections.

The oil of its seeds is sued in place of Coconut oil and in ointment preparation, the butter of the tree is employed as substitute for Cod liver oil.

Folk Uses

Its young leaves tied up tin a plantain leaf and steeped in hot ashes and rubbed in cold milk are given as a remedy for dysentery. Besides, the juice of its fruits is made into a syrup which is taken as a cooling drink in dysenteric fever.

Species: *Garcinia Morella* (Gaertn.) Desr.

Description of Plant

It is a small or medium-sized evergreen tree with 9.0-15.0 m height. The wood is hard, molted and yellow. Leaves are thinly coriaceous, sagittate or ovate-lanceolate and 10.0-15.0 cm long. The flowers are unisexual, greenish-white, sessile in the axills of fallen leaves; females larger than male. The fruits are globose, 4-lobed, 2-celled and about 1.8cm across. Seeds are 4, ovoid or reniform, slightly compressed and dark brown.

Flowering and Fruiting

During October-April.

Distribution

It is found in the forests of Assam, Khasi and Jaintia hills,West Bengal and in western ghats upto 900m.

Plant Parts Used

Gum-resin, obtained from stem-bark is usable.

Uses

The gum-resin is strong purgative and anthelmintic and is used in amenorrhoea, dropsy and obstinate constipation and in ascites. Externally, it is applied to relieve pain and swellings.

Folk Uses

The resionous exudates obtained from its stem bark is applied externally in pains and swellings by the native people.

Species: *Garcinia xanthochymus* Hook. F. and T. and.

Description of Plant

This is a moderate-sized bushy evergreen tree with dense dark shining foliage and spreading angular branches arising in tiers. Leaves are thickly coriaceous, oblong or elliptic-oblong, acute, bright green and 25.5-38.0 x 3.5-15.0cm. The flowers are axillary, thick, tough and white coloured; male having stamens in 5 broad bundles of 3-5 on a fleshy lobed disk and bisexual with 5-celled ovary. Fruits are dark yellow, globular and about 5,.0-7.5 cm in diameter with a conspicuous beak somewhat on one side. Seeds are 1-4 and oblong.

Flowering and Fruiting

Flowering: During spring and Fruiting: In summer season.

Distribution

This is reported to be found wild in the lower hill forests of Eastern Himalayas, West Bengal. Western ghats, Nilgiris and Andamans upto 1200 m.

Plant Parts Uses

Fruits, seeds and bark of the plant are usable.

Uses

This fruits are very acidic and sweetish (Nandkarni, 1954) and a syrup made from their juice, is used as a cooling drink in biliouness. An oil is obtained from its seeds, which is considered as antiscorbutic and is used as a potent remedy for dysentery and diarrhea. Besides, the bark of the plant is known as astringent and is used successfully in dysentery.

Folk Uses

The ripe fruits are eaten and a 'sharbat' is made from their juice with a little rock salt, pepper, ginger, cumin and sugar, which is taken as a cooling drink in biliousness etc. in the different parts of country. Tribal people of Tripura, chew semi-dried pulp of its mature fruits in dysentery.

Genus: *Mesua* Linn.

The generic name has been assigned in the honour of Johannes Mesu (77-857) of Damascus, a celebrated Arabian Physician.

The genus comprises of trees or shrubs. Leaves are closely and finely penninerved. Flowers are hermaphrodite or polygamous, axillary, solitary and large. Sepals are 4. Petal are 4. Stamens are indefinite; filaments free or connate at the base and anthers erect, oblong and 2-celled. Ovary is 2-celled with 2 ovules in each cell. The fruits are fleshy and sub-woody, dehiscing by 4 valves. Seeds are 4 with fragile testa.

About, 40 species are found distributed in Indomalayan region. Only one of which has been reported from India.

Species: *Mesua ferrea* Linn.

Description of Plant

It is a medium-sized or large evergreen tree with short trunk, often buttressd at the base and dark red heartwood. Leaves are 5.0-15.0 x 3.5-4.5 cm, coriaceous and lanceolate; upper side shining and lower covered with a white waxy powder. Flowers are white, large and solitary or in clusters and fragrant. The fruits are ovoid,2.5-5.0 cm long with persistent calyx. Seeds are 1-4, shining, dark brown with oily and fleshy cotyledons.

Flowering and Fruiting

Flowering: In February–April and Fruiting: During September-October.

Distribution

It is found in the mountains of Eastern Himalayas and East Bengal, Assam and Eastern and Western ghats as well as in Andamans ascending upto 1500m.

Plant Parts Used

Flowers, sees, fruits, leaves and bark are the parts, used medicinally.

Uses

Its flowers and stamens form a major drug Nagapushpa and Nagkesar in indigenous system of medicine. There are considered astringent, stomachic and expectorant and are used in cough and their paste made in ghee (butte fat) is applied in bleeding piles and burning of the feet and buds in dysentery. Its seed oil is used in rheumatism and cutaneous affections. The unripe fruits of the plant are aromatic and sudofiric while ripe ones are astringent and both are used for gastric troubles. The bark of the plant is also aromatic and astringent and its decoction combined with ginger is used a sudorific. Leaves as poultice are applied on forehead in severe colds.

Folk Uses

Its dried flowers are powdered and after mixing with ghee or their paste with butter and sugar is given in given in bleeding piles and dysentery indifferent parts of India. About 25 gm seeds of *'methi'* (*Trigonella foenum-graceum*), 50 gm wood of *'beeja'* (Pterocarpus marsupium) and 10 gm leaves of 'tejpat' (*Cinnamomum tamala*) are mixed and ground to make powder; of which 20 gm is given to women twice daily for 5 days with rice water or cow's milk to clear left over blood after mensuration or delivery in Jharkhand. Besides, the tribal people in Tripur, take a spoonful of its flower paste mixed with honey twice daily regularly in bleeding piles till the disease is cured.

Preparation

It is much used a flavouring agent in various ayurvedic preparation like, Chavyanprash, Amritarista, Lavanbhaskar churna, Pippalasav, Kanakasav and Khadirarista.

Family: TAMARICACEAE

Genus: *Tamarix* Linn.

The generic name has been derived from Latin name Tamaricis = a tamarisk plant, which grow on Tamaris river on the border of Pyreness.

The genus comprises of shrubs or small trees. Leaves are alternate, scale like and amplexicaul or sheating. The flowers are regular and bisexual in simple or panicled spikes or racemes. Sepals and petals are usually 5, imbricate and free or connate below. The stamens are as many as or twice as the petals and are inserted on a glandular disk which is more or less lobed. Ovary is of 3-5 carpels. The capsules are dehiscing into 3-5 valves. Seeds are hairy or winged.

About 54 species are known to occur in Europe, Africa and Asia. Out of which 6 species have been recorded from India.

Species: *Tamarix dioica* Roxb.

Description of Plant

it is a gregarious shrub with reticulately cracked bark reaching a height upto 2 m. Leaves are minute, closely adpressed, acuminate, obliquely truncate and green with white margins. Flowers are light rose coloured and dioecious in dense peduncled spikes. The fruits are conical or oblong capsules which are about 1.2 cm long. Some abnormal growths are also found on the plants, known as galls and in ayurveda by the name *'Choti ami'*. These are caused by some insects and are globular in shape.

Flowering and Fruiting

Flowering: During May-June and Fruiting: In cold season.

Distribution

It grows throughout Northern India along the Ganges and near the sea-coast in Tamil nadu ascending upto 750 m.

Plant Parts Used

Twigs and galls are used.

Uses

Its twigs and galls are used as astringent. The galls are also used as adulterant and substitute for galls of *T. troupii*.

Folk Uses

The fumigations of the leaves are taken in wound and piles by the inhabitants of northern India.

Species: *Tamarix troupii* Hole (*T. gallica auct.* Non Linn.)

Description of Plant

It is a gregarious bushy shrub attaining a height upto 9 m. Bark is brownish, smooth when young and rough when mature. The wood is hard and tough, white and tinged with red usually. Leaves are minute, persistent, at first imbricated, afterwards distant, not sheathing, subulate and usually green. The flowers are very small, numerous, pink or white and are borne in slender panicled racemes. Capsules are conical, somewhat trigonous, tapering pale- pink coloured and about 0.6 cm long. Seeds are with a plume of white hairs.

Flowering and Fruiting

Flowering: During July-August and Fruiting: In winter.

Distribution

This is found in Punjab, Uttar Pradesh, Jharkhand and West Bengal, particularly along marshy land and river sides.

Plant Parts Used

Galls and manna, gummy exudation of the plant are used.

Uses

The galls formed on the leaves of the plant, are considered astringent and are given internally in dysentery and diarrhea. Pulverised galls mixed with vaselline are successfully applied in piles and fissure and their decoction is applied to foul and sloughing ulcers and infusion as a gargle for sores throat. Manna, an exudates gum of the plant is sued as a laxative and expectorant.

Folk Uses

Its pulverized galls are mixed with vaselline to form a paste, which is applied in piles and anal fissure by the native people. A decoction of its root nodules is given in sprue and diarrhea by the Rabari people in Barda Hills area in Gujarat.

Family: DIPTEROCARPACEAE

Genus: *Dipterocarpus* C.F. Gaertn

The generic name has been derived from Greek words, Diptero = two winged and Korpos = fruit, referring to the nature of the fruits.

The genus comprises of tall trees. Leaves are renewed at the end of hot season, when the old leaves fall. Flowers are large, pink in short fewfid axillary panicles. Calyx tube is with 5 segments, 2 of which are larger. Ovary is generally hairy, continued into a conical fleshy stylopodium. The fruits are globose or ovoid, smooth or with 5 ribs or wings.

There are 76 species found distributed in Sri Lanka and India to West Malaysia and Bali. 10 of which have been reported from India, especially from Assam and the Andamans.

Species: *Diptercarpus turbinatus* Gaertn. F.

Description of Plant

It is a lofty tree, attaining a heigh5t of 35-45 m. Wood is dull red or reddish brown, coarse textured. Young branches are cylindric and canescent. Leaves are ovate or ovate-lanceolate; blade 1.25-30.0 cm long and petiole about half of blade. Flowers are borne on 5.0 cm long pedicels; calyx-tube in fruits neither ribbed nor winged.

Flowering and Fruiting

Flowering: During November-December and Fruiting: In April-May.

Distribution

This is found in the semi-evergreen or evergreen moist tropical forests of Assam, Tripura and Andamans.

Plant Parts Used

Oleo-resin, obtained from the stem in used.

Uses

Oleo-resin, a balsamic exudation from the trunk commercially known as 'Garjan balsam' is considered diuretic and is used in tuberculoid leprosy, psoriasis, indolent ulcers and gonorrhoea. Balsam in the form of an emulsionor ointment mad ewith three parts of lime water to one of the oil, is rubbed over affected parts twice a day for about 2 hours daily in leprosy and internally is given in a mixture containing 80 gm oil and mucilage with about 320 gm lime water twice daily.

The oleo-resin tapped from the trunk of the plant is used as an adulterant to Capaiba balsm (*Copaifera* species) and as a substitute for Kapur (*Dryobalanops* species).

Family: MALVACEAE

Genus: *Abutilon* Mill

The generic name has been derived from Abutilun, Arabic name for the plant.

The genus comprises of tomentose herbs or shrubs. Leaves are usually cordate and long petioled. Flowers are solitary on axillary oeduncles, jointed near the top and are yellow or orange. The bracteoles are absent. Calyx in 5-cleft. Corolla is of 5 imbricate petals, connate below and adnate to the staminal tube which is divided at the apex into numerous antheriferous filaments. Carpels are 5 to many and styles as many as the carpels. Ripe carpels are found separating from the short central axis and are dehiscent. Seeds are reniform; upper ascending and lower descending.

About 100 species occur in tropics and subtropics of the world, however, Babu (1977), mentioned 150 species; Out of which 11 species are reported from India.

Species: *Abutilon indicum* (Linn.) Sweet (*A. asiaticum* (Linn.) Sweet; *Sida guineenis* Schumach)

Description of Plant

It is a perennial shrub, covered with a minutely hoary tomentum and upto 3 m in height. Stems are round and frequently tinged with purple. Leaves are broadly ovate, cordate, acuminate, irregularly and coarsely toothed and 1.7-2.5 cm long. Flowers are solitary on joined peduncles, orange yellow or yellow and about 2.5cm in diameter. Capsules are hispid, longer than the calyx and with a distinct, small and acute point. Seeds are 3-5, reniform, tubercled or minutely stellate-hairy and black or dark brown.

Flowering and Fruiting

Throughout the year, chiefly during August-December.

Distribution

It is found as a weed in the Sub-Himalayan tracts and hills upto 1200m, in hotter parts of India and throughout the tropics.

Plant Parts Used

The seeds, leaves, bark, flowers and roots are used.

Uses

The seeds are known as laxative and expectorant and are used in cough, gonorrhoea, gleet and chronic cystitis. Leaves are demulcent and are applied to boils and ulcers and as a fomentation to painful parts of the body. These are cooked and eaten in bleeding piles and their decoction is used in toothache and tender gums and is given also for enema and vaginal infections. The bark is used as astringent and diuretic. The flowers are applied to boils and ulcers and their powder is eaten in ghee for blood vomiting and cough. The roots are nervine tonic and are used in piles and leucoderma and also in strangury, haematuria, stones of bladder and as a wash in eye disease. The powdered ones are used in cough and leprosy.

Folk Uses

The leaves are used as a poultice and their juice is given to children in spleen and liver enlargement. Leaf-paste mixed with water is taken orally about 150 ml twice a day in stomachache Its seven raw leaves are eaten to check diabetes for seven days by the tribals in Western Maharashtra.

Preparation

Bala taila, Chavyanprash, Atibalaghrit, Mahanarayan taila and Mahavishgarbh tails.

Genus: *Malvastrum* A. Gray

The generic name has been derived from Latin word Malva=mallow and Aster, suffix to noun stems to form diminutives, indicating its similarity to the genus Malva.

The genus comprises of herbs or undershrubs. Leaves are entire or divided. The flowers are in axillary or terminal inflorescence. The bracteoles are 3 and narrow. Calyx is 5-parted. Petals are longer than the sepals. The staminal tube is antheriferous to the summit. Ovary is 5-more celled. The fruits are composed of ripe carpels separating from a short torus, indehiscent and 1-seeded. Seeds are ascending.

According to Airy Shaw (1973), its 12 species are known to occur in tropical and substropical America; 3 species are found in India.

Species: *Malvastrum coromandelianum* (Linn.) Garcke (*M. tricuspidatum* (R.Br.) A. Gray; *M. coromandeliana* Linn.)

Description of Plant

The plant is a decumbent-ascending or erect undershrub about 6-90 cm high. Leaves are 2.0-10.0x1.5-5.0 cm, ovate-lanceolate to oblong with a subcordate-truncate or cuneate base, acute or obtuse; dentate-serrae and strigose with simple and appressed hairs. Flowers are yellow and mostly solitary in the axils on peduncles shorter than the petioles. Bracteoles are 3 and linera. Calyx is campanulate and 5-lobed. Petals are obliquely obcordate and as long as the calyx. Carpels are 1-12 and reniform. Mericarps are with dorsal sharp edges, laterally rugose, hairy dorsally at the top and 3-aristate.

Flowering and Fruiting

Almost throughout the year, chiefly during summer.

Distribution

It is a native of America, now has been naturalized and found throughout the India.

Plant Parts Used

Whole plant, leaves and its flowers are used.

Uses

The whole plant is considered as emollient, resolvent and bechic and its decoction is given in dysentery. Its leaves are applied on wounds and inflamed sores as a cooling and healing salve. Besides, flowers are used as pectoral and diaphoretic (Ambasta, 1986).

Folk Uses

The natives of the country apply its bruised leaves on wounds for early healing.

Genus: Pavonia Cav.

The generic name has been assigned in the honour of Don Antonio Pavon (1754-1840), Spanish botanist and Plant explorer.

The genus comprises of herbs or undershrubs. Flowers are axillary, solitary or clustered at the end of the branches. Sepals are 5, connate at base; petals 5, connate below or adnate to the tube of the stamens. Ovary is 5-celled with single ovule in each cell.

About 200 species are found distributed in tropical and subtropical parts of the world. Only 7 of them have been reported from India.

Species: *Pavonia odorata* Willd.

Description of Plant

This is an erect branching and pubesecent annual herb with root having a musk-like odour. Leaves are 2.5-12.5 cm long, cordateovate, shallowly 3-5 lobed, dendate or lower ones entire. Flowers are pink or white and fragrant and are borne axillary. Fruits are 2-chambered; each with single seed. The seeds are brown and oily.

Flowering and Fruiting

During October-December.

Distribution

The herb is known to be distributed in waste places in the Deccan, Rajasthan, Uttar Pradesh, Madhya Pradesh, Orissa and West Bengal.

Plant Parts Used

Root are used medicinally.

Uses

Its roots are astringent, cooling, demulcent, carminative and febrifuge and are recommended in dysentery and intestinal haemorrhage. Besides, the entire herb is bruised well with ghee and as poultice, is applied to erysipelas and its paste with dehusked rice water is used as antiemetic.

Folk Uses

Its fruits are eaten much by the people and leaves are rubbed over body in rheumatism.

Genus: *Sida* Linn.

The generic name has been adopted by Carrolus Linnaeus from classical authors.

The genus comprises of herbs or undershrubs with stellate pubescence. Leaves are simple and often lobed. The flowers are pedicellate, axillary, solitary or clustered; the pedicel is jointed. The bracteoles are absent. Sepals are 5, valvate and connate below into a broad tube. Corolla are 5, free above, yellow or white and connate below and adnate to staminal tube. The staminal tube is divided at the summit into numerous anther bearing filaments. The carpels are 5-10, 1-ovuled and styles are as many as carpels. Ripe carpels are separating from the axis and are generally 2-awned at the apex. Seed is solitary and pendulous.

About its 200 species occur in warmer parts of the world, especially in America, 9 species are found in India, mostly in hotter parts.

Species: *Sida acuta Burm.* F. ssp. *acuta* Borssum (*S. carpinifolia* sensu Masters)

Description of Plant

It is an erect, glabrate or thinly stellate hairy, perennial undershrub, about 60-90 cm high. Leaves are lanceolate-oblong with a subcordate or roundedbase, subacute, serrate-dendate and 2.5-5.0 cm long. Flowers are yellow; calyx lobes ovate-triangular, sharply acute or acuminate; petals obovate-cuneate. The capsules are about 0.5 cm across and glabrous.

Flowering and Fruiting

During September-March.

Distribution

It is found in waste places throughout the hotter parts of India.

Plant Parts Used

Roots and leaves are used.

Uses

The roots of the plant are astringent, cooling and tonic and are used innervious and urinary diseases and in disorders of the blood and bile; as bitter, febrifuge and stomachic is used in bowel complaints. Its infusion with little ginger in the doses of small teacups twice a day is recommended in intermittent fever and expressed juice of the roots I the form of electuary is given for the removal of intestinal worms. Warm leaves alone or moistened with gingil oil are employed to hasten the suppuration.

Folk Uses

A decoction of the shrub is given to remove calculi from the urinary tracts in different parts of India. Besides, in Alwar (Rajasthan), the seed powder of the plant is given in leucorrhoea, gonorrhoea and spermatrorrhoea.

Preparation

Baladikwath, Baladya ghrit, Baladya arista, Chandanbala lakshadi taila, Sudarshan churna and Kukuvadi churna.

Species: *Sida cordata* (Burm. F.)

Description of Plant

It is a prostrate or decumbent-ascending and some-what hairy herb. Leaves are ovate-cordate, acuminate, serrate with alternate and short teeth and glabrate or thinly stellate-hairy. The followers are yellow and axillary and solitary or borne in pairs or in small cymes. Fruits are subglobose and 0.35-0.38 cm in diameter. The seeds are brown.

Flowering and Fruiting

Almost throughout the year.

Distribution

the plant is found in the hotter parts of India upto 1500m.

Plant Parts Used

Flowers, fruits, leaves, root-bark and whole plant are used.

Uses

It flowers and unripe fruits with sugar are given for burning sensation in micnutrition. The juice of its leaves is given indiarrhoea during pregnancy and as a poultice, leaves are applied to cuts and bruises. The root-bark is used in leucorrhoea, micnutrition and gonorrhoea. The whole plant as cooling, astringent and tonic and it is used in fevers and urinary complaints. A decoction of entire herb is given to prevent joint swellings in arthritis.

Folk Uses

An aqueous extract of plant together with sugar is given twice a day for two days in dysentery and for 7-15 days in spermatorrhoea by the Tharus people in Uttar Pradesh .

Preparation

Baladikwath, Baladya ghrit, Baladya arista, Chandanbala lakshadi taila, Sudarshan churna and Kukuvadi churna.

Species: *Sida cordifolia* Linn.

Description of Plant

It is a small erect, branched and stellate-tomentose herb or shrub. Leaves are ovate-oblong to cordate and very downy on both surfaces. Petioles are as long as the blade. The flowers are yellow and small and are borne solitary or in fascicles. Calyx is 0.7cm long and divided halfway down, lobes ovate-triangular and ciliate. Corolla is 0.7-0.8cm long and ciliate at the tops. Staminal column is 0.3 cm long and hairy below. The carpels are 10. The capsules are 0.7-0.8 cm in diameter.

Flowering and Fruiting

During August-December.

Distribution

It grows wild throughout the tropical and subtropical India, ascending upto 1050m.

Plant Parts Used

The whole plant, its sees, leaves, roots and root-bark are used.

Uses

The juice of entire plant mixed with water is given in rheumatism, gonorrhoea and spermatorrhoea. Its seeds are considered aphrodisiac and are useful in colic, tenesmus and gonorrhoea. An infusion of its leaves is prescribed as febrifuge and also a vegetable to patients suffering from piles. Its roots as astringent, diuretic and tonic and an infusion of its roots is given in nervous and urinary diseases, cystitis, strangury, bleeding piles, haematuria, leucorrhoea, chronic dysentery and in asthma as cardiac tonic. A strong decoction of its roots is used as diaphoretic, antipyretic, stomachic and tonic. The roots with asffoetida and rock salt is given in hemiplagia, facial paralysis, stiff necks and noises in the ear with headache. A paste of its roots made with juice of Palmyra palm is used in elephantiasis.

The root bark with sesame oil and milk is given facial paralysis and sciatica. The bark of the root in the form of powder is given with milk and sugar for the relief in micnutrition and leucorrhoea.

Folk Uses

The juice of the whole plant with water in the doses of 250 gm is used for spermatorrhoea and gonorrhoea. A poultice of its leaves is applied to boils to promote suppuration. An infusion of the plant mixed with table salt and their filtrate is taken orally in asthma. In excessive menstrual flow, a past of its shoots with black pepper in equal parts by weight made into pills of about 5.0 gm each and is given thrice daily for two weeks from the first day of menses by Assamese (Borthakur, 1993). Besides, stembark of this plant is crushed with the roots of *Aristolochia indica*, tubers of *Cyperus scarious* and whole plant of *Selaginella bryopteris* and *Phyllanthus fraternus* and made into pills. 3 of which are taken for 5 days in early morning to cure epileptic fits by the tribals of Santhal Pargana in Jharkhand.

Preparation

Baladi kwath, Baladya ghirt, Baladyarista, Chandanbala lakshadi taila, Sudarshan churna and Kukuvadi churna.

Species: *Sida rhombifolia* Linn.

Description of Plant

It is an erect, much-branched and glabrate to stellate-hairy, annual or perennial herbs. Leaves are lanceolate to rhomboid acute, cuneate or rounded at base and glabrous above and hoary or grey pubescent beneath. The flowers are yellow and small and are borne axillary, solitary or in pairs. Fruits are small capsules, 0.35-0.4 cm across and stellate-hairy I the upper part. Seeds are 1-2, smooth and black.

Flowering and Fruiting

Flowering: During rainy season and Fruiting: In winter season.

Distribution

It is found throughout India, particularly in moist region ascending upto 1800 m in the Himalayas.

Plant Parts Used

Whole plant, its leaves, stem and roots are used.

Uses

The whole plant is used in pulmonary tuberculosis and rheumatism. The leaves are pounded and are applied on swellings. Its mucilaginous stem both externally and internally is used as demulcent and emollient. The roots are considered valuable in the treatment of rheumatismand their paste in milk with honey is given to relieve pain of leucorrhoea.

Its leaves are used as a substitute for tea.

Folk Uses

Two teaspoonfuls of the plant juice is given two times a day in headache by the people in different parts of the country. Besides, an aqueous extract of the plant is also taken thrice a day for two weeks in diabetes.

Preparation

Baladi kwath, Baladya ghrit,Baladyarista, Chandanbala lakshadi taila, Sudarshan churna and Kukuvadi churna.

Genus: *Urena* Linn.

The generic name has been derived from Malabarian name Uren, a kind of mallow.

The genus comprises of herbs or undershrubs, covered with stellate hairs, Leaves are angled or lobed. The flowers are clustered. The bracteoles are 5 which are adnate to the calyx. Calyx is 5-cleft. Petals are 5, connated below and united to the base of staminal tube. Anthers are nearly sessile. Ovary is 5-celled. The fruits are covered with hooked sbristles or smooth and indehiscent. The seeds are ascending.

About 6 species occur in tropical and subtropical countries; of these 2 species are reported from the hotter parts of India.

Species: *Urena tobata* Linn

Description of Plant

It is an erect and stellate-pubescent to stellate tomentose herb or under shrub, upto 2.0m in height, Leaves are variable, angled or shallowly lobed, roundish or ovate and 5.0-7.5 cm long. The flowers are pinkish and are borne solitary or in axillary cluster of 2-3. The capsules are 0.8-1.0 cm across, indehiscent, densely pubescent, echinate and are covered with hooked bristles, therefore, easily stick to the clothes and hair/fur of animals. The seeds are smooth, rounded on the back and wedge-shaped o the innerside.

Flowering and Fruiting

Flowering: In rainy season and fruiting in winter.

Distribution

The plant is found throughout the tropical regions of India as a weed in waste places, forest clearings and along road-sides.

Plant Parts Used

the roots, stem, flowers, seeds and leaves are used.

Uses

Its roots as diuretic and these are externally used as embrocation in lumbago and rheumatism. A decoction of its stem is sued in cases of flatulent colic. The flowers are expectorant and pectoral (Agarwal, 1986) and are given in cough and their infusion is used as a gargle in sore throat and aphthesis. Its seeds are also used in gonorrhoea.

Its leaves are used as a substitute of Patchouli adulterant.

Folk Uses

Its leaves are crushed in water and then are applied on head for hair loss by the natives of Andamann and Nicobar and pounded leaves boiled in coconut oil are applied to treat wounds; also their juice is dropped in eye conjunctivitis and leaves mashed between palms are applied on cuts to clot the blood. Whole plant is boiled in sesame oil and is applied externally on rheumatic pains by the tribal people in Western Maharashtra.

Family: STERCULIACEAE

Genus: *Helicteres* Linn.

The genetic name has been taken from the Greek word Heliktos meaning a twisted bracelet, referring to the fruits of some species.

The genus comprises of trees or shrubs. Leaves are simple. The flowers are axillary and ar borne solitary or in fascicled. The follicles are spirally twisted or straight. Seeds are tubercled with scanty albumen.

About 60 species are found in the tropical Asia and America; 4 species have been reported from India.

Species: *Helicteres isora* Linn.

Description of Plant

It is a sub-deciduous small tree or shrub of about 1.5-3.0 m height. Young branches are rough with scattered stellate hairs. Leaves are serrate, obliquely cordate or obovate, shortly acuminate and rough above and pubescent beneath. The flowers are solitary or in spare clusters with red reflexed petals, becoming pale blue when old. The fruits are 5.0 cm long, greenish-brown, beaked and cylindrical with 5 spirally twisted carpels. Seeds are tubercled and many.

Flowering and Fruiting

Flowering: During rainy season and Fruiting: Later in August-September.

Distribution

It is found distributed in dry forests throughout Central and Western India, from Bihar as far West as Jammu and Western Peninsula.

Plant Parts Used

Fruits, seeds, bark and roots are used.

Uses

The fruits are demulcent and astringent and are useful in the gripping of bowels and flatulence of children. The bark is used in dysentery and diarrhea. Its seeded are powdered and mixed with pure castor oil or coconut oil and are prescribed in otorrhoea. Its roots are considered astringent, demulcent, diuretic and antigalactogogue and their juice or decoction is prescribed in diarrhea, dysentery, griping pain in the bowels, flatulence and in diabetes. Besides, its pods are fried and are given to children to kill intestinal worms.

The decoction of its fruit with *Achyranthes aspera* plant is also given in fever.

Folk Uses

One teaspoonful aqeous extract of its roots is given for 3-7 days for the treatment of dogbite in Uttar Pradesh. Tribals of Singhbhum district of Jharkhand tie its fruits as an amulet in neck to treat disease of malnautrition, which is locally known as Dubli disease among children.

Preparation

Gandharva churna and Sidha praneshwar ras.

Genus: *Pterospermum* Schreb.

The generic name has been derived from Greek words Pteridion = wing and Sperma=seeds, referring to its winged seeds.

The generic name has been derived from Greek words Pteridion= wing and Sperma = seeds, referring to its winged seeds.

The genus comprises of tree or shrubs with stellate hairs. Leaves are bifarious, simple or lobed, penninerved, unequal-sided and leathery. The flowers are bisexual and bracteolate. Calyx is 5-cleft and deciduous. Petals are 5. Staminal coloumn is short and anthers are in triplets, opposite the sepals and alternating with 5 liguilate staminodes. Ovary is inserted within the top of the staminal coloumn and is 3-5 celled with many ovules in each cell. The fruits are woody or coriaceous capsules which are terete or angled and loculicidally 5-valved. Seeds are winged.

About, 40 species are known to occur in Eastern Himalayas, South East Asia and Western Malaysia. Out of these, 12 species have been reported from India.

Species: *Pterospermum acerifolium* Willd.

Description of Plant

It is a large evergreen tree with thin grey and smooth bark upto 24 m in height; young branches are covered with ferruginous tomentum. Leaves are 15.0-30.0 x 12.0-25.0 cm, orbicular, peltate, lobed or entire and deep green above. The flowers are axillary, solitary or in 2-3 flowered cymes, fragrant and yellowish. The fruits are 5-angled, 10.0-15.0 cm long woody capsules. Seeds are winged and brown.

Flowering and Fruiting

Flowering: During March-June and fruiting in cold season.

Distribution

The plant is found distributed in the Sub Himalayan tracts upto an elevation of 1200 m and extends Southwards to Western ghats. Also cultivated along road-sides.

Plant Parts Used

Flowers, bark and leaves are used.

Medicinal Uses

Its flowers are considered as tonic and are useful in leucorrhoea, inflammation, ulcer, tumour and leprosy. The flowers and bark charred and mixed with Kamela (*Mallotus philippinensis*) are applied in suppurating small-pox. A plaster made out of its calyx is used in glandular swellings of the neck and ears. Besides, the leaves are haemostatic and are used for thatching.

Folk Uses

Its leaves are useful binder and are used to stop bleeding from the wounds hence their paste is also applied in headache. Halwa, a preparation is made from its flowers by mixing these with sugar and ghee which is eaten in bleeding piles (Singh, 1983). In migraine, its flowers are ground in rice-water and are applied as a paste on forehead. In Madhya Pradesh, a paste of its flowers is applied on glandular swellings around neck by the Sahariya tribals.

Preparation

Himanshu taila.

Genus: *Ptergota* Schott and Endl.

The generic name has been derived from Greek word Pterygotos=winged, referring to the winged seeds.

The genus comprises of the trees. Leaves are simple, cordate or lobed and wholly glabrous. The flowers are polygamous and are borne in panicles. Calyx is campanulate and their segments scarcely exceeding the tube. Petals are absent, Staminal columns are bearing at apex a ring of sessile anthers. Ovary is of 5 carpels. The fruits are radiating follicles. Seeds are many and parominently winged.

About, 20 species are known to occur in the tropical parts, chiefly in the old world. Only one species is reported from India.

Species: *Pterygota alata* (Roxb.) R. Br. (*Sterculia alata* Roxb.)

Description of Plant

it is a large tree with smooth and grey bark. Leaves are 10.0-25.0 x 7.5–20.0 cm, broadly ovate, cordate, entire and glabrous. The flowers are brownish-yellow and are borne in rusty tomentose racemes. The follicles are 1-5, woody, early globose, shortly beaked and 1.2 cm across. Seeds are many, tightly packed, elliptic, compressed and winged.

Flowering and Fruiting

Flowering: During February-March and Fruiting: In November-January.

Distribution

The plant occurs wild in Western Peninsula, Chittagong, Sylhet and in the Andamanns.

Plant Parts Used

Seeds are used.

Uses

According to Nadkarni (1976), its seeds are used as a substitute for opium as narcotics.

Genus: *Sterculia* Linn

The generic name is based on Latin word Stercus = dung., in allusion to the foetid smell of the flowers.

The genus comprises of trees. Leaves are undivided, lobed or digitate. Inflorescenes are paniculate and usually axillary, rarely racemosoe; flowers unisexual or polygamous. Calayx is 4-5 fid or partite, usually coloured; petals absent. Stamens column is bearing 10-30 anthers at its apex which are arranged in a ring or without order. Carpels of the ovary are 5 and subdistinct. Ovules are 2 many in each carpel; style connate at the base. Seeds are 1- many, naked or rately winged, sometimes arillate with flat or undulate cotyledons.

About, its 300 species are known to occur in tropical parts of the word. Of them 12 have been reported from India.

Species: *Sterculia foetida* Linn.

Description of Plant

This is a moderate-sized, deciduous tree attaining a height of 30 m with whitish, flaking and soft bark. Leaves are digitate, crowded at the end of branches; leaflets 5-9, subsessile, oblong-lanceolate and 10.0-18.0 x 4.0-5.0 cm. Flowers are foetid smelling, red and yellow or dull purple, 2.5-5.0 cm across and are borne in panicles. The follicles are woody, thick, boat-shaped and 10.0-12.0 cm long; possess foul smelling when ripe. Seeds are 10-15, ovoid-oblong, black and 2.0cm long.

Flowering and Fruiting

Flowering: During April- May and Fruiting: In November-December.

Distribution

This is found along the Western coast from Konkan southwards at low elevations; also cultivated on road-sides.

Plant Parts Used

Seeds, leaves, fruits and bark of the plant are usable.

Uses

An oil obtained from its seeds is used as laxative and carminative. Leaves and bark are used as diuretic and aphrodisiac. Besides, decoction of its fruits is taken as an astringent.

Its seeds are used as an adulterant to Cocoa (*Tehobroma cacao*).

Folk Uses

Its seeds are roasted and are eaten as a laxative by the people in the southern parts of the country. According to Joshi (1989), Bhills of Rajasthan apply a paste of its juvenile leaves on chapped and cracked skin of children.

Species: *Sterculia urens* Roxb.

Description of Plant

It is a large deciduous tree with smooth, white or greenish-grey bark. Leaves are crowded at the ends of branches, glabrous above and tomentose, beneath, shallowly

5-lobed; lobes entire, acuminate and blade 20.0-30.0 cm long. Flowers are bisexual and are crowded in erect, more less pyramidal panicles. The fruits are re-coloured, covered with stiff stinging bristles. Seeds are 3-6 in each carpel and are oblong and are dark brown.

Flowering and Fruiting

Flowering: During January-March and Fruiting: Later.

Distribution

This is found in the forests of Gujarat, Konkan, Deccan, Chennai, Chhota Nagpur and in central India.

Plant Parts Used

Gum and leaves are usable.

Uses

The gum obtained from this plant is commercially known as 'Gum karaya or Indian tragacanth' and is used in throat affection. Besides, its leaves and tender branches are steeped in water to yield a mucilaginous extract which is useful in pleuro-pneumonia of cattle.

Folk Uses

Sometimes, seeds of the plant are roasted and are eaten in the different parts of the country.

Family: TILIACEAE

Genus: *Triumfetta* Linn.

The generic name has been assigned in the honour of G.B. Triumfetti, an Italian botanical author of the 17th century.

The genus comprises of herbs or under shrubs with stellate pubescence. Leaves are serrate and entire or lobed. The flowers are yellow and small and are borne axillary or in leaf-opposed dense cymes. Sepals are 5, distinct and frequently mucronate at the apex. Petals are 5 and distinct. Stamens are 5-25, free and inserted on a glandular torus. Ovary is 2-5 celled with 2 ovules in each cell. The fruits are oblong or globose, spiny or bristly, and indehiscent or tardily dehiscent. Seeds are pendulous and 1-2 in each cell.

About 150 species occur in the tropical parts of the world. Of which, 8 species grow in India, mostly in tropical parts with 3-4 species extending to the Himalayas.

Species: *Triumfetta rhomboidea* Jacq. (*T. bartramia* Linn.)

Description of Plant

It is an annual or perennial, erect and simple or branched hirsute herb with about 10-90 cm height. Leaves are cordate, rhomboid or ovate, irregularly serrate and usually 3-lobed. The flowers are yellow, small and about 1.9 cm across and are borne in dense cymes. Sepals are lanceolate-narrowly oblong, mucronate and stellate-hairy outside. Petals are oblong and shorter than sepals. The fruits are ovoid or globose capsules with smooth hooked spines. Seeds are 0.3 x 0.2 cm.

Flowering and Fruiting

During October-December.

Distribution

The herb is found throughout tropical and sub-tropical parts of India; upto an elevation of 1200 m in the Himalayas.

Plant Parts Used

The leaves, bark, roots, flowers and fruits are used.

Uses

Its leaves, flowers and fruits are mucilaginous, demulcent and astringent and these are given to promote parturition when it is delayed. Its flowers are rubbed with sugar and water and are given in gonorrhoea to stop burning sensation caused by urine. The bark and its fresh leaves are used in diarrhea and dysentery and the leaves and flowers in leprosy. Its roots are considered diuretic and a hot infusion of them is given to facilitate the child birth. Besides, the roots are pounded and are given for intestinal ulcer.

Folk Uses

A hot infusion of its roots is given to facilitate the child birth or to hasten the inception of parturition when it is delayed by the women in different parts of the country.

Family: ZYGOPHYLLACEAE

Genus: *Peganum* Linn.

The generic name is derived from Greek work Peganon meaning rue, solid and this old Greek name used by Theophrastus for the Rue plant.

The genus comprises of branching, glabrous or pubescent, perennial, rooted herbs, Leaves are alternate, entire or multifid. Flowers are white and are borne solitary in sub-terminal leaf-opposed peduncles. Sepals are 4-5, persistent; petals 4-5, imbricate; stamens 12-15, inserted at the base of disk, some antherless and ovary is globose and deeply-3-lobed. The fruits are globose, 3-4 celled, dry and 3-valved or fleshy and indehiscent with many seeds.

About 5 or 6 species are found from Mediterranean to Mongolia, South U.S. and Mexico. Only one of which is reported from India.

Species: *Peganum harmola* Linn.

Description of Plant

It is a bush-like, glabrous, perennial, rooted herb upto 30-90 cm height. Leaves are alternate, 5.0-8.0 cm long, pennatifidly cut into linear, very narrow and acute with spreading lobes. Flowers are white and are borne solitary in the axils of the branches; calyx-lobes narrow, much exceeding the corolia and persistent. The capsules are globose and depressed at the apex.

Flowering and Fruiting

Flowering: In June-July and Fruiting: During September-October.

Plant Parts Used

Whole plant, its seeds and leaves are used .

Uses

The whole plant is used as an abortifacient, alterrative, antiperiodic, aphrodisiac, emmenagogue and stimulant. Its sees are anodyne, anthelminitc, antipyretic, antispasmodic, hypnotic and narcotic and their decoction is given for abortionand in amenorrhoea and colic; also used as a gargle in laryngitis. The smoke of its seeds is believed to have antiseptic properties, hence the fumigation is applied in palsy and lumbago. Besides, its leaves are useful in asthma, cholilithiashis, colic, dysmenorrhoea, hiccup, thiasis, neuralgia and rheumatism and their decoction or infusion is used as anthelminitc, mild emmenagogue and sudorific.

Folk Uses

A paste of its roots is applied on head to kill lice by the people in different parts of India.

Genus: *Tribulus* Linn.

The generic name has been derived from Greek words Tri = three and Balls = to project, referring to the carpels provided with three prickly points.

The genus comprises of prostate and branched silky herbs. Leaves are stipulate, opposite and abruptly pinnate; leaflets are usually unequal. The flowers are white or yellow and are borne solitary on pseudoaxillary peduncles. The fruits are 5-angled and are composed of 5-12 winged or spinous or tuberculate indehiscent cocci. Seeds are 2-more in each cell without albumin.

About its 20 species occur in tropical and subtropical regions of the world. Of which, 3 species are found in India, one in Eastern and Southern parts, one in Punjab and one is throughout India.

Species: *Tribulus terrestris* Linn.

Description of Plant

It is a small prostate or decumbent and hairy or silky annual herb. Leaves are in unequal pairs, stipulate, opposite and abruptly pinnate; leaflets are in 4-7 pairs, oblong and mucronate. The flowers are yellow,1.0-1.5 cm across, solitaryand are axillary or leaf opposed. The fruits are globose and hairy 5-angled cocci with 2 long and 2 short spines. Seeds are obliquely pendulous.

Flowering and Fruiting

During rainy season.

Distribution

It is found in waste places and dry habitats throughout the warmer regions of India ascending upto an altitude of 3000 m including West Rajasthan and Gujarat.

Plant Parts Used

The fruits, leaves, stems and the roots are used.

Uses

Its fruits a cooling, diuretic, tonic and aphrodisiac and their decoction or infusion is used in painful micnutrition, kidney diseases, chronic cystitis, gonorrhoea, gout, gravel and in impotence. Its dry fruits are powered and are given in doses of 18 gm with sugar and black pepper in gleet, spermatorrhoea and impotence. Besides, its fruits are also prescribed in Bright's diseases. The leaves are used in affections of urinary calculi and as stomachic and the stem is considered astringent and its infusion is given in gonorrhoea. Its roots are also used as aperient, demulcent and as tonic.

Folk Uses

A decoction of the whole plant is taken as a house hold remedy to remove gravels by the people in different parts of the country.

Preparation

Dashmula kwath, Gokshuradi churna, Gokshuradi kwath, Gokshuradi guggulu, Gokshuradi avaleha, Abhayarista, Chavyanprash, Rasna saoptak kwath. Haritakayadi kwath, Vriharvarunadi kwath, Ark mukkab mussafikhoon, Sharbat bazuri motadil, Sharbat mudir, Lulab-al-asrar, Majun Zangibil and Sufuf kalan.

Family: OXALIDACEAE

Genus: *Oxalis* Linn.

The genus comprises of annual or perennial herbs or rarely of shrubs with acid juice. Leaves are tri-foliate with or without stipules. The flowers are in axillary 1-or more flowered peduncles and regular. Sepals are 5 and imbricate. Petals are 5 and hypogynous. Stamens are 10, free or united at base and all are anther-bearing. Ovary is lobed, 5-celled and with one or more ovules in each cell. Fruit is a loculicidal capsule. The seeds are with an outer fleshy coat which bursts elastically.

About 800 species are known to be cosmopolitan in distribution, chiefly in Central and South America and South Africa. Of them, 6 species are found in Himalayas and warmer parts of India.

Species: *Oxalis corniculata* Linn.

Description of Plant

It is an appressed-pubescent, diffuse and creeping perennial herb with ascending or suberect branches. Leaves are digitately trifoliate, long petioled and stipulate; leaflets are obcordate. The flowers are yellow and are borne in 2-8 flowered umbeliform infolorescence. The fruits are oblong capsules, narrowed to the apex, 5-angled, 1.5-2.0cm long and pubescent. Seeds are many, transversely ribbed, 0.1 cm across and dark brown.

Flowering and Fruiting

Almost throughout the year, mainly during July-October.

Distribution

This is found throughout the warmer parts of India and in the Himalayas upto 2438 m. Very common in moist and cultivated places, open lands and surrounding rice field.

Plant Parts Used

The whole plant and its leaves are used.

Uses

The entire plant is considered antiscorbutic and its fresh juice is used in anaemic, dyspepsia, piles and tymparities. Its leaves are a good source of vitamin C and are antidiarrhoeal, antipyretic, antiscorbutic, antidyspeptic, astringent, appetizing, digestive and refrigerant. These are applied externally on inflammatory swellings as poultice. An infusion of the leaves is used to cure opacity of the cornea and their juice is given to counteract poisoning of Datura.

Folk Uses

Expressed juice of its leaves is made into a Sharbat with sugar and it prescribed to allay thirst in dysentery. The whole plant is rubbed down with water, boiled and juice of onions is tehn added to this mixture which is applied to the head in bilious headache. Its leaves are frequently eaten by the children. In Pauri Garwhal, the plant is used in 'Aankh ka phaulu' disease in which a white spot is seen in eye lens with redness in outsphere. Though, the juice of the leaves is dropped in to the eyes but if the right eye is affected, the extract juice of the leaves is rubbed on the left shoulder of the patient and if the leaf eye is affected, the juice is rubbed on the right shoulder of the patient. In Jammu and Kashmir, people use, its chopped leaves mixed with egg albumen in dysentery. Whole plant mixed with black pepper is used for skin eruptions, alopecia and wounds by the natives of Garhwal.

Preparation

Changeri ghrit.

Family: BURSERACEAE

Genus: *Boswellia Roxb.* Ex Colebr.

The genetic name has been assigned in the honour of Dr. Boswell (1740-1795) of Edinburgh.

The genus comprises of tree with papery bark and abounding in resin, Leaves are alternate, exstipulate, imparipinnate; leaflets opposite, sessile and usually serrate. Flowers are white, small and bisexual in racemes or panicles; calyx small, 5-toothed; petals 5, distinct, narrowed at the base and imbricate; disk annular, crenate; stamens 10 and ovary sessile and 3-celled. The drupes are trigonous containing one-seeded pyrenes. Seeds are compressed and pendulous.

About 24 species are known to be distributed in tropical Africa, Madagascar and tropical Asia. Only 3 species are reported from India.

Species: *Boswellia serrata* Roxb. Ex Coleb (*B. serrata var. glabra* (Roxb.) Bennet; B.*glabra* Roxb.)

Description of Plant

It is a moderate-sized deciduous a tree with smooth, greenish or ash-coloured papyraceous bark. Leaves are alternate, imparipinnate, crowded towards the ends of branches; leaflets 17-31, opposite, sessile, lanceolate or ovate, crenate and pubescent. Flowers are small and white in axillary racemes or panicles. The drupes are about 1.2 cm long and trigonous, splitting into 3 valves and substended by the woody disk. Seeds are compressed and pendulous.

Flowering and Fruiting

Flowering: During March-April and Fruiting: In winter season.

Distribution

It is found in the forests of Western and Central India, extending from Bihar to Rajasthan and southwards into Deccan peninsula.

Plant Parts Used

Gum, obtained from the trunk and bark of the plant are used.

Uses

Gum, obtained from the plant commercially known as Indian oilbanum or Indian frankincense or *Salai guggulu*, is considered antiseptic, astringent, diaphoretic, diuretic, ecbolic, emmenagogue and expectorant and is used beneficially in cutaneous and nervous diseases, cystic breast, chronic diarrhea and dysentery, gout, goiter, piles, rheumatism, tumours and ulcers and is also applied locally on boboes. That gum mixed with that of acacia is used to correct breath and also taken for a long time to reduce obesity in the doses of about 30 gm. Besides, the bark is useful indiarrhoea, piles and cutaneous diseases; mixed with butter is applied as a poultice on bleeding.

Folk Uses

Gum of the plant mixed with coconut oil is applied to sores and to stimulate the growth of hairs by the people in different parts of the country. In gastric pain, bark of the plant is inhaled through 'chilam' (earthen censor) by the natives of Madhya Pradesh and in Bihar, about 100 gm of fresh bark is ground in water which is given twice only for a day to check dysentery. A tribal people in Jharkhand pound its stem bark together with *Curcuma longa* in kerosine and apply in rheumatic pains.

Genus: *Commiphora* Jacq.

The generic name is derived from Greek words Kommi = gum and Phero=to bear, reffering to the rich gum exudation from the trunk of the plants.

The genus comprises of balsamiferous trees or shrubs with often spiny branches. Leaves are usually trifoliate; lateral leaflet sometimes absent or small. Flowers are small, fascicled and polygamous. Calyx is urceolate or tubular; petals usually 4, inserted on the edge of disk and valvate. Stamens are 8-10, on the disk's edge, alternately long and short and ovary sessile, 2-4-celled with 2 ovules in each cell. The drupes are ovoid and resinous containing 1-3-celled nut or 2-3 nuts.

About 185 species are found in the warmer parts of Africa, Madagascar and from Arabia to Western India; out of which only 6 species have been reported from India.

Species: *Commiphora wightii* (Arnott) Bhandari [*C. mukul* (Hook. Ex Stocks) Engl.; *Balsmodendron wightii* Arnott; *B. roxburghii* Stocks; *B. mukul* Hook. Ex Stocks]

Description of Plant

It is a small tree of more usually a stunted bush with spinescent branches. Leaves are usually unifoliate, alternate or crowded at end of short branches, cuneate-obovate, rhomboidal or oval, acute, deply serrate and smooth and shining. Flowers are small, nearly sessile, few together and unisexual; male with short and barren ovary and females with short stamens and imperfect anthers. Calyx are cylindrical and petals 4-5, brownish red, strap-shaped. The drupes are red, ovate-acuminate containing 2-celled stones, rarely 4-valved.

Flowering and Fruiting

Flowering: During March-April and Fruiting: Later.

Distribution

It is reported to found wild in the arid and rocky zones in some parts of South-west and North-western regions of India including Mysore and Rajasthan.

Plant Parts Used

Gum, obtained from the plant is used medicinally.

Uses

Gum, obtained from this plant commercially known as Indian bdellium is alterative, antiseptic, anti-inflammatory, antispasmodic, antisuppurative, aphrodisiac, aperient, appetizing, ecbolic, emmenagogue and expectorant and is used in amenorrhoea, anaemia, encometritis, leucorrhoea, menorrhagia, nervous diseases, rheumatism, scrofulous affections and cutaneous diseases, specially applied in bad wounds and indolent ulcer and recommended in treatment of urinary diseases, obesity, in marasmus of children and rheumatoid arthritis. The fumes of its burnt gum are also inhaled inchronic bronchitis, chronic nasal catarrh, laryngitis, and tuberculosis. In indolent ulcer, it is applied as a plaster after mixing with lime juice or coconut oil and a lotion, prepared after mixing 80gm of its tincture (20 per cent in 90 per cent of alcohol) in 280 ml of water, is used as a gargle in pyrrhoea, weak tonsillitis and pharyngitis and ulcerated throat.

The gum obtained from this tree is used as a substitute or adulterant to Harbol myrrh (*Commiphora myrrha*).

Folk Uses

the gum is burnt as incense on the holy occasions throughout the country (Dhiman, 2003); also its fumes are inhaled in bronchitis, catarrh, laryngitis and tuberculosis by the native people. Triphala guggulu, prepared by taking of 5 parts of guggulu, 3 parts of triphala, one of pippali and sufficient honey to make pills, is a simple household remedy which is used in gonorrhoea, dropsy, foul, ulcers, fistula and syphilis in different parts of the country. Besides, Rabari people, pound its stem gum in calcium carbonate and apply externally on tumour of spleen and liver in Barda Hills in Gujarat.

Preparation

There are five different forms of guggulu, but in preparations, only Kanka guggulu and Mahisksha guggulu are used. The main preparations are Yograj guggulu, Kaisar guggulu, Gokshuradi guggulu. Trayodshang guggulu, Adityapaka guggulu, Sadana guggulu, Amrita guggulu, Kachnar guggulu Triphala guggulu, Vatari ras, Chandraprabha vati and Aroygyvardhini vati.

Family: RUTACEAE

Genus: *Aegle* Corres

The generic name has been derived from Greek word Aegle, mythologically refers to one of the naiads who presided over river and springs.

The genus comprises of trees with spines. Leaves are alternate and 3- foliate; leaflets are membranous and subcrenulate. The flowers are large and white and are borne in axillary panicles. Calyx is small, 4-5 toothed and deciduous. Petals are 4-5 and spreading. Stamens are many, filaments are short and anothers elongated. Ovary is ovoid with broad axis, style short and stigma is capitate, oblong or fusiform and deciduous. The ovules are many. Fruits are large, globose, ovoid or renmiform and many seeded with woody rind. Seeds are embedded in aromatic pulp and are oblong and compressed.

About 3 species are found in Indomalayan region. Only 1 species has been reported to grow in India.

Species: *Aegle marmelos* Correa ex Roxb.

Description of Plant

This is a small or medium-sized deciduous and aromatic tree about 10m, high with straight sharp, axillary and 2.5 cm long spines. Leaves are trifoliolate, petiole 2.5–6.0 cm long and terete; leaflets are ovate-lanceolate, lateral sessile, terminal long and acuminate, 5.0-10.0 x 2.5-6.0 cm, cuneate to obtuse at the base. The flowers are greenish-white, sweet scented and about 2.5 cm in diameter and are borne in axillary panicles. Fruits are 5.0-10.0 cm across, globose, grey or grayish-yellow, hard with orange coloured sweet pulp. The seeds are numerous, oblong and compressed and are embedded in aromatic pulp; testa woolly and mucilaginous.

Flowering and Fruiting

Flowering: During April-May and Fruiting: In March-April in next year.

Distribution

The plant is common throughout India in dry hilly areas, gardens and along road-sides; also cultivated in various places in India.

Plant Parts Used

Fruits, seeds, flowers, leaves, bark and roots are usable.

Uses

The ripe fruits of the plant are alterative, cooling, laxative and nutritive and are useful inhabitual constipation, chronic dysentery and dyspepsia. The unripe fruits are used as antidiarrhoeal, astringent, demulcent, digestive and stomachic. Seeds are used as laxative and the flowers as antidiarrhoeala and antitemetic. The leaves are expectorant, febrifuge and the fresh ones are used in dropsy and effective in bronchial asthma. Its fresh leaf-extract is reported to reduce the period of convalescence in patients suffering from cholera or cholearic diarrhea. The leaf-juice is applied externally in abscess and ash of the leaves is to kill worms and injuries caused by animals. The bark of roots and stem is beneficial in intermittent fever, melancholia, palpitation of heart and in stomach pain. The root is also supposed to be used as anthelmintic.

The pulp of its fresh fruits mixed with milk when administered with Cubeb powder acts as diuretic and astringent on the mucous membranes of the generative organs, therefore, useful in gonorrhoea.

Folk Uses

The pulp of its ripe fruits is used for the preparation of a tasteful aromatic *sharbat* which is a popular drink in India in hot weather. Besides this, its fresh leaves are also chewed by the diabetic patients to control the blood sugar throughout the country.

Preparation

Bilvapanchak kwath, Bilvadichurna, Bilvadighrit, Bilvaa taila, Bilvadileh, Brhidgangadhar churna and Majaun gbawasir, It is also one of the ingredients of 'Dashmula', a common ayurvedic formation, particularly useful in the loss of appetite and puerperal diseases.

Family: MELIACEAE

Genus: *Toona (Endl.)* M. Roem.

The generic name has been adopted from its Hindi name Toon.

The genus comprises of deciduous trees. Leaves are small and pinnate; leaflets are opposite or subopposite. The flowers are bisexual in terminal panicles. Calyx is short and 5-cleft. Petals are free, imbricated and suberect. Stamens are 4-6, free are inserted at the top of the disk, sometimes alternating with staminodes; the filaments are subulate and anthers oblong and versatile. Disk is thickly, fleshy and 5-loboed or cylindrical. Ovary is sessile on the top of the disk, 5-celled. The ovules are 8-12 in each cell, pendulous and 2-seriate. Fruits are 5-celled and coriaceous capsules. Seeds are numerous, compressed and winged.

About its 15 species occur in tropical Asia, America and Australia. Out of which, 4 species have been reported from India.

Species: *Toona ciliata* Roes. (*Cedrela toona* Roxb. Ex Rottl.)

Description of Plant

It is a large tree with a dense spreading crown and thin dark grey bark. Leaves are 25-45cm long and pinnate; leaflets are 8-30, usually opposite, obliquely ovate or lanceolate, acutely acuminate, glabrous, shining, entire and usually wavy. The flowers are white and are borne in terminal panicles. Calyx is short and lobes ciliate. Petals are free, oblong or ovate and ciliate. Stamens are 5 and are inserted on the lobes of the disk. Stigma is capitate and with a large depression at the apex. The fruits are oblong capsules, coriaceous, 5-celled and septifragally 5-valved. Seeds are compressed and winged.

Flowering and Fruiting

Flowering: During March-April and Fruiting: In June-July.

Distribution

This plant is found in Sub-Himalayan tracts from the Indus eastwards, Chittagong, Assam, Chotga Nagpur, Western ghats of Mumbai to Nilgiris and Annamalais and other hills of Western Peninsula.

Plant Parts Used

Bark and flowers are used.

Uses

That its bark as astringent, tonic and antiperiodic and its infusion is given in chronic dysentery. Externally, powder of the bark is applied to ulcers. The flowers are emmenagogue and are used inmenstrual disorders.

Folk Uses

That its young leaves are crushed and are pasted on external injuries in Pauri Garhwal by the local people.

Family: CELASTRACEAE

Genus: *Celastrus* Linn.

The genetic name has been derived from Greek word Kelas: the latter season, referring to an evergreen tree having fruits remaining of the tree throughout winter, the name was given by Theophrastus.

The genus comprises of climbing shrubs or small trees which are often spinous. Leaves are alternate and stipulaes are minute, decidyous or sometimes none. The flowers are small, 4-5 merous and polygamous in terminal or axillary panicles or racemes. The fruit is globose or obovoid and loculicidal capsule. Seeds are 1-2 in each cell, completely or partially enclosed in an aril and with foliaceous cotyledons.

About 30 species are known to occur in tropical and subtropical parts of the world. Out of which 4 species have been recorded from India.

Species: *Celastrus paniculatus* Willd.

Description of Plant

It is large deciduous unarmed climber or scrambling shrub with terete branches; the young shoots and branches are pendulous. Leaves are alternate, glabrous, broadly ovate or obovate, acuminate or acute and 6.3-10.0 x 3.5-7.5cm. The flowers are unisexual, yellowish-green, 0.38 cm in diameter and are borne in terminal pendulous panicles. The fruit is a capsule, which is globose, usually 3-celled, bright yellow when ripe and 3-6 seeded. Seeds are enclosed in a complete red arillus and black.

Flowering and Fruiting

Flowering: During April-June and Fruiting: In cold season.

Distribution

It is found in tropical and sub-tropical Himalayas ascending upto 1400 m from Punjab to Assam, Bihar, Uttar Pradesh and South India.

Plant Parts Used

The seeds, bark, leaves and the roots are used.

Uses

The seeds of the plant are considered alterative, antirheumatic, aphrodisiac, emetic, laxative and nervine tonic and their decoction is used ingout, leprosy and paralysis. The oil of the seeds is rubifacient and stimulant and is considered effective in beriberi and oedema and is also used to improve memory. The oil in doses of one or two drops internally is used for pneumonia of childredn and externally in scabies, Besides, the seeds are crushed and are used to treat ulcers and leucoderma, its bark is used as abortifacient and the roots and leaves as a poultice in headache.

Folk Uses

About 5.0cm long root of the plant and 7 black peppers are made into a paste in *'Chawaldhua'*, water in which rice is washed and is given twich a day to cure spermatorrhoea, leucorrhoea and piles (Brahmam and Sexena, 1990). The natives of Southern Uttar Pradesh, take its one teaspoonful root powder twice a day for a long time in tumour cancer.

Preparation

Jyotishmati taila, Mall taila, Marichyadi taila,Karanjadi yog, Laghu-vis-garva-taila and Rogan malkangni.

Family: RHAMNACEAE

Genus: *Zizyphus* Mill.

The generic name has been adopted from Zizyphan, Greek name of Mediterranean jujuaabe which is derived from Arabic name Zigouf of the lotus.

The genus comprises of trees or shrubs and armed with stipulary prickles. Leaves are alternate, distichous and sub-coriaceous. The flowers are small in axillary fascicles or cymes. Calyx is 5-fid and spreading and lobes are keeled within, Petals are 5 or 9 and deflexed. Disk is 5-10 lobed with free margins. Stamens are 5, Ovary is immersed in the disk and adnate to it at the base and 2-4-celled and the styles are 2-4 and free or partly connate. The fruits are fleshy drupes with a hard 1-4-celled with 1-4 seed-stones.

About 100 species occur in tropical Asia. America, Mediterranean parts, Indomalayan region and in Australia. Out of which, 17 species have been reported form India.

Species: *Zizyphus mauritiana* Lamk. (*Z. jujube* (Linn.) Gaertn.

Description of Plant

It is a small tree or large shrub, almost evergreen and usually armed with dark-grey or nearly black bark. Its young branches are rusty tomentose. Leaves are elliptic-ovate or suborbicular, dark green and glabrous above and densely woolly-tomentose beneath. Prickles are in unequal pairs; one straight, the other is recurved. The flowers are in axillary clusters or shortly peduncled cymes. The fruits are oblong or obovoid or globose drupes and are 2-celled, fleshy, glabrous, red when ripe and edible. Seeds are plano-convex.

Flowering and Fruiting

Flowering: During September-October and Fruiting: In October-March.

Distribution

It is common in hotter parts of India, cultivated in gardens, villages and road-sides and also found wild in waste places and tropical forests.

Plant Parts Used

The leaves, twigs, bark, roots, fruits and seeds are used.

Uses

The leaves are astringent and diaphoretic and their infusion is used as an eye lotion in conjunctivitis. A paste of its leaves and that of twigs is applied to abscess, boils, and carbuncles to promote suppuration and to strangury. The bark of the stem is considered astringent and its powder or decoction is used in diarrhea. It is also used in dysentery, colic and inflammation of the gums. The juice of the root bark is sued as purgative and externally, is applied to gout and rheumatism.

The decoction of the roots is given in fever and as a powder is applied to old wounds and ulcers. Its fruits are mucilaginous, pectoral and styptic and are eaten to purify the blood and to aid the digestion. The dried fruits are used a smild laxative and expectorant. The seeds are antidiarrhoeal and their kernels are used for abdominal pain in pregnancy, as an antidote to aconite poisoning, antiemetic, sedative and soporific. Its seeds are given with butter milk in bilious affections. An ointment made of its seeds with some bland oil is used as a liniment in rheumatism.

Folk Uses

The fruits of this plant are eaten as cooling and tonic by the people in different parts of the country.

Preparation

Sharbat zufah murrakab, Sharbat sadar, Sharbat unnab, Sharbat murakkab musaffikhun and Arq murkkab musaffikhun.

Species: *Zizyphus oenoplia* Mill. (*Rhamnus oenoplia Linn.*)

Description of Plant

It is a thorny shrub with sarmentose branches and dark-grey and rough bark. Leaves are 3.7 x 2.5 cm, obliquely ovate-lanceolate and crenate-serrate. The flowers are 10-20 and are borne in sessile or sub-sessile, axillary, condensed and paniculate cymes. Calyx is tomentose outside. Petals are ob-triangular. Ovary is 2-celled and styles are 2, united above the middle. The fruits are obovoid or globose drupes, black and 1 or 2 celled with scanty pulp.

Flowering and Fruiting

Flowering: During April- July and Fruiting: In winter.

Distribution

It is common in village thickets and hedges throughout the hotter parts of India.

Plant Parts Used

Fruits, roots and the bark are the parts used medicinally.

Uses

Its fruits are used as stomachic. The roots are used in hyperacidity and ascaris infections. The decoction of its root-bark is used for healing of wounds and stem bark as digestive, febrifuge and tonic. Bhakuni *et al*. (1971), reported that the extract of the plan also possesses hypotensive and diuretic properties.

Folk Uses

The fruits of the plant are eaten in stomachache.

Preparation

Fruits are used as an ingredient in the preparation of stomachache pills

Family: VITACEAE

Genus: *Cissus* Linn

The generic name has been taken from the Greek word Kissos = ivy, referring to the habit of the plant.

The genus comprises of climbing shrubs with leaf-opposed tendrils. Leaves are simple or digitately or pedately 3-9 foliate and rarely pinnate. The flowers are in umbellate, paniculate, racemose or spicate cymes and are usually without bracts and sometimes polygamous. The berries are ovoid or sub-globose and 1-2 seeded.

About its 350 species are found in torpical and rarely insub-tropical parts of the world. Out of which, 7 species have been reported from India.

Species: *Cissus quadrangular* Linn. (*Vitis quadrangularis* Wall ex W.& A.)

Description of Plant

It is a climbing shrub or herb. The stem is glabrous, 4-angled, fleshy and contracted at nodes with simple tendrils, sometimes leafless. Leaves are cordate or reniform, opposite, 2.5-3.7cm long, sometimes divided into 3-5 lobes, crenate-serrate and glabrous. Flowers are small and are borne in short-peduncle cymes and are greenish-white incolour. Fruits are globose, succulent, apiculate berries, 0.6 cm across and red when ripe.

Flowering and Fruiting

During May-August.

Distribution

The plant occurs throughout the hotter parts in India, frequently in shrubberies; also planted in gardens.

Plant Parts Used

Leaves, young shoots, stem and roots are used.

Uses

The leaves and young shoots are considered alterative and stomachic and their powder is given indigestive problems. The stem is antiscorbutic and stomachic and its paste is given in asthma and the juice internally and externally both for the union of fractures of bones. This is also beneficial in epistaxis when dropped in nostril, in otorrhoea as an eardrop and in irregular menstruation.

The powder of the ashes of its shoots is administered internally in cases of dyspepsia. The root-powder is used to unite bone fractures and is considered as much effective as the plaster. The root-powder is also used orally in constipation, *vatarakta* (gout) in piles.

Folk Uses

Its pounded fresh shoots are applied over burns and wounds throughout the country. In Meghalaya, a mixture of the plant along with the fruit walls of *Moringa oleifera*, ginger juice and leaf-paste of *Justicia gendarussa* is tied in cases of fractures and dislocated bones.

Preparation

Asthisanhar taila.

Family: SAPINDACEAE

Genus: *Cardiospermum* Linn

The generic name is derived from Greek words, Kardia = heart and Sperma = seed, referring to the shape of the seeds.

The genus comprises of climbing herbs with wiry stems and branches. Leaves are alternate, 2-ternate and exstipulate; leaflets coarsely dendate. Flowers are borne in axillary racemes. Sepals are 4, outer two smaller; petals 4, 2 larger, lateral ones usually adhering to the sepals. Disk is unilateral, wavy, almost reduced to 2 round or linera glands opposite petals. Stamens are 8; ovary is 3-celled with single ovules. Capsules are 3-celled and 3-valved with globose and exalbuminous seeds.

About 12 species are known to be distributed in tropical parts of the world, especially in America. Out of which only 2 have been reported from India.

Species: *Cardiospermum halicacabum* Linn.

Description of Plant

This is an annual or perennial thinly pubescent or glabrous herb with striate branches. Leaves are 3.7-7.5cm long, alternately compound, deltoid or ovaote; leaflets deeply cut. Flowers are small, white, few in umbellate eymes, The fruits are broadly pyriform, depressed and trigonous capsules which are expanded like balloons, about 1.2-1.8 cm across. Seeds are smooth, black and globose.

Flowering and Fruiting

During December-May.

Distribution

It is commonly found throughout India in waste places and a shedges and thickets.

Plant Parts Used

Whole plant is usable.

Medicinal Uses

The herb is used both externally and internally in rheumatism and lumbago. Its roots are considered aperient, diaphoretic, diuretic, emmenagogue, laxatie and rubifacient and are used for lumbago and nervous diseases; decoction beneficially, in piles. Besides, its leaves are applied as a poultice in rheumatism and their juice is dropped in earache.

Folk Uses

The leaves of the plant after frying are applied to the pubes to increase the menstrual flow in amenorrhoea in different parts of the country. The leaves boiled in castor oil are also applied in rheumatic pains, swellings and tumours by the native people. Besides, Rabari tribal people in Barda Hills in Gujarat, crush the whole plant in Ground nut oil and apply all over the body in fever, jaundice and swellings (Jadeja, 1999).

Genus: *Sapoindus* Linn.

The generic name has been derived from Latin words Sopo=soap and Indicus=Indian, as the drupes of some species contain a pulp which lathers with hot water and is used for washing.

The genus comprises of trees or shrubs. Leaves are alternate or subverticillate and usually paripinate; leaflets are coriaceous and entire. The flowers are regular and polygamous and are borne in terminal or axillary panicles. The fruits are of 1-3 indehiscent cocci. Seeds are globose and are with or without aril.

About 13 species are found in the tropical and subtropical Asia, Pacific (excluding Australia) and America. Out of which 7 species have been reported from India.

Species: *Sapindus emarginatus* Vahl. (*S. trifoliatus* Linn.; *S. laurifolius* Vahl. *Var. emarginatus* (Vahl.) Cooke)

Description of Plant

It is a large handsome tree. Leaves are paripinnate; leaflets elliptic or oblong, acuminate or emarginated, glabrous or pubescent beneath with short, curved or stellate hairs. The flowers are white and are borne in terminal panicles; calyx rusty pubescent; petals 4 or 5, oblong or lancelolate without scales or with two tufts of white hairs and ovary is 3-lobed and ferruginous-tomentose. The drupes are saponaceous or coriaceous, fleshy and 1-2 coccus; cocci oblong or globose and indehiscent.

Flowering and Fruiting

Flowering: During October-December and fruiting in March-April.

Distribution

it is common on the Aravalli hills and in the Western Penninsula; also cultivated in Bengal and North India.

Plant Parts Used

Fruits, seeds and roots are used.

Uses

The fruits of the plant are considered alexipharmic and are used internally as emetic, expectorant, purgative and nauseant and as an errhine used in epilepsy, asthma, haysteria and hemicrania. A thick aqueous solution is dropped into nostril to give relief in hysteria, epilepsy and migraine. Its seeds as narcotic and their oil as emetic, expectorant, purgative and tonic; their kernels are used as abortifacient. Besides, its roots are used successfully in gout, rheumatism and paralysis. Besides, its roots are used successfully in goat, rheumatism and paralysis.

Its fruits are also used as fish poison.

Folk Uses

Pessaries made of the kernels of its seeds are used to stimulate the uterus during childbirth and in amenorrhoea by the native people. Besides, seeds, pounded up with water are introduced in mouth to cut short the paroxysm of epilepsy and fragrant leaves are used for baths in painful joints.

Species: *Sapindus mukorossi* Gaertn.

Description of Plant

it is a small deciduous trees. Leaves are pinnate; leaflets are 10-16, shortly stalked, 5.0-15.0 cm long, lanceolate-oblong, alternate or subopposite, acuminate or obtuse and glabrous. Flowers are white or purple, 2.5cm across and are borne in terminal panicles. The fruits are 1.7cm across, subglobose, fleshy, glaucescent and saponaceous. Seeds are 0.8-1.2 cm in diameter, globose and black.

Flowering and Fruiting

Flowering: In May-June and Fruiting During cold season.

Distribution

It occurs wild in Dehradun and in the valleys of the Himalayas up to 1219 m; also cultivated throughout North West India, Bengal and Assam.

Plant Parts Used

Fruits, roots, leaves and the seeds are used.

Uses

That its fruits as expectorant and emetic and Dastur (1962), considered these as emetic, detergent, astringent, anthelmintic,m tonic and antidotal. These are used in salivation, chlorosis and epilepsy. The pulp of fruits is rubbed in water until it soaks and then strained and is given by mouth to the people, those bitten by venomous reptiles and also to those suffering from severe diarrhea or cholera. The pulp is given to relieve colic with sherbet and is also used as emetic in small doses and in larger doses as purgative. A new drops of thick solution of the pulp are placed in the nose for relief in hemicrania, hysteria and epilepsy. A paste of its fruits with vinegar is used for reducing scrofulous swellings and also for killing lice. Besides, the paste of the fruits is also externally applied to the bites of reptiles, scorpion and centipedes etc.

Its roots are used as expectorant and also in gout, rheumatism and paralysis. The fragrant leaves are used for baths in painful joints. The pessaries made of the kernel of its seeds are used to stimulate the uterus in child birth and in amenorrhoea. Its seeds are pound up with water and are introduced in mouth cut short the paroxysm of epilepsy.

Folk Uses

The fruits of this plant are used as a substitute for soap for washing hair and to killlice. In hysteria and melancholia, the fumigations of the fruits are smoked by the people.

Preparation

Tiriyak afuyum.

Genus: *Semecarpus* Linn. F.

The generic name is derived from Greek words, Senna = token and Karpos = fruits, reffering to poisonous nature of the fruits as it is only a token fruit.

The genus comprises of trees. Leaves are alternate, simple, entire and coriaceous. Flowers are small, polygamous or dioecious, in terminal axillary panicles. Calyx are 5-6 fid, petals 5-6, imbricate. Disk is broad and annular. Stamens are 5-6 and are inserted at the base of disk and ovary is 1-celled with single pendulous ovule. The drupes are fleshy, obliquely oblong or subglobose. Seeds are pendulous with coriaceous testa and inner fleshy coat.

About 50 species are found in Indomalayan region, Micron and Solomon. 6 species have been reported from India.

Species: *Semecarpus anacardium* Linn,

Description of Plant

It is a medium-sized, dioecious and deciduous tree with rough dark-coloured bark. Its young branches, inflorescences and undersides of leaves are clothed with find pale pubescent. Leaves are crowed towards the extremities of the branches and are large, oblong, obovate, rounded at apex, cartilaginous at margins and coriaceous. Flowers are greenish-yellow, fasciculate and are arranged in erect, compound and terminal panicles. The drupes are about 2.5 cm long, obliquely oval or oblong, smooth, shining and purplish-black and are seated in an orange cup.

Flowering and Fruiting

Flowering: During May-June and Fruiting: In November-February.

Distribution

It is reported to occur throughout the hotter parts of India including tropical outer Himalayas.

Plant Parts Used

Fruits and bark are the parts, used medicinally.

Uses

That its fruits are used in asthma, ascites, epilepsy, neuralgia, psoriasis and rheumatism and also as abortifacient and vermifuge. A decoction of its fruits mixed with milk and butte fat is used efficaciously in asthma, gout, hemiplegia, sciatica and syphilis. The fruits are bruised and are applied to os uteri to procure abortion and are also given as vermifuge. Oil obtained from its nuts is used as vesicant and externally applied in rheumatism and leprous nodules. A paste containing equal parts of the juice of its nuts, *Plumbago zeylanica, Balaiospermum montamum, Euphorbia ligulartia, Calotropis gigantean,* sulphate of iron and molasses is applied over scrofulous glands of the neck. Besides, its ripe fruits are boiled with cow's dung, washed and mixed with butter to purify and then used internally in dyspepsia, nervous debility and skin diseases; also given to relieve asthmatic attacks. A brownish gum exudates obtained from its bark is found useful in nervous debility and in leprous, scrofulous and venereal diseases.

Folk Uses

In Bihar, people take about 125 gm of seeds of this plant and boiled them in 3 litre of buffalo's milk and then 250 gm fresh herb Phyllanthus niruri is added to it and milk is again boiled till it reduced to 500 ml. This preparation is taken once daily for three consecutive days to check bleeding in urine (Singh and Khan, 1990).Besides, a warm oil obtained from its mature fruits is applied over cuts and old wounds 2-3 times daily till recovery by the native people.

Preparation

Amritabhallatak, Bhallatak taila, Bhallatak kshripak and Sanjivani vati.

Family: PISTACACEAE

Genus: *Pistiacia* Linn.

The generic name is based on its Greek name, Pistake which has been derived from its Arabic name Foustaq.

The genus comprises of trees or shrubs. Leavs are alternate, exstipulate and pinnate or 3-foliate. Flowers are small, dioecious and are borne in axillary racemes or panicles. Male flowers: calyx 3-5 fid; stamens 3-7 and disk small. Female flowers: sepals 3-4; stamens and disk absent and ovary sessile with pendulous ovule. The drupes are dry with bony stones.

About 10 species are found distributed from Mediterranean to Afghanistan, East Asia to Malayasia and in the warmer parts of America. Only 2 of which have been recorded from India.

Species: *Pistacia chinenis* Bunge ssp. *integerrima* (Stewart) Rech.f. (*P. integerrima* Stewart)

Description of Plant

It is a moderate-sized, glabrous tree with dark grey or blackish bark and reaching a height of 16m. Leaves are 15.0-25.0 cm long and impair-or pari-pinnate; leaflets in 4-5 pairs, lanceolate and coriaceous with oblique base. Flowers are borne in lateral panicles. Male panaicles are compact and pubescent; stamens 5,6 or 7 and anthers large, oblong and deep red coloured. Female flowers are borne in long and lax panicles; style 3-fid nearly to the base with broad and recurved stigmas.The drupes are globose, dry, somewhat broader, rugose and glabrous. Seeds are with a membranous testa.

Flowering and Fruiting

During March-May.

Distribution

It is known to occur wild in North west Himalayas from Indus to Kumaon between 350-2500m.

Plant Parts Used

The galls are used medicinally.

Uses

According to Kirtikar and Basu (1935), the galls of the plant are known for their acrid, anthelmintic, bitter, expectorant and hot properties and this drug has a great reputation both in Hindu and Mohammedan medicine, as a tonic and expectorant. The galls are ground and after frying in ghee are given internally in dysentery and their paste is applied externally in psoriasis. Besides, that a decoction or lotion of its galls is found very effective as a gargle to suppress haemorrhage from gums and it is also used in bleeding from nose and discharge from mucous membranes such as gleet and leucorrhoea etc.

Folk Uses

In Uttar Pradesh, people take 250 gm each of galls of this plant, fruits of *Solanum surratttense*, roots of *Barleria prionit* is and leaf of Piper betel and add water to make the volume 4 litre. Then, it is boiled to reduce ½ litre and 10 ml. of this is given once daily for two months to cure asthma .

Preparation

Balchatur-bhadra, Karkatadi churna and Sringyadi churna.

Family: MORINGACEAE

Genus: *Moringa* Adans.

The generic name has been derived from Malyalam name Muringa of the plant.

These are soft-wooded deciduous tree. Leaves are alternate, imparibi or tripinnate; pinnae and leaflets are opposite, quite entire, glandular at base and caduceus. Flowers are large and bisexual in axillery panicles. Calyx is cup-shaped. Petals are 5 and unequal. Stamens are 5, fertile and opposite to petals. Disk is lining the calyx-tube. Ovary is stipitate and 1-celled and ovules numerous. The fruits are long, 3-6 angled and beaked capsules. Seeds are numerous, winged or not, with planoconvex cotyledons.

About its 12 species are found in North-east and South-west Africa, Madagascar, Arabia, 2 species have been reported from India.

Species: *Moringa oleifera* Lamk. (*M. pterygosperma* Gaertn.)

Description of Plant

It is fairly a large tree with thick corky bark upto 10 m in height. Leaves are usually tripinnate and 30.0-60.0 cm long; leaflets are elliptic and opposite and their glands are linerar and hairy. The flowers are white, bisexual and irregular and are borne in large panicles. Capsule are 20.0-50.0 cm long, triangular, ribbed and greenish. Seeds are 3-cornered, winged and about 2.5cm long.

Flowering and Fruiting

Flowering: During February-April and Fruiting: In May-June.

Distribution

It is indigenous to North West India; also cultivated throughout India.

Plant Parts Used

Root, bark, leaves, flowers, fruits, gum and the seed-oil are used.

Uses

That its roots are used as stimulant in paralysis and intermittent fever, as rubifacient in palsy and chronic rheumatism, as cardiac and circulatory tonic and in form of a compound spirit are useful in fainting giddiness, nervous debility, hysteria and flatulence. The root-juice is used internally in calculus affection in combination with milk and the decoction of the roots as gargle in hoarseness and throat sore. Its stem bark is used as abortifacient and root-bark as fomentation to relieve the spasm. The bark-juice mixed with molasses is administered in headache.

The leaf-juice is useful in hiccough and in higher doses is used as emetic. This is combination with honey is applied to eyelids in eye diseases. Its flowers are used as diuretic, cholagogue and stimulant and the fruits are considered antipyretic and anthelmintic and are eaten. The gum exudates of the plant is used for dental caries and after mixing with sesame oil is poured into ears for relief in otalgia (Chopra *et al.*, 1956). Besides, the seed oil in combination with equal parts of ground nut oil is applied locally to relieve pain in rheumatism.

Its root-powder is used as a good substitute of mustard powder.

Folk Uses

It leaves are cooked as vegetable curry and are eaten in influenza and catarrhal affections by the people and the crushed bark boiled in mustard oil is used as a balm in acute rheumatic pain.

Bhat tribals use a decoction of its fruits with equal quantities of the fruits of *Clerodendrum indicum. Sesamum orientale* and *Pipe longum* and *gur* daily for 20 days as strong sterilizer for regulating fertility Chenchus people in Andhra Pradesh, mix its stem-bark powder with little quantity of *methi* and leaves of *Murraya koenigii* and pound to make pills. These pills are taken orally to get relief from backache.

Preparation

Shobhanjndi lep, Syamdi churna, Sudarshan churn, Sarv-jwar-har louha and Shobhanjanadi kwath.

Family: FABACEAE

Genus: *Abrus* Adans.

The generic name has been derived from Greek word Habros meaning soft, referring to softness of the leaves.

This genus comprises of climbing shrubs. Leaves are with numerous deciduous leaflets. The flowers are small in dense racemes on axillary peduncles or short branches. Pods are oblong or linear-oblong, turgid or flat, moderately firm and thinly septate. The seeds are polished.

About 12 species occur in tropical parts of the world; 2 species are reported from India.

Species: *Abrus precatorius* Linn.

Description of Plant

It is much branched and climbing undershrub with woody stem. Leaves are paripinnate, 5.0-7.5 cm long and 16-40 folialate; leaflets are oblong to elliptic, deciduous, 0.6-2.5x0.8-1.2cm, membranous and glabrous or thinly silky beneath. Inflorescence is racemose, axillary or terminal and shorter than leaves. Flowers are reddish or white. Pods are 2.5-4.2cm x 0.5-0.8cm, oblong, slightly inflated and fulvopuberulent with 3-5 seeds. Seeds are globose or ellipsoid, 1.2cm long lustrous and commonly scarlet with a black spot surrounding the hilum or rarely white, or black with a white spot.

Flowering and Fruiting

Flowering: In August-September and Fruiting: During winter.

Distribution

It is common throughout India, ascending upto an elevation of about 1060 m in outer Himalayas; also found in Andamann and Nicobar islands.

Plant Parts Used

The seeds, leaves and roots are used.

Uses

The plant ahs been used for medicinal purpose by the Hindus from very early times and its is mentioned in Sushruta. Its seeds are purgative, emetic, tonic, antiphlogistic and aphrodisiac and are used in nervous disorders and as abortifacient. A paste of its seeds is used as local application in stiffness of shoulder joint, sciatica and paralysis. The leaves have sweetish taste and are used in biliousness, leucoderma, itching and other skin diseases. The fresh juice of its leaves mixed with a bland non irritating oil is applied on painful swellings and admixed with Chitrak roots (*Plumbago zeylanica*) in leucoderma. The juice of leaves is also applied to bare skin of alopecia for regrowth of the hairs. The decoction of its leaves is given incough and cold. The roots are described as diuretic, tonic, emetic and alexeteric and are used in preparations for gonorrhoea, jaundice and haemoglobinuric bile. An infusion and the paste of its seeds is included in British pharmacopoeia for medicinal use.

Besides, various parts of the plant are used in night blindness, inflammation of gums, muscular pain, convulsions, mucus in urine, gravel, diarrhea, and been fracture in cattles.

Its roots and leaves, sometimes are used as a poor substitute for liquorice.

Folk Uses

The paste of its seeds is applied on skin in leucoderma and other skin diseases. A decoction of its leaves and roots is widely used for coughs, colds and colic. Besides, the people in North-Eastern Karnataka, use a watery extract of its leaves for 15 days for blood purification.

Preparation

Gunjabhadra Ras. An extract of roots is regarded as official in Indian pharmacopoeia, dose ad libitum.

Genus: *Acacia* Mill.

The generic has been adopted from Greek word Akis= a sharp point, referring to the thorns of the plant.

The genus comprises of erect or climbing spinose or prickly shrubs or trees, armed with stipular spines and prickles. Leaves are bipinnate with minute leaflets. The flowers are in globose heads or cylindrical spikes and are hermaphrodite or polygamous and usually 5-merous. Calyx is funnel-shaped or campanulate. The pods are either dry or fleshy, dehiscent or indehiscent and rarely turgid or subcylindric. Seeds are compressed.

About 750-800 species are found in the tropical and sub tropical parts of the world. Of them, 25 species have been reported from India.

Species: *Acacia catechu* (Linn.) Willd.

Description of Plant

It is a moderate-sized tree with thorny branches. Bark is rough, dark, grey or grayish-brown and about 1.2cm thick, exploiting in long narrow strips. Leaves are 10.0-17.5 cm long, pinnate with a pair of recurved prickles at base of rachis and pinnae 40-80; leaflets are 60-100 small and ligulate. Flowers are pale-yellow in axillary cylindrical spikes. The pods are thin, blabrous, lustrous, straight, star-shaped, dark brown and 5-6 seeded. Seeds are flat, dark-brown and orbicular.

Flowering and Fruiting

Flowering: In rainy season and Fruiting: During November-January.

Distribution

It occurs in drier regions of Sub-Himalayan tract upto an altitude of 1200 m from Punjab to north-Eastern states, Madhya Pradesh, Gujarat, Andhra Pradesh and Tamil Nadu.

Plant Parts Used

The bark and the heartwood are the parts used generally.

Medicinal Uses

The bark of the plant is astringent and is useful in passive diarrhea usually in combination with cinnamon or opium. The decoction of its bark is given internally in leprosy. The *katha*, obtained from heartwood of 20-30 years old trees is described as astringent, cooling and digestive and is used in relaxed condition of throat, mouth, gums, cough and in diarrhea. This is also applied externally to ulcers, boils, and eruptions of the skins. In cracked nipples and sores of various etiologi, a decoction of its bark is used as compresses. For leprosy, the decoction of its *panchang* is given as drink with meal along with washing of the affected part with it and application of ointment of catchu to ulcers.

Kheersal, a product met within some older trees in the form of white crystalline deposit of catechism is used for treatment of cough and sore throat. The decoction of *kheersal* and *chirayata* is used in spleen enlargement during chronic fever.

Folk Uses

A powder of its resinous exudates is dusted over wounds in different parts of the country.

Preparation

Khadirarist, Khadir sar, Khadiradi kwath, Khadirastak, Kadiradi vati, Kutajarista, Lavangadi vati, Chandanadi vati, Vrihtkhadiradi vati, Jarurkhula and Marham kharish jadin.

Species: *Acacia farnesiana* (Linn.) Willd.

Description of Plant

It is a small sized bushor about 3.0-4.5m high with dark brown and smooth bark. The branches are spreaded out from the main trunk which also bear brown streaks. Leaves are bipinnate with stipular spines and 0.7-1.2 cm long; leaflets are 20-40 on each pinna, very small and rigid. The flowrs are borne in axillary heads and are gragrant and bright yellow. The pods are dull-brown, 5.0-7.5 x 1.2cm, cylindrical and turgid with a double row of seeds.

Flowering and Fruiting

Flowering: In winter and Fruiting: During rainy season.

Distribution

It occurs wild throughout India, often planted in gardens.

Plant Parts Used

The leaves and bark are used.

Uses

Its bark is described as astringent and demulcent and is useful is polyuria. An infusion of its tender leaves is given internally in cases of gonorrhoea. Besides, various plant are parts used in madness, carbuncle, epilepsy, rabies, convulsions, delirium, sores, cholera, snake-bite, rinderpest and in sterility of women.

The gum exudate of the plant is used as a substitute for Arabic gum.

Folk Uses

In Punjab, the plant is supposed to be obnoxious to rats and snakes and is accordingly planted as a protection against the injury caused by these animals burrowing in embarkments.

Species: *Acacia nilotica* Delile ssp. *indica* (Benth.) Brenan *(A. Arabica* (L.amk.) willd. Var., *indica* Benth.; A. *nilotica* var. *indica* (Benth.) Mill A. *arabica sensu* Baker)

Description of Plant

It is a moderate-sized tree with a short trunk, spreading crown and feathery foliage usually with a height of 15m. The bark is dark-brown and much fissured. Leaves are 2.5-5.0 cm long, bipinnate with spinescent stipules; leaflets are in 10-20 pairs, long, linera and glabrous. The flowers are yellow and fragrant in axillary globose heads forming axillary cluster of 2-5 heads. The pods are generally solitary, 8-12 seeded, white, flat and about 7.0-15.0 cm long.

Flowering and Fruiting

Flowering: In rainy season and Fruiting: During cold season.

Distribution

It is found throughout the drier regions of India, particularly in Andhra Pradesh and Maharashtra.

Plant Parts Used

The parts used are the bark, pods, leaves and gum.

Uses

Its bark is used as astringent and demulcent. The powder of its bark is applied externally in ulcers and the decoction is used as a gargle in sore throat and toothache and also in chronic dysentery and diarrhea as an astringent enemata. The decoction of its pods is said to be beneficial in urinogenital diseases. An infusion of its tender leaves is given as an astringent and as a remedy for diarrhea and dysentery. Its gum is described as astringent and styptic and is useful indiarrhoea, dysentery and diabetes mellitus.

That its gum is used as a substitute for arabic gum.

Folk Uses

The twigs of the plant are used as datum (natural tooth brush) to strengthen the gums and teeth.

Preparation

Babularisht, Lavangadi vati, Akakia, Hab awazkusha and Sunun poast mughlian.

Species: *Acacia Polycantha* Willd. [*A suma* (Roxb.) Kurz; *A. suma* Buch.-Ham. Ex Voigt]

Description of Plant

This is a medium-sized or large tree with white bark exfoliating in papery flakes and marked at intervals by horizontal patches of dark colour. Pinnae are 10-20 paired; leaflets numerous linear, grey or grayish-green when dry. The flowers are pale-yellow or white in lax spikes. Pods are 7.5-10.0 x 1.3-3.5 cm and pubescent when young with 6-8 seeds.

Flowering and Fruiting

Flowering: during rainy season and Fruiting: In winter.

Distribution

It is reported to meet wild in West Bengal, Bihar and Maharashtra.

Plant Parts Used

Bark, leaves and the pods are usable.

Uses

The bark of the plant as astringent and its decoction is used as a gargle in sore throat and toothache and dry powder is applied in ulcers. An infusion of its tender leaves is also used as an astringent and a remedy for diarrhea and dysentery. Besides a decoction of its pods is also used successfully in urino-genital diseases.

Species: *Acacia Senegal* Willd.

Description of Plant

It is a small thorny tree with smooth and shining bark. Pinnae are in 3-5 pair; common petiole oftern armed with minute prickles; leaflets 8-12 paired. Grey, cillate.

The flowers are white and fragrant in lax pedunclulate spikes. Pods are thin, straight and pubescent when young, tardily dehiscent and 7.5 x 1.8cm.

Flowering and Fruiting

Flowering: During August-December and Fruiting: Later.

Distribution

It is found wild in Punjab and in the forests in Aravalli range.

Plant Parts Used

Gum, obtained from the tree is used medicinally.

Uses

A gum of the tree is known as demulcent and emollient and externally is applied over burns, sore nipples and other inflamed surfaces and internally used in the inflammation of intestinal mucosa.

Besides, it is also used a s substitute for amylaceous food in diabetes, since it is not cohered into sugar.

Folk Uses

The powdered gum of this plant is blown up into nostrils to check severe epixtaxis by the people in different parts of the country.

Species: *Acacia tomentosa* Willd.

Description of Plant

This is a small tree. Its branchelts and leaf-rachis are densely covered with fine grey hairs and leaf-rachis also with glands. Bark is grey-black and cracked into quadrangular pieces. Leaves are about 10.0cm long and are crowded on arresting axillary branches; leaflets 8-16 paired, sessile, oblong-linear and glabrescent. The flowers are white and are borne in axillary pedunculate heads. Pods are 10.0-15.0 x 1.0-1.5 cm, thin, flat, ligulate, short-stalked and slightly falcate with 6-10, brownish-green seeds.

Flowering and Fruiting

Flowering: During summer season and Fruiting: In rainy season.

Distribution

It is found wild in Sunderbans and Western parts of India.

Plant Parts Used

Pods, leaves and bark are the parts, used medicinally.

Uses

A decoction of its pods is beneficial in urino-genital diseases. An infusion of its tender leaves is taken as aremedy for diarrhea and dysentery. Besides, decoction its bark is also used as a gargle in sore throat and in toothache; dry powder applied externally in ulcers.

Folk Uses

Native people use a decoction of its bark as a gargle in teeth and gum problems in Western parts of the country.

Genus: *Bauhinia* Linn.

The generic name has been assigned in the honour of John and Casper Bauhin, German botanists of the 16th century.

The genus comprises of tree, shrubs or climbers with circinate tendrils. Leaves are simple, more as less distinctly bi-lobed and palmately curved. The flower are usually showy and white, pink or purple in simple or panicled racemes. The pods are linear or oblong, flat and dehiscent or indehiscent. Sees are albuminous.

According to Airy Shaw (1973), its 300 species are found distributed in the warmer parts of the world. Only 8 species have been reported from India.

Species: *Bauhinia malabarica* Roxb.

Description of Plant

It is a moderate-sized deciduous tree with dark brown bark, exfoliating in thin long strips. Leaves are broader than longer with 2 obtuse lobes, slightly cordate, glabrous, glaucous beneath and 3.7-10.0 x 5.0-12.5 cm and 6-9 nerved. Inflorescence is a raceme, about 3.7-5.0 cm long, dense and often with 2-3 together flowers. The flowers are 1.2 cm long and reddish-white and are borne on 2.5cm long slender pedicels; the buds are obovoid and rounded at tip. Pods are stalked upto 17.5-30 x 1.7-2.5, acuminate, firm, glabrous and nearly straight with regular reticulate veins which start diagonally from both sutures and meet in middle, Seeds are 20-30.

Flowering and Fruiting

Flowering: During September-November and Fruiting: In December-January.

Distribution

The plant is found wild in Sub-Himalayan tracts, Bengal, Assam and South India.

Plant Parts Used

The flowers and leaves are used.

Uses

The plant is said to be used as medicine in skin diseases. An infusion of its new flowers is given in dysentery. Besides, the decoction of its leaves is used in piles externally, where the patients are advised to sit in a tub having the decoction which pay relief in piles.

Folk Uses

An infusion of its new flowers is taken in the dysentery in some parts of the country.

Species: *Bauhinia vahlii* Wight and Arn.

Description of Plant

The plant is a large climber with densely pubescent branchlets and copius circinate tendrils. Stem may be upto 30m long. Leaves are broader than long, rigidly subcoriaceous, deeply cordate, 11-13 nerved with obtuse loboes and densely pubescent beneath; petiole is stout and 7.5-15.0 cm long and stipules obtuse and falcate. Flowers are white and 1.8cm across and are borne in dense subcorymbose racemes. Calyxtube is elongated and limb splitting into 2-lobes. Stamens are 3. The pods are pendulous, flat, 15.0-35.0 x 5.0-7.5cm, woody and clothed with dense rusty coloured tomentum and dehiscent.

Flowering and Fruiting

Flowering: In April-June and Fruiting: During winter.

Distribution

The plant is found distributed in Sub-Himalayan tract upto 915 m and in Assam,a Bihar and Madhya Pradesh.

Plant Parts Used

Leaves, seeds, roots pods and bark are used.

Uses

The leaves of the plant are described as demulcent and mucilaginous and are used in malaria. Its seeds are used as tonic and aphrodisiac. The bark is astringent and is used in dysentery and the roots as demulcent. The root fibres and leaves are said to be very useful in malaria.

The gum exudates of the plant known as Bauhinia gum is used as a substitute for Arabic gum.

Folk Uses

One teaspoonful aqueous extract of its roots is given thrice a day orally for the treatment of syphilis for three days indifferent parts of the country.

Species: *Bauhinia variegate* Linn.

Description of Plant

The plant is a medium-sized deciduous tree. The bark is grey or brownish with longitudinal cracks and light pink inside. Leaves are 10.0-12.5 cm long, 1-foliate,2-lobed, not deeply cleft, rigidly subcoriaceous and deeply cordate. Flowers are pink coloured and sessile or borne in short peduncled corymbs. Pods are 15.0-30.0x1.2-2.0cm long, flat , hard, compressed and dehiscent. Seeds are 10-15 in a pod.

Flowering and Fruiting

Flowering: During November-December and Fruiting in Janbuary-February.

Distribution

It is met in the Sub-Himalayan tract from the Indus eastwards; also found in dry forests of Central, Eastern and Southern India.

Plant Parts Used

Flowers, bark and roots are used.

Uses

The flowers and dried buds of the plant are anthelmintic and are used in diarrhea, piles and dysentery. The bark is tonic, astringent and anthelmintic and an emulsion of the powder of its bark with rice water mixed with ginger is used in scrofula and cutaneous affections. The decoction of its root-bark is carminative and is used in dyspepsia and flatulence. Besides, its flowers with sugar are used a gentle laxative.

Folk Uses

The decoction of its bark is used indiarrhoea and that of its roots as antifat remedy by the native people.

Preparation

Kachnar guggulu, Kachnaradi kwath, Kachnar gudika, Hab mussafi khum, Sufuf kalan and Majun suparipak.

Genus: *Butea Roxb.* Ex Willd.

The generic name has been assigned in the honour of John Earl of Bute, a botanical author and patron of botany in the 18th century.

The genus comprises of trees or climbing shrubs. Leaves are 3-foliate and stipellate. The flowers are large, densely fascicled and showy in axillary racemes or terminal racemes or panicles. Calyx is broadly campanulate. Corolla is much exerted. Stamens are 2-adelphous and ovary is sessile or stalked and 2-ovuled. The pods are ligulate, firm, splitting round the single apaical seed and indehiscent below.

About 30 species are distributed in Indo Malayan region and China. Out of which 3 have been recorded from India.

Species: *Butea monosperma* (Laamka.)Taub. (*B. frondosa* Koen, ex Roxb.; *Erythrina monosperma* Lamk.)

Description of Plant

It is a moderate sized deciduous tree upto 15m high and 1.5-1.8m girth. The bark is fibrous, bluish-grey or light brown and exuding a ruby red viteous gum. Leaves are pinnately 3-foliate, large, unequal and 10.2-20.4 cm. The flowers are 3.8-5.0 cm long and orange red and are fasicled on rigid axillary and terminal racemes. The pods are silvery-white and broad. Seeds are flat, elliptic, reddish-grey and 3.2 cm across.

Flowering and Fruiting

Flowering: In February-March and Fruiting: During April-May.

Distribution

It is common throughout India upto 1219 m except in very arid parts.

Plant Parts Used

The seeds, flowers, bark, roots and the leaves are used.

Uses

Its seeds are described as anthelmintic, acrid, bitter, aperient and rubifacient and are used in flatulence and piles. A decoction of its seeds is given in gravel and paste of powdered sees with lemon juice is applied in herpes and as a cure of ringworms and also for cooling effect. Fine powder of its seeds along with *Acorus calamus* rhizomes or mixed with juice of *Cyperus rotundus* rhizomes is used as a cure for delirium. Its flowers are diuretic, depurative, astringent and aphrodisiac and are used as emmenagogue and as poultice in orchitis and to reduce swellings in bruises and sprain also. A decoction of its flowers is given in diarrhea and to puerperal women. A lotionis prepared after distilling its flowers which is used for some eye diseases. The flowers with leaves and rots of *Hygrophila auriculata* are given with milk in leucorrhoea and their juice is given to induce sterility in women. The bark is astringent, pungent, alterative, aphrodisiac and anthelmintic and is useful in tumours, bleeding piles, and ulcers. The decoction of its bark is used in cold, cough, fever, haematuria and menstrual disorders and in bloody diarrhea. The gum is astringent and is used in diarrhea and dysentery. Fresh gum is applied on ulcers and septic sore throat and its infusion is used as local application in leucorrhoea. The gum-solution is also applied to bruises, ring worms and erysipelatous inflammation.

The leaves of the plant are described as diuretic and aphrodisiac and are used to cure pimples, boils, tumours and haemorrhage; also given in flatulence, colic, worms and piles. The juice of leaves is used in skin diseases and with cow's milk is used as slow sterilizer. The roots are used in night blindness and cause temporary sterility in women. The root-bark is considered as aphrodisiac and is used in elephantiasis, as analgesic and anthelmintic and is also applied in sprue, piles, ulcers, tumours and dropsy. The bark is given in conjunction with ginger in cases of snake-bite. The boiled flowers are tied over abdomen in pain and swelling of kidney and relieve urine and roots rubbed with water are dropped in nostrils in epilepsy fits.

The gum of the plant is used as a substitute for the gum of Pterocarpus marsupium.

Folk Uses

Its warmed and dried flowers are tied over abdomen and swellings to testicles throughout the different parts of the country. The people in Pauri Garwhal, use its seeds as antidote for snake-bite and with lime juice to remove dhoby's itches.

Preparation

Palashbijadi churna, Palashkshar ghirt, Palashadi kwath, Calcury (used against ureteric calculus), Majun zanzgbil and Mujun supari pak.

Genus: *Caesalpinia* Linn.

The genetic name has been assigned in honour of Andreas Caesalpoini, a Professor of medicine at Pisa in the 16th century.

The genus comprises of erect trees, shrubs or prickly climbers. Leaves are large and abruptly pinnate. The flowers are showy and yellow and are borne in axillary or terminal racemes, the latter often corymbose. Calyx lobes are 5, imbricate and hood-shaped, the lowest is largest. Petals are 5 and clawed, the upper most is smaller.

Stamens are 10, free and declinate. Ovary is sessile and few ovuled and style is filiform. The pods are oblong or ligulate, thin and flat or thick and subturgid, dehiscent or indehiscent and smooth or armed all over with spines.

About 100 species occur widely distributed in the tropics and subtropics. Out of which 9 species have been reported from India.

Species: *Caesalpinia bonduc* (Linn.) Roxb. (C.*crista* Linn.; C. *bounducella* (Linn.) Fleming; *Guilandina bonduc* Linn.)

Description of Plant

It is a large scandent shrub with small and yellow branches and is covered with downy pickles. Leaves are about 30cm long and bipinnate; leaflets are 12-16, elongated, upper part thick, 2.0-2.5 x 1.0-1.5 cm, oblong or elliptic and obtuse. The peduncles are long and bear dense flowers at the top which are yellow in colour. The pods are shortly stalked, prickly wingless, swollen in the middle and slightly flat, Seeds are 2-3, lead coloured, 2.5 cm long and bean shaped.

Flowering and Fruiting

Flowering: During July-September and Fruiting: In December-January.

Distribution

It is found throughout the hotter parts of India particularly along the sea coast and upto 830 m. Common in West Bengal and South India. Also cultivated as hedge plant in the villages.

Plant Parts Used

The seeds, leaves, roots and bark are used.

Uses

It seeds are described as antiperiodic, tonic and ferrifuge and in powder form they are used in fever asthma and colic in the doses of 0.7-2.0 gm with equal parts of black pepper and externally in inflammation. The oil of the seeds is emollient and efficacious for stopping discharges from ear and other skin diseases. This oil is used as embrocation to remove freckles from the face. Its tender leaves are used for disorders of liver. These are boiled with castor oil or butter fat and are applied externally on painful and swollen testicles. The leaves and bark are used as emmenagogue, anthelmentic and febrifuge. Besides, adecoction of its roots is said to be useful in calculus and with honey is used for leucorrhoea.

Folk Uses

In Andhra Pradesh, leaf-paste is bandaged on the swellings and tumours and bruised leaves are bandaged on the testicles to cure the inflammation due to hydrocele and hernia.

Preparation

Bijai churna, Karanjadi ghrit, Karanjadi churna and Visham javarghani vat.

Species: *Caesalpinia sappan* Linn.

Description of Plant

This is a small tree or shrub with yellow wood which turns red quickly when cut freshly. Prickles on branches are thinly dispersed and small. The flowers are yellow; peduncle ad along as the rachis. Pods are 7.5-10.0 x 3.5-4.0 cm, prickly and slightly flated.

Flowering and Fruiting

Flowering: During summer and Fruiting: In winter season.

Distribution

It is found wild in South India and Bengal; also cultivated in different parts of the country.

Plant Parts Used

Wood of the plant is used.

Uses

The wood decoction of the plant is considered emmenagogue and is used in mild cases of dysentery and diarrhea. In the form of a paste, it is also employed in some of skin diseases, especially in lichen.

It is also used as a substitute for logwood.

Genus: *Cassia* Linn.

The generic name is the classical name of some trees with aromatic bark.

This genus comprises of the shrub or trees and rarely of herbs. Leaves are abruptly pinnate. The flowers are large in axillary racemes and terminal panicles. Calyx tube is very short. Sepals are broad or narrow. The pods are variable, flat or terete usually septate, dehiscent or indehiscent and dry. The seeds are sometimes parallel to the valves and sometimes to the septa.

There about 500-600 species occur in tropical and warm temperate regions of the world excluding Europe; of which, 24 species are found in India, mainly in tropical Himalayas and Western Peninsular India.

Species: *Cassia alata* Linn.

Description of Plant

This is a shrub with very thick downy branches. Leaves are simple, pinnate, subsessile, 30.0-60.0 cm long; leaflets oblong, obtuse, minutely mucronate, subcoriaceous and glaucocus or obscurely downy beneath. Flowering are large, yellow in peduncled racemes. Pods are about 10.0-20.0 cm long, straight, membranous, dehiscent and glabrous with numerous seeds.

Flowering and Fruiting

Flowering: During October-November and Fruiting: In December-January.

Distribution

It is reported to meet wild in lower Bengal, also cultivated in other parts of India.

Plant Parts Used

Leaves and flower are used.

Uses

That a decoction of its leaves and flowers is used internally in bronchitis and asthma; also for washing eczematous patches. An ointment prepared by mixing its leaf paste with the equal weight of simple ointment or borax is beneficially applied in various skin diseases. Besides, its flower extract along is also given internally in bronchitis.

Folk Uses

A decoction of its leaves is used to cure dermatitis and indigestion by the native people in Rajasthan.

Species: *Cassia auriculata* Linn.

Description of Plant

This is a tall shrub reaching a height upto 2.5 m with finely downy spreading branches. Leaves are nearly sessile, 7.5-10.0 cm long, rachis grooved, pubescent and furnished with a single linear gland between the leaflets of each pair; leaflets 8-12 paired, ovate-oblong, obtuse or emarginated, mucronate, subsoriaceous and downy. Flowers are about 5.0 cm across and are borne in corymbose racemes. The pods are 10.0-12.5 cm long, straight, ligulate, obtuse, flexible and dark brown with 10-12 seeds.

Flowering and Fruiting

Flowering: During cold season and Fruiting: Later.

Distribution

It is found wild in dry regions of Madhya Pradesh, Chennai, Rajasthan, Andhra Pradesh and Karnataka.

Plant Parts Used

Its bark, roots and seeds are used medicinally.

Uses

That bark and roots of the shrub are considered astringent and are used in cutaneous diseases. Besides, seeds are used in opthalmia and conjunctivitis, in diabetes and chylous urine, seeds of plant with their testa and kernels are finely powdered and are blown into the eyes or its powder mixed with Coconut or gingelly oil is applied to sore eyes.

Its seeds are sued as good substitute of Coffee (*Coffea aravica*) and are prescribed in giddiness due to heart diseases.

Folk Uses

In Gujarat, pessaries made from its flowers are used to check excessive menstrual flow. Bhills of Rajasthan take stem bark of the plant orally in cough and colds. Besides, the twigs of the plant are also used as tooth brush to strengthen the gums.

Species: *Cassia fistula* Linn.

Description of Plant

It is a moderate sized and wholly glabrous tree with greenish-grey bark and about 3.7-4.8m in height. Leaves are paripinnate and 5.0-12.0 cm long; leaflets are 8-16, acuminate and ovate-lanceolate. The flowers are yellow in lax pendulous racemes and fragrant. The pods are about 30-40 cm long and 2.0-2.5 cm across, cylindrical, smooth, hard, indehiscent and yellowish-grey. Seeds are about 0.6 cm in diameter, compressed and albuminous and are parallel with septa.

Flowering and Fruiting

Flowering: In April-July and Fruiting: During August-September.

Distribution

It grows wild in many parts of India; also cultivated as an ornamental plant for attractive yellow blossoms in pendant racemes.

Plant Parts Used

The pods, leaves and root bark are used.

Uses

An ethanol extract of its pods and of stem-bark is considered hypoglycaemic, antiviral and anticancer. It is used in cancer, epilepsy, convulsions and delirium febris, pimples, burns, syphilis, dysuria, haematuria, gravel and diarrhea. The pulp of its fruits is used as a tonic and in chest infection and with sugar is taken in constipation; also useful in heart and chest diseases. Its leaves are ground into a paste which is applied to the ringworm lesions and after mixing with bark and rubbed with oil are used efficaciously in Chilblains ringworm, pyoderma, insect bites, eczema, scabies and psoriasis. The seeds are emetic and are used in jaundice and the bark paste internally in blindness. Besides, the fruit paste is said to be applied in snake-bite.

The gum of the seeds is used a a substitute for Gaur gum obtained from Cyamposis tetragonoloba.

Folk Uses

Native people take its leaf-juice for purification of the blood. In Phulbani (Orissa), a paste of some young leaves is kept inside the genitals of women once daily tocure amenorrhoea for a week.

Preparation

Aragvadhadi-kshar, Aragvadhadi taila, Aragvadhadi leh, Aragvadharishta, Aragvadhadisutravarti and Aragvadha kwatham.

Species: *Cassia obtusifolia* Linn.
(*C. tora* Linn.)

Description of Plant

It is an annual bushy foetid herb or under shrub about 30-150 cm high with flexous stem. Leaves are paripinnate and 2.5-3.8 cm long; leaflets are ovate-oblong, obtuse rounded and glabrous above. Petiole is 3-4 cm long and stipules 1.0cm long and fugacious. The flowers are small, yellow and 5-6 together. Pods are slender, sub-4-angled and 12.5-15.0 x 0.8 cm. Seeds are many, dark brown and broadly ovoid.

Flowering and Fruiting

Throughout the year, chiefly during July-December.

Distribution

It is found distributed throughout the tropical parts of India.

Plant Parts Used

Leaves, seeds and roots are used.

Uses

The decoction of its leaves is used as laxative. Its leaves are pounded and when applied on cuts, act like tincture of iodine; also applied on eczema and other skin diseases. Tender leaves are taken internally to prevent skin diseases. The leaves are also prescribed in decoction form in 60gm doses for children suffering from feverish attacks while teething and after boiling in caster oil they are applied to foul ulcers and also on inflammation caused by any irritant.

The seeds boiled with tea are taken for cold and their powder is given in abnormal delivery; with turmeric and mustard oil are used against eczema. Water soluble senna extract is used in constipations. Its seeds are steeped in the juice of Euphorbia nerlifolia and are made into a past with cow's urine, which is used an application to cheloid tumours. These are also useful in leprosy and psoriasis, when ground with sour butter milk or lime juice and applied to ease the irritation of itch or skin eruptions. Its roots are also rubbed into a paste with lime juice and are used for ring worm and applied for buboes in plaque. An oil called *Chakramarda* containing *Cassia tora* and *Eclipta* prostrate is very useful application in obstinate skin diseases such as ring worm etc. Besides, ethanol (50 per cent) extract of the plant is considered antiviral, spasmolytic and diuretic and is used in epilepsy, scabies and sores.

The seed gum, known as Chakunda gum is used as a substitute for Guar gum.

Folk Uses

The seeds of the plant and haldi are taken in equal quantities and made into a paste which is applied as poultice in gonorrhoea by the Kondh tribals in Eastern India.

Preparation

Dadrughnivati and Pamari taila.

Species: *Cassia occidentalis* Linn.

Description of Plant

it is diffuse under shrub or herb with 60-120cm height. The stem is angular, striated with reddish purple and glabrous except short gland-hairy young parts. Leaves are paripinnate with foetid smell. Peduncle is small bearing a few yellow flowers having reddish ringe. The pods are laterally compressed, slightly upcurved, appressed-short hairy, 12.0-15.0 x 0.8-1.0 cm and dehiscent. Seeds are many and 0.4-0.42 x 0.25-0.26 cm with an aerole of 0.28-0.3 x 1.5cm.

Flowering and Fruiting

Throughout the year, chiefly during July-December.

Distribution

It is found scattered from the Himalayas to South India.

Plant Parts Used

The seeds, leaves and roots are used.

Uses

The whole plant is said to be febrifuge, purgative, diuretic and tonic. Its leaves and seeds are used as purgative and the root as diuretic, purgative and antiperiodic. The seeds and leaves are used externally in skin diseases and as antiperiodic. Its root juice is used in ringworm. Its leaves are pounded and made into a paste which applied to fresh wounds to bring their healing by first intension. Besides, the seeds are used externally in the treatment of scabies.

Its seed gum is used a substitute for Guar gum obtained from Cyamposis tetragonoloba.

Folk Uses

Its seeds are used as a substitute of coffee and the decoction of the plant is as food diapahoretic in various parts of the country.

Preparation

Parval bhasm.

Species: *Cassia sophera* Linn.

Description of Plant

it is diffuse, subglabrous bushy undershrub with green branches reaching a height upto 2.5-3.0m. Leaves are paripinnate; leaflets in 6-7 pairs and are oblong-lanceolate. Flowers are small, yellow in axillary distinctly peduncled corymbs. The pods are 7.5-10.0 cm long, slightly turgid, not impressed between seeds. Seeds are 30-40 and dark brown in colour.

Flowering and Fruiting

During November-July.

Distribution

It is very common throughout the tropical parts of India.

Plant Parts Used

Whole plant is used.

Uses

The whole plant, its seeds, stem bark and roots are used to cure cough and its decoction as expectorant in acute bronchitis. Juice of the leaves is mixed with Sandal wood to form a paste or a paste made from its roots with Conjee is applied on '*dhobhi itch*' and a decoction of its leaves is given in asthma and hiccup. Besides, an ointment prepared from its bruised seeds, leaves and sulphur or from its root bark powder after mixing with honey, is applied over ringworm and patches of pityriasis and psoriasis.

Its roots are used as a substitute and adulterant to Cassia occidentalis roots.

Folk Uses

The juice of young leaves of the plant is applied in ringworm by the native people in different parts of Uttar Pradesh.

Genus: *Dalbergia* Linn. f.

The generic name has been assigned in the honour of Nicholas Dalberg, a Swedish botanist who died in 1820.

The genus comprises of trees and rarely of climbing shrubs. Leaves are imparipinnate with alternate, sub-coriaceous leaflets. The flowers are copious and small in terminal or lateral panicles. The bracts and bracteoles are small. The pods are thin, flat, coriaceous, 1-4 seeded and indehiscent and are not thickened or winged at the sutures.

About 300 species are distributed in the tropical and subtropical parts of the world and South Africa. 25 species have been reported from India.

Species: *Dalbergia sissoo* Roxb.

Description of Plant

It is fairly a large deciduous tree with grey bark upto 15-18 m in height. The branches are downy, grey and spreading. Leaves are 3-5, alternate, imparipinnate, bifarious, roundish, acuminate and 3.5-6.0x3.0-5.4 cm. The flowers are sessile in axillary panicles shorter than leaves and yellowish white. The pods are 3.5-4.5 cm long, thin, strap shaped and pale grey. Seeds are 1-4, kidney shaped and flat.

Flowering and Fruiting

Flowering: In summer and fruiting during winter.

Distribution

It occurs in the Western Himalays upto 1500 m Sikkim to upper Assam; extensively planted along road sides in different parts of India.

Plant Parts Used

The leaves, bark, wood and roots are used.

Uses

That its leaves are considered bitter and stimulant and their decoction is used in gonorrhoea. The bark is alterative and dried bark is haemostatic and is effectively used in bleeding piles, menorrhagia and in varicose veins. The wood is alterative and is used in leprosy, boils and eruptions and to allay vomiting. Besides, the roots are used as astringent.

The plant is also used in Unani system of medicine. The drug in Unani called, *Shisham katel* which is used as counter-irritant, antisyphilitic and blood purifier. The mucilage of its leaves mixed with sweet oil is used in excoriations. The raspings of its wood are boiled in water until these become half, then mixed with 'sharbat' of shishamare advised to drink about for forty days in leprosy.

Folk Uses

Its leaves are warmed and tied over breast of women in swellings and their juice mixed with sugar and cured is given to cure blood dysentery (Singh and Maheshwari, 1983).

Genus: *Derris* Lour.

The generic name has been adopted from Greek work Derris = a skin or leathery cover, referring to the nature of pods.

The are climbers or three. Leaves are oddpinnate; leaflets are opposite and usually exstipellate. The flowers are copious in axillary or terminal racemes or panicles. Calyxis campanulate and truncate . Corolla in much exerted, standard board, silky, keel obtuse and the petals are cohering slightly. Stamens are monoadelphous and the anthers versatile. Ovary is oblong and versatile. The pods are oblong and indehiscent.

About 80 species are found in the tropical parts of the world. Out of which 25 species have been reported from India.

Species: *Derris indica* (Lamk.) Bennet (Syn. *Galedupa indica* Lamk.; *Pongamia glabra* Vent.; *P. pinnata* (Linn.) Pierre)

Description of Plant

It is a moderate-sized glabrous tree with a short bole and spreading crown and soft grayish-green bark. Leaves are imparipinnate; leaflets are 5-7, elliptic or ovate and 5.0-17.5 cm long. The flowers are liliac or white, tinged with pink or violet and fragrant and are borne in axillary racemes. The pods are woody, compressed, glabrous, indehiscent, elliptic to obliquely oblong and 4.0-7.5 x 1.7-3.5 cm. Seeds are 1-2 white, elliptic or reniform, wrinkled and 1.7-2.0 x 1.2-1.8 cm with reddish-brown leathery testa.

Flowering and Fruiting

During May-June.

Distribution

It is found throughout India upto 1200 m; also in the Andamanns.

Plant Part used: The seeds, flowers, leaves, bark and roots are used.

Uses

The plant is used both in ayurvedic and Unani system of medicines. Its seeds are efficacious in whooping coughs and irritation and their pulp is applied on leprosy. The seed oil is applied to skin diseases including herpes, scabies and sores. A mixture of its oil and zinc oxide (3.8 gm: 31.1 gm) is applied usefully in eczema. He further stated that sometimes, the oil is used internally as a stomachic and cholagogue in cases of dyspepsia with sluggish liver. Anembrocation made of equal parts of the oil with lemon juice forms and application in psoriasis, pityriasis and pruigo capitis. Its flowers are used as a remedy for diabetes. The leaf-juice is used in cough, flatulency, dyspepsia and diarrhea. A paste of its leaves is used internally in ulcers infested with worms. In leprosy, the leaves with that of *chitraka* are made into a powder and are given with curd (Nadkarni, 1954). Its bark is used externally in bleeding piles and decoction of the same is given is beri-beri (Ambasta, 1986). The juice of the roots by itself or which *Neem* or *Nirgundi* juice is effective to destroy worms of foul ulcers and fistulous sores and also used for cleaning teeth and strengthening the gums.

Folk Uses

Locally, a decoction of its bark is taken in beri-beri diseases. It leaves boiled with broken-rice and leaves of *Ach* (*Morinda citrifolia*) are made into a salt gruel, which is used as a good nutritive food for children. Besides, its seeds are ground into a paste and applied externally for knee and hip joints for rheumatic diseases by the Irula tribes in South India.

Preparation

Karanjadi churna, Karanjadi ghrit, Karanjadi taila, Karanjadi yog, Vidangadi let and Habb karanjaba

Species: *Derris trifoliat* Lour.
(D. *uliginosa* Benth.; *Robnia uliginosa* Willd.)

Description of Plant

It is a large climbing shrub with glabrous branches and dark grey bark. Leaves are alternate, odd-pinnate, 12.5-20.0 cm long; leaflets 3.7-10.0 x 3.5-5.0 cm, subcoriaceous, ovate or ovate-oblong, acute or acuminate and glabrous with rounded or cordate base. The flowers are red coloured, 0.9 cm long in axillary racemes. The pods are obtuse, suborbicular, glabrous and winged only at upper suture. Seeds are 1-2, reniform, compressed and about 3.8 cm long.

Flowering and Fruiting

Flowering: In rainy season and fruiting during winter.

Distribution

It is known to occur wild along the sea-coasts in Maharashtra, Tamil Nadu, Orissa, Sunderbans and the Andamans.

Plant Parts Used

Root bark of the plant is usable.

Uses

The root bark of the shrub possesses alterative properties and is successfully used in rheumatism and dysmenorrhoea.

Besides, the bark is also used as a fish poison.

Genus: *Mucuna* Adans.

The generic name, Mucuna is the Brazilian name of the plant.

The genus comprises of climbing annual or perennial herbs or undershrubs. Leaves are simple, 3-foliate and stipellate. The flowers are large, showy and purple turning black when ripe. Calyx-tube is campanulate. Corolla is exerted. Stamens are 2-adelphous and anthers dimorphous. Ovary is sessile and many–ovuled. The pods are variable in size and are usually covered with needle-like bristles.

About 120 species occur in tropics and subtropics of the world, 15 species are reported from India, mostly in Eastern Himalayas and Western Peninsula.

Species: *Mucuna pruriens* (Linn.) DC. (*M. prurita* Hook.; *Dolichos pruriesns* Linn.; *Stizolobium pruriens* (Linn) Medicus)

Description of Plant

It is an annual herbaceous twiner. Leaves are grey-silky below and 3-foliate; leaflets ovate–oblong or ovate–rhomboid, unequal at base and 12.0-15.0 x 6.0-7.0 cm. The flowers are dark violet in pendulous racemes. The pods are curved, turgid, longitudinally ribbed, 5.0-8.0 cm long and clothed with grey bristles. Seeds are 4-6 and ovoid.

Flowering and Fruiting

Flowering: During September-October and Fruiting: In December- March.

Distribution

It occurs wild almost throughout India; also in Andamann and Nicobar islands.

Plant Parts Used

Pods, seeds, roots, and leaves are used.

Uses

The pods are anthelmintic and considered most active against *Taenia caunina* and *T. paraphistonum*. Its seeds are aphrodisiac and nervine tonic and are used in scorpionstings. Experimentally showed that an aqueous extract of its seed powder increase the sexual activity. The roots are diuretic and nervine tonic and their decoction is beneficial in urinary problems and delirium in fevers. A strong infusion of its roots mixed with honey is given in cholera. The roots are powdered and made into a paste that is applied to the body in dropsy.

A paste of its leaves with water which is used to heal ulcers. The hairs after prolonged heating are used in form of an infusion in cases of gall and liver disorders. Besides, an ointment in oil is prepared from its root-paste which is used in elephantiasis, as a local stimulant and mild vesicant.

Folk Uses

A piece of cloth dipped in decoction of its roots is inserted in vagina as *'yoni sanchochan yog'* to make vagina contractive. Besides, the piece of roots is also tied over wrist in dropsy and seeds are applied over scorpion stings.

Preparation

Vanari gutika, Masabaladi pachan, Kaunch pak and Itriphal kabir

Genus: *Ougeinia* Benth.

The generic name has been taken from Ujjain, a town in India where seeds of the plant were sent in 1795 to Dr. Roxburgh of Calcutta Botanical Gardens.

This monotypic is confined to India only in its distribution.

Species: *Ougeinia oojeinensis* (Roxb.) Hochr. (*O. dalbergioides* Benth.)

Description of Plant

It is medium-sized deciduous tree with a short and often crooked trunk and about 12 m height. Bark is deeply cracked, thin and grey. Leaves are pinnately trifoliate; leaflets 7.5-15.0 cm long, roundish or obovate, entire or obscurely crenate and coriaceous. The flowers are many, fragrant and pink or white in short fascicled racemes. The pods are flat, light brown and 5.0-10.0 cm long. Seeds are 2-5, flat, smooth, 1.2 cm long and brown when mature.

Flowering and Fruiting

Flowering: During March–April and Fruiting: In May–June.

Distribution

The plant is found throughout the outer Himalayas from Jammu to Sikkim and Deccan India; also planted ornamentally.

Plant Parts Used

The bark is used.

Uses

The bark is used as febrifuge. A decoction of its bark is prescribed when urine is highly coloured. A gum-like exudation from the incised bark is useful in diarrhea and dysentery.

The gum-resin exudates of the plant, Saidan gum is used as a substitute for Kino gum.

Folk Uses

A decoction of its bark is used as a fomentation to subside the boils and body swelling. An aqueous extract of its roots is given for 2 months for treating menstrual disorder. The juice of stem bark is dropped in the eye for the treatment of cataract two times a day for one month and stem bark juice along with butter milk is given two times a day for 2 days in diarrhea

Genus: *Psoralea* Linn.

The generic name is derived from Greek word, Psoraleou = scurfy or warted, in allusion to the many glandular dots covering the leaves.

The genus comprises of herbs or under shrubs punctuate with black or pellucid glands. Leaves are simple or imparipinnate and conspicuously gland-dotted. Flowers are capitate, spicate, subracemose or fasciculate. Calyx tube is connate and ovary sessile or shortly stalked and with one ovules. The pods are ovoid or oblong indehiscent and 1-seeded.

There are 130 species known to be distributed in tropical and subtropical parts of the world. Only 3 of them have been reported from India.

Species: *Psoralea corylifolia* Linn.

Description of Plant

It is an erect, annual herb about 0.5-1.2 m high with grooved stem and branches. Leaves are simple, 3.5-7.5 x 2.5-5.0 cm, broadly elliptic and inciso-dentate. Flowers are bluish purple or yellow in axillary long–peduncled heads. Pods are indehiscent, ovate, mucronate, somewhat compressed and about 0.5 cm long. Seed is one, smooth and adhering to the pericarp.

Flowering and Fruiting

During April-June.

Distribution

It is found wild on road sides and waste places almost throughout the India.

Plant Parts Used

Seeds of the plant are used.

Uses

That its seeds are considered as deobstruent, anthelmintic, stomachic, diuretic and diaphoretic and are used in febrile conditions, leprosy, leucoderma and other cutaneous diseases. The powder of its seeds is recommended much by the *Vaudyas* in leprosy and leucoderma internally and in the form of a paste, are applied externally. Besides, an ointment prepared by mixing one part of an alcoholic extract of the seeds with two parts of *Choaulmogra* oil and two parts of lanoline, is rubbed once or twice daily in *leucoderma,* white leprosy, psoriasis and other skin diseases and febrile condition.

Folk Uses

In Uttar Pradesh, people take the seeds of *Babchi* and *Panwar* (*Cassia brachycarpa*) in equal quantities and ground them to make a fine powder. A little sulphur is added to this and the paste thus obtained, is applied externally on the affected parts of the body with mustard oil and after one hour take bath with lukewarm water in scabies. Besides, about 3.0 gm seeds of the plant are ground and are taken with honey twice daily for a month to purify blood by the blood by the native people of Madhya Pradesh.

Genus: *Pterocarpus* Jacq.

The generic name has been derived from Greek words, Pteron = a wing and Karops = a fruits, referring to the nature of the fruits.

The genus comprises of erect unarmed tree. Leaves are alternate, imparipinnate; leaflets alternate and exstipellate. Flowers are usually yellow and are borne in axillary and terminal racemes or panicles. Calyx is turbinate, usually curved before expansion;

corolla exerted, petals with long claws and glabrous. Staminal sheath is slited both above and below generally; ovary stalked and with two ovules. Pods are orbicular or broadly ovate with broad rigid wing. Seeds are 1 or 2, oblong or subreniform.

About 100 species are found distributed in the tropical parts of the world. Only 4 of them have been reported India..

Species: *Pterocarpus marsupium* Roxb.

Description of Plant

It is a moderate-sized or large deciduous tree with a stout crooked stem and widely spreading branches, attaining a height upto 30 m and an girth of 2.5 m. Leaves are imparipinnate; leaflets 5-7, coriaceous, 6.5-10.0 x 3.8–5.0 cm, oblong, obtuse, rounded, truncate or more or les retuse at apex and glabrous. Flowers are pale-yellow, and are borne in short lateral and terminal fusco-pubescent paniculate racemes. Pods are flat, orbicular, glabrous or nearly so, winged and about 5.0 cm across. Seeds are 1-3, bony, convex and separated by hard dissepiments.

Flowering and Fruiting

Flowering: During rainy season and Fruiting: In summer.

Distribution

It commonly occurs in the hilly regions thoughout Southern India, extending to Gujarat, Madhya Pradesh, Uttar Pradesh, Bihar and Orissa.

Plant Parts Used

Bark, leaves and the gum, obtained from its bark are used.

Uses

The bark is astringent and its decoction is used successfully to treat diarrhea. A paste of its leaves is applied externally to abscess, sores and other cutaneous diseases. Besides, the gum obtained from the bark of plant commercially known as *'kino gum'* is good astringent and is used in pyrosis, toothache and in diarrhea.

Folk Uses

Its leaves are bruised and are applied externally to boils, sores and some skin diseases by the natives in different parts of the country. That stem bark is boiled and its copious vapour bath is taken as restorative and tonic to overcome mother's general weakness and fatigue by the Bhills in Rajasthan.

Species: *Pterocarpus santalinus* Linn. f.

Description of Plant

It is a small–sized deciduous tree attaining a height of 11 m with extremely hard and dark purple heart wood. Leaves are generally imparipinnate; leaflets 3 or 5, broad-elliptic, obtuse and 3.5-7.5 cm long. Flowers are yellow and are borne in simple or sparingly branched racemes. The pods are woolly and about 3.8 cm across. Seeds smooth, coriaceous and reddish-brown.

Flowering and Fruiting

During summer.

Distribution

This is known to found wild in Andhra Pradesh, particularly in Cuddapah district and neighbouring areas of Karnataka and in Tamil Nadu.

Plant Parts Used

Wood of the plant is usable.

Uses

The wood of the plant is known to possess astringent and cooling properties and is beneficial in fever, dysentery and haemorrhage; also applied externally in the from of a paste in headache, inflammation, boils, piles and opthalmia.

Folk Uses

A powder of its wood with milk is taken by the people in bleeding piles. Besides, wood rubbed on a piece of stone with water is applied on the skin as a cooling agent and purifier of skin after bathing like white sandal wood.

Genus: *Saraca* Linn.

Saraca, the generic name is a corruption of India name Asoka.

These are unarmed trees. Leaves are pinnatifid and stipules are intrapetiolar and more or less united. The flowers are in panicles with coloured bracts and bracteoles. Calyx-tube is long and funnel-shaped and limb 4-cleft. Petals are absent. Stamens are 2-8, exserted, oblong and versatile and anthers are on long slender filaments. The pods are flat, dehiscent and coriaceous. Seeds are exalbuminous.

About 20 species are known to occur in tropical Asia. Out of which 2 species have been reported from India.

Species: *Saraca asoca* (Roxb.) De Wild (*S. indica* auct. non Linn.)

Description of Plant

It is small evergreen tree about 5-8 m in height. The bark is with warty surface and dark brown to grey or almost black. Leaves are equally pinnate and 15.20 cm long; leaflets 6-12, oblong-lanceolat, glabrous and 7.5-20.5 x 1.3 cm. The flowers are orange or orange yellow and very fragrant and are borne in exillary corymbs. Pods are flat, oblong, woody and 7.5-25.0 x 3.8-5.0 cm. Seeds are ellipsoid-oblong and compressed.

Flowering and Fruiting

Flowering: In March-April and Fruiting: During August-September.

Distribution

It grows wild along streams or in shady evergreen forest upto an altitude of 750 m in Central and Eastern Himalayas as well as in the Khasi, Garo and Lushai hills.

Plant Parts Used

Bark, flowers and seeds are used.

Uses

The stem-bark issued to cure colic, dysentery, dyspepsia, piles, ulcers and uterine problems, particularly in menorrahagia due to uterine fibroids, leucorrhoea and menstrual pain. Its flowers are pounded and used in blood dysentery, diabetes, as excellent uterine tonic and in syphilis. Besides, the seeds are considered diuretic and their powder is given in gravels and strangury.

It bark is adulterated with that of bark *Polyalthia longifolia* and the fruits are chewed as substitute of areca fruits.

Folk Uses

A decoction of its bark in milk is given during the course of the days in menorrhagia from the fourth day of the monthly period till the bleeding ceases. In Orissa, the bark is used as astringent in cases of internal haemorrhoids.

Preparation

Ashokarishta and Ashokaghrit.

Genus: *Tephrosia* Pers.

The generic name has been derived from Greek word Tephrous = ashcoloured, referring to the pubsescence of most species.

The genus comprises of herbs, undershrubs or shrubs. Leaves are odd pinnate or rarely simple; leaflets are opposite and sub-coriaceous. The flowers are in terminal and leave-opposed racemes or solitary or in pairs in the leaf-axis. Calyx is campanulate. Petals are clawed. Stamens are 2-adelphous and anthers are uniform and obtus. Ovary is sessile, linear and many ovuled and style is incurved, filiform or flattened and glabrous or beared. The pods are linear, flattened any many-seeded.

That species occur in the tropical and subtropical parts of the world; 25 species are found in India, chiefly in Western Peninsula.

Species: *Tephrosia purpurea* (Linn.) Pers. (*T. hamiltonii Drumn* ex Gamble., *Cracca purpurea* Linn.)

Description of Plant

It is copiously branched perennial herb about 30-60 cm in height. Leaves are imparipinnate and 5.0-15.0 cm long; leaflets 9-21, glabrous above and obscurely silky beneath, narrow and oblanceolate. The flowers are red or pink in leaf-opposed racemes. The pods are linear, slightly curved and glabrescent and 2.5-5.0 x 0.4-0.5 cm. Seeds are 5-10, smooth and greenish-grey.

Flowering and Fruiting

Flowering: During rainy season and Fruiting: In winter.

Distribution

It is found all over India ascending upto an elevation of 1850 m in the Himalayas.

Plant Parts Used

Whole plant, its seeds, roots, root-bark and fruits are usable.

Uses

The whole plant as laxative, anthelmintic, and tonic and it is used internally as a purifier of the blood. The extract of herb is useful in insufficiency of liver. An infusion of its seeds issued against worm infestation in children and oil is used externally in scabies and dermatitis.

The roots are bitter and are used in dyspepsia and chronic diarrhea. A liniment prepared from the roots is used in elephantiasis. The pills made from its root-bark with little black pepper are given in refractory colic. Besides, the decoction of the fruits is also used as a vermifuge and to stop vomiting.

Folk Uses

A decoction of its roots with ginger is consumed to relieve headache. The root extracts, one teaspoonful one a day along with seed paste is given for 15 days in spermatorrhoea. The powdered root bark along with black pepper is given as antidote for snake-bite. In snake bite, an aqueous extract of the plant together with that of *Calotropis procera* is also given as an antidote. Its stem is used as toothbrush for pyorrhea. The root of the plant is introduced through vaginal passage into uterus and kept for half an hour to produce abortion and to start monthly bleeding in women by the tribals in Arunachal Pradesh.

Genus: *Teramus* P. Br.

The generic name has been derived from Greek work, Teramos, meaning soft, referring to the nature of plants.

The genus comprises of twining herbs. Calyx-tube is companulate with distinct and subequal teeth. Corolla is little exerted; petal about equal in length. Stamens are 1-adelphous. Ovary is sessile with many ovules. The pods are linear, flattish with septa between the seeds.

There are 15 species known to be distributed in tropical parts of the world. Only 2 of them are reported from India.

Species: *Teramus labials* Spreng

Description of Plant

It is a wide spreading slender climbing herb with a few appressed hairs on the stems. Leaves are trifoliate and above 6.0–12.0 cm long; leaflets 3, membranous or subcoriaceous, ovate-oblong with rounded base, subacute or obtuse, glabrescent above, sparsely appressed hairy beneath and pink, purple or white coloured in lax axillary racemes. The pods are narrow, elongated, glabrous and recurved with 8-10 seeds. Seeds are smooth oblong and slightly rounded or truncate.

Flowering and Fruiting

Flowering: During September-October and Fruiting in November-December.

Distribution

This is known to found almost throughout India.

Plant Parts Used

Its fruits are used medicinally.

Uses

In the nighantus, the plant is described as cooling and astringent and dry producing semen, strength and blood in consumption and fever; also used in disorders of *vayu, pitta* and blood. Its fruits are astringent, febrifuge and stomachic and are used in rheumatism, paralysis, tuberculosis and catarrh.

Folk Uses

In Madhya Pradesh, for treating hotness of head with giddiness, people pour leaf-lice of the herb into the water which becomes thick. This is rubbed on the skull during hot sensation in head for quick relief.

Genus: *Uraria* Desv.

The generic name is derived from Greek word, Oura = the tail referring to the shape of the bracts.

The genus comprises of suffruticose perennial. Leaflets are 1-9 and stipellate. Flowers are minute and racemose; calyx tube very short, 2 upper teeth short and 3 lower elongated; stamens 2- adelphous and ovary sessile or shortly stalked with few ovules. The pods are 2-6, small turgid and with single seed.

About 20 species are known to distributed in tropical, Africa, South east Asia, Formosa, Indomalayan region, North Australia and in Pacific. Of them 10 species have been reported from India.

Species: *Uraria picta* (Jacq.) Desv. Ex DC (*Hedysarum pictum* Jacq.)

Description of Plant

This is an erect about 0.8-1.2 m high shrubby perennial herb or under shrub. Leaves are 25.0-30.0 cm long and 1.3–foliate; leaflets 4-6, lanceolate or linear-oblong, mucronate and obtuse. Flowers are purple coloured and are borne in dense cylindrical racemes. The pods are glabrous, pale lead-coloured or whitish with 3-6 joints. Seeds are reddish, reniform and 1-12.

Flowering and Fruiting

Flowering: During rainy and Fruiting in winter.

Distribution

It is known to be found in dry grasslands, waste places and forests throughout India.

Plant Parts Used

Fruits and roots of the plant are used.

Uses

That its fruits and roots are used for prolapse of anus of infants and fruits alone in sore mouth. Besides, a decoction of its roots is also given in cough, chills and in fever.

Folk Uses

In some parts of India, decoction of entire plant is given to the persons suffering from cough and fever. In Bastar, (Madhya Pradesh), tribal people give 5.0–10.0 gm of its root paste maxing in water as an antidote for snake–bite (Hemadri and Rao, 1989). If the patient is unconscious, then the paste is blown into the nose of the patient.

Genus: *Vigna* Savi

The generic name has been assigned in honour of Dominico Vigna, Professor of Botany at Pisa and author of a commentary on Theophrastus in 1625.

The genus comprises of twining or prostrate, rarely of suberect herbs. Leaves are pinnately tri-foliate; leaflets stipellate. Flowers are in racemes at the upper part of axillary peduncles; calyx campanulate, teeth short or long corolla much excerted; stamens 2-adelphous and ovary is sessile with numerous ovules. The pods are linear, straight and incurved, subterete with septa between the seeds.

About 80-100 species are found in tropics, abundantly in Africa and Asia; out of which 9 have been reported from India.

Species: *Vigna trilobota* (Linn) Verdcourt. (*Phaseolus trilobatus* (Linn) Schreb; *P. trilobus* sensu Ait.; *Dolichos trilobatus* (Linn.) Maut)

Description of Plant

This is a annual or perennial herb with woody root stocks about 30-60 cm long. Leaves are tri-foliate; leaflets 1.3-2.5 long, membranous, ovate or rhomboid, glabrous or with a few hairs; the central division spathulate, very rarely entire. Flowers are yellow in subcapitate few-flowered racemes. Pods are glabrous, recurved, subcylindircal and 2.5-3.0 cm long. Seeds are 6-12 nearly flat, flat, cylindrical–oblong, truncate are each end and brown coloured.

Flowering and Fruiting

Flowering and Fruiting is during winter.

Distribution

It is known to occur throughout India, ascending upto and elevation of 2100 m in the North-east.

Plant Parts Used

Whole plant, its fruits and leaves are used.

Uses

Whole plant is considered febrifuge and its juice is given in rat-bite fever. Its fruits are cooling, anthelmintic, aphrodisiac, astringent, bitter and styptic and are used to cure consumption, inflammations, fever, burning sensation, piles, dysentery, cough, got and biliousness. Besides, leaves are known to possess cooling and sedative properties and they are used as a cataplasm for weak eyes and their decoction in intermittent fever.

Folk Uses

In Jharkhand a decoction of the leaves in administered in irregular fever by the native people..

Family: COMBRETACEAE

Genus: *Anogeissus* (DC) Wall.

The generic name has been derived from the Greek words Ano = above and Geisson = a covering roof, referring to the compression of 2-winged nuts placed horizontally into a dense head, having the appearance of tiled roof.

The genus comprises of trees or shrubs. Leaves are alternate or opposite, petioled and entire. The flowers are in dense globose heads, on axillary peduncles and usually shorter than leaves. Calyx-tube is compressed at base, prolonged beyond ovary into a slender tube, limb small, with 5 lobes and deciduous and the petals are absent. Stamens are 10 and 2-seriate. Ovary is inferior and 1-celled with 2 ovules. Fruits are small, coriaceous, compressed and broadly 2-wingh. The seed is only one.

That species occur in tropical Africa, Arabia and South East Asia. 5 of which have been reported from India.

Species: *Anogeissus latifolia* (Roxb. Ex DC.) Wall ex Guill. (*Conocarpus latifolia Roxb.* Ex DC.)

Description of Plant

It is tall tree with smooth, white-grey bark and drooping branchlets. Leaves are alternate or inconstantly opposite, 5.0-10.0 cm x 2.5-6.75 cm, elliptic or obtuse and coriaceous. Flowers are small in dense and globes heads or borne in short exillary racemes. Fruit is a dry drupe with glabrous and orbicular wings. Seed is only one.

Flowering and Fruiting

Flowering: In September-January and Fruiting in November-February.

Distribution

It is common and after gregarious in the deciduous forests of South and Central India ascending upto 1000 m in hilly places in India.

Plant Parts Used

The flowers, bark and gum are often used.

Uses

That its flowers are used as collyrium in eye diseases. The plant is said to be used in toothache. Spleen enlargement, dysuria, excessive perspiration, diarrhea, cholera, cough and cold, rinder pest and urinary disorders. The gum '*Ghatti*' obtained from the plant is used as a tonic after delivery. Besides, the bark is used an astringent

Folk Uses

Korku tribals in Melghat, take 250 gm gum of the plant with groundnut oil, 100 gm of kaju, 100 gm pista and 100 gm khajoor and mixed them with flour and from this mixture, ladoos are prepared which are eaten to cure '*sandhivat*'. In paralysis, people of Dadra and Nagar Havely, apply warm bark paste of the plant on the affected parts of the body.

Genus: *Calycopteris* Lamk.

The generic name is derived from Greek words, Lalyx=calys and Pteron = wing, referring to the spreading winged and persistent calyx lobes.

This monotypic genus is known to be distributed in Indomalayan region.

Species: *Calycopteris floribunda* Lamk.

Description of Plant

This is a dense and gregarious shrub about 0.7–1.0 m high with drooping bracnches. Leaves are 7.5–12.5 x 3.0–5.5 cm, apposite, ovate-lanceolate, acute or acuminate, glabrous and entire with 0.6–1.0 cm long petioles. Flowers are greenish in fulvous-pubescent terminal panicles. Fruits a re about 0.6-0.8 cm long, oblongor ellipsoik, 5- ribbed and crowned by the persistent calyx.

Flowering and Fruiting

During March-May.

Distribution

It is distributed in deciduous forests in South India., Orissa, Assam and Chittagon.

Plant Parts Used

Leaves of the plant are used.

Uses

That its leaves are considered anthelmintic, astringent, bitter and laxative and their juice is given in colic and with butter in dysentery and malarial fever. Externally in paste form, these are applied on ulceric wounds.

Folk Uses

A paste of its leaves and roots of *Grewia pilosa* with honey is applied on ulcerous wounds in some parts of the country.

Genus: *Citrullus* Schrad.

The generic name has been adopted from Latin name Citrus for the Citron, alluding to the similarity in appearance of citrus fruits.

These are climbing or creeping herbs with 2-3–fid tendrils. Leaves are petioled and palmately 3-7 lobed. The flowers are monoecious, all solitary and rather large. Male flowers: calyx is capanulate, lobes 5, stamens 3 and short and anthers scarcely cohering one 1- celled, two 2- celled. Female flowers: calyx and corolla are as in male, ovary ovoid, style short, stigmas 3 and reniform and ovules many and horizontal. The fruits are globes or ellipsoid, smooth, fleshy and indehiscent. Seeds are many, oblong, compressed and smooth.

About 3 species occur in Africa, Mediterranean and tropical Asia; 2 species are found in India, out of which one is cultivated throughout India for its edible fruits and another in hotter parts of India for the drug.

Species: *Citrullus colocynthis* Schrad.

Description of Plant

It is a diffuse or creeping, slender, branched and hirsute or scabrid perennial herb with simple or 2-fid slender hairy tendrils. Leaves are 3.8-6.0 x 2.5-5.0 cm, pale–green above, ashy beneath, 5-7 lobed or commonly 3-lobed; middle one is largest. Male flowers in 0.6-1.2 cm long peduncles; calyx hairy, campanulate; corolla 0.6 cm long. Segments obovate apiculate. Female flowers are with ellipsoid ovary. The fruits are globular, slightly depressed and about 5.0-7.5 cm across with spongy bitter pulp. Seeds are pale brown and 0.4-0.6 cm in diameter.

Flowering and Fruiting

During July-September.

Distribution

It occurs wild throughout India, particularly in North-west, central and south India and on the sea shores in Gujarat and other parts of Western India; also cultivated.

Plant Parts Used

Fruits, seeds and roots of the plant are used medicinally.

Uses

Its fruits and seeds as a strong purgative. A spongy internal pulp of its dried fruits finds application in British Pharmacopocia and is useful in biliousness, fever, intestinal parasites, constipation, hepatic and abdominal, visceral and cerebral congestion and dropsy. Mohammedan Physicians, use this drug extensively as a drastic purgative in ascites and jaundice andin various uterine diseases, especially in amenorrhoea. Its roots are also useful in jaundice, ascites, urinary diseases and rheumatism and are also given in abdominal enlargements and in cough and asthmatic attacks to children.

Folk Uses

The fresh roots of the plant re used a tooth brush to strengthen the gums. A poultice of its roots is applied inn inflammation of the breast by women. Besides, its fruits or roots with nux-vomica are rubbed into a paste with water and are applied to boils and pimples in different parts of the country.

Preparation

Abhyarista, Narayan Churna and Phalatrikayadi kwath.

Genus: *Combretum* Loefl.

The generic name is the Latin name for climbing plant.

This genus comprises of large shrubs with long, pendent or scandent branches; rarely spinescent. Leaves are entire, opposite or alternate or ternate. Flowers are polygamo-dioecious and are borne in spikes or racemes. Calyx tube is constricted above ovary, funnel-shaped or tubular; petals equal to calyx lobes, rarely absent. Stamens are twice the number of petals and ovary is inferior and 1- celled with 2-5 ovules. Fruits are dry, generally indehiscent and with 4-5 wings with only one seed.

About 250 species are known to be distributed in tropical parts of the world excluding Australia. Only 10 of them have been reported India.

Species: *Combretum pilosum* Roxb.

Description of Plant

It is large scandent shrub. Leaves are opposite and 10.0-20.0 cm long, usually with some rusty hairs on the lower surface and about 0.3-0.6 cm long petiole, Calyx-tube is 5-fluted upwards; petals narrow-obovate and disk low in the funnel of calyx; ovary sessile. Fruits are elliptic or circular, usually pubescent and about 2.5 cm in diameter.

Flowering and Fruiting

During January-May.

Distribution

It is known to found wild in East Bengal and Assam.

Plant Part Used

Leaves are used.

Uses

Its leaves are considered anthelmintic and their decoction is given for round worms, specially for *Ascaris lumbricoides* and Oxyuris vermicularis.

Folk Uses

In Eastern parts of India, people take decoction of leaves of the plant to remove round worms from the body as anthelmintic.

Genus: *Terminalia* Linn.

The generic name has been derived from Latin word Terminus=end, referring to the leaves borne at the ends of branchlets.

The genus comprises of large trees. Leaves are alternate or sub-opposite, entire or slightly crenate, exstipulate and often with glands on the petiole. The flowers are small, sessile an usually bisexual; bracts are deciduous. Calyx-tube is produced above the ovary with constricted mouth, lobes 4-5 and valvate. Petals are absent. Stamens are 10 and inserted on the calyx-tube. Ovary is 1-celled, style long, simple and ovules 2-or 3 and pendulous. The fruits are oblong or ovoid, ellipsoidal or globose, smooth or angular or with 25 wings, indehiscent and coriaceous. The seeds are solitary and exalbuminous.

About 250 species occur in the tropical parts of the world. Only 12 species of them have been reported from India.

Species: *Terminalia arjuna* (Roxb.) W. and A.

Description of Plant

It is large tree often with buttressed trunk, smooth grey bark and dropping branchlets about 20-25 m high. Leaves are usually sub-opposite, oblong orelliptic-long, pale dull green above and pale-brown beneath, 10.0-20.0 cm long and hard. The flowers are yellowish-white and are borne in shortly panicled-spikes. The fruits are 2.5-5.0 cm long drupes, ovoid or obovoid-oblong, fibrous woody, glaborous and with 5-7 equal, hard, coriaceous and thick narrow wings.

Flowering and Fruiting

Flowering: In April-May and Fruiting: During September-November.

Distribution

It is common on the banks of rivers, streams and dry watercourses in Sub-Himalayan-tract and Central and South India and in West Bengal.

Plant Parts Used

Fruits, leaves and the bark are used.

Uses

The fruits are used as tonic and deobstruent. Externally, the leaves are used as a cover on sores and ulcers and their juice is dropped in earache. The bark is antidysenteric, antipyretic, astringent, cardiotonic, lithotriptic, styptic and tonic and its powder acts as a diuretic in cirrhosisofliver and gives relief in symptomtic hypertension. A decoction of its thick bark made with milk is given every morning on an empty stomach or its powder with milk and gur as a cardiotonic. About 500 mg brk powder can be used three times daily, but it may cause mild gastric intolerance and constipation in some persons. They further remarked that the drug appears quite safe for prolonged use. The powder of the bark is also given with honey in fractures and contusions with ecchymosis. The extract of the barkisused for cleaning sores, ulcers and cancers etc. as astringent. An ointment made from its bark mixed with honey is used to cure acne. The ashes of its bark are also prescribed in scorpion-stings. A linctus made from the bark soaked and dried seven times successively in the leaf juice of Bansa (*Justicia adhatoda*) and well mixed with honey, sugar-candy and cow's ghee is prescribed in cases of phthisis; it stops the blood in the sputum and clears up and cures the sores.

Folk Uses

Its twigs are used to clean mouth and teeth; also to remove mouth sores. A powder of the bark with '*Khir*')a preparation of rice in milk), is given before Sunrise by the tribals of Santhal pargnas for allergies. The khir should be prepared an earlier day and exposed to full moon. Generally, its single dose is said to be sufficient. Besides, 1-2 teaspoonful of its bark powder are taken with milk as a cardiotonic by the people throughout the country.

Preparation

Arjunarista, Arjunaghirt, Parthodhyarista, Arjunatwak churna, Kukubhadi churna, Hridayarnava ras and Arju ksheerpak.

Species: *Terminalia bellirica* (Gaertn.) Roxb. (*T. bellirica* var. *laurinoides* Cl.; *Myrobalanus bellirica* Gaertn.)

Description of Plant

It is a large deciduous tree with bluish or ashy-grey bark and about 20-25 m high. Leaves are clustered at the ends of the branchlets and are coriaceous, pale beneath, broad elliptic or obovate elliptic, 8.0-20.0 cm long and often with unequal base; petiole 2.5-7.5 cm long and usually glandular. Spikes are slender, interrupted and 8.0-15.0 cm long. The flowers are about 1.25 cm across, dirty-grey or greenish-yellow and with offensive smell. The fruits are 2.5 cm long, ovoid, ellipsoidal or globose and grey velvety with 5, more or less indistinct furrows. The nuts are thick walled and hard.

Flowering and Fruiting

Flowering: During April-June and Fruiting in December-January.

Distribution

It is found distributed throughout the forests of India upto 914 m elevations and along the foot of the Himalayas.

Plant Parts Used

Fruits, bark and the gum are used.

Uses

Its fruits are bitter, astringent, tonic, laxative, antipyretic and antidiarrhoeal and they are used in piles, dyspepsia, leprosy, billousness, headache, hoarseness and eye diseases. The ripe fruits are used as an astringent in combination with chebulic myrobalan and half ripen fruits are used as a purgative. The oil obtained from its fruits is applied to hairs and rheumatic swellings. The fruits pulp mixed with honey is used in opthalmia. However, excessive eating of the seed pulp may cause intoxication. Besides, it is an ingredient of *triphala*, prepared locally and used daily for constipation by thepeople.

The kernels are used as narcotics. The bark is used as diuretic and gum yielded by the tree as demulcent and purgative.

Folk Uses

Its fruits are used locally, in coughs, hoarseness, sore throat and dyspepsia. A peel of the fruit with a little pulp is given to chew to children in coughs and colds (Joshi 1989). Besides, its fruits in equal combination of *T. chebula* and *Embelica officinalis* forms *triphala* which is used daily for constipation by the people throughout the country.

Preparation

Bibhitak tails, Triphala churna, Phalatrikadi kwath, Talisadichurna, Lavangadi vati, Sudarshan churna, Sarvajwar louha, Agnitundi vati, Sanjivani vati, Punarnava mandur, Navayas louha, Sringadi churna, Karanjadi yog, Laghuvis- garva taila, Arogyavardhani vati, Khadirarista, Gokshuradi guggulu, Chandraprabha vati and Vidangadi louha

Species: *Terminalia chebula* Retz.

Description of Plant

It is a large deciduous tree with dark–brown bark; young branchlets, leaf-buds and leaves are long and soft and with shining rust coloured, sometimes silvery hairy. Leaves are mostly sub-opposite, distant, 7.5-15.0 cm long, ovate or oblong-ovate, acuminate and deciduous in the cold season. The flowers are dull white or yellowish with offensive smell and bisexual and are borne in spikes from the upper axils and in small terminal panicles. The fruits are 2.5-5.0 cm long, obovoid, ellipsoid or ovoid glabrous and more or less 5-ribbed when dry. The nuts are hard and with a rough grooved surface.

Flowering and Fruiting

Flowering: During April-June and Fruiting in January-March.

Distribution

It is abundantly found in Northern India from Kangra to Bengal and South wards to the Deccan tablelands at 304-914 m and upto 1828 m in Travancore; higher forests of the Bombay ghats, Satpuras, Belgaum and Kanara.

Plant Parts Used

The fruits, bark, gum and leaves are used.

Uses

The fruits as laxative, astringent, alternative and stomachic. A cold infusion of its fruits is used as a garglein stomatitis and in chronic ulcers, carious teeth, in cough, asthma and urinary diseases. Finely powdered fruits are used in bleeding and ulceration of the gums. Internally, these are efficacious in chronic diarrhea, dysentery and flatulence A fruit extract issued as wash in watering eyes and fruit paste in burns. The bark is diuretic and cardiotonic and the gum obtained from the tree is eaten as tonic and also in colitis. Besides, it is an ingredient of *triphala*, prepared locally and used daily for constipation by the people.

Folk Uses

Its root paste issued for conjuctivitis by the tribalsin Maghalaya. Besides, a paste of its fruits is taken orally for abortion by the tribal women in Arunachal Pradesh.

Preparation

Abhayamodak, Abhayarista, Agastiharitaki, Agnitundivati, Amrita haritaki, Amlakyadi churna, Arogyavardhini vati, Agragbadhyadi kwath, Byaghri haritaki, Chittrak haritaki, Chandraprabha vati, Gokshuradi guffulu, Danti haritaki, Haritaki Khand, Haritakadi kwath, Khadirarista Kukuvadi churna, Karanjadi yog. Leghu-vis-garva taila, Navayaslouha, Pathyadi kwath, Pathyadi churna, Panchaskar churna, Phaltrikadi kwath, Punarnava mandur, Sanjivani vati, Sarvajwar louha, Sringayadi churna, Sudhrshan churna, Dashmulaharitki, Triphala Vasadi kwath, Vidangadi lep,Itriphalsagir, Sharbat murrakale musaffi khun and Majun mochra.

Famly: CUCURBITACEAE

Geneus: *Coccinia* W. and A.

The generic name has been taken from Latin work Coccineus = scarlet, referring to the colour of the fruits.

These are annual or perennial and prostrate or climbing herbs with simple tendrils which are rarely 2 fid. Leaves are petioled, 5 lobed or 5- angular and toothed. The flowers are dioecious, large, solitary and white without bracts. The fruits are fleshy, indehiscent, cylindric and smooth. Seeds are many, ovoid, compressed and margined.

About 30 species occur in tropics and South Africa, One species is reported to found throughout India.

Species: *Coccinia grandis* (Linn.) Voigt (*C. cordifolia* Cogn; C. Indica. W. and A.; *Cephalandra indica* Naud)

Description of Plant

It is an annual or perennial, scandent or prostrate and much branched climbing herb. Stems are grooved and slender with slender tendrils. Leaves are 55-8 cm long, ovate, deeply cordate at base, 5-angularpartite, with entire to pinnatipartite lobes or segments, denticulate, pellucid-dotted above and gland-dotted on the lower surface at the base and between the main nerves; the petioles main nerves; the petioles 2.0-3.0 cm long. Male flowers are in I-flowered peduncles which is 2.0-3.5 cm long and female flowers show 1.3-2.5 cm long peduncles. The fruits are ellipsoid-oblong, cylindrical and with rounded ends and about 5.0 cm long. Seeds are obovoid, rounded at the apex and yellowish grey.

Flowering and Fruiting

Flowering: In December–February and Fruiting during April-September.

Distribution

It is found distributed throughout India.

Plant Parts Used

Roots, leaves and fruits are used.

Uses

Its ethanol (50 per cent) extract is antiprotozoal and further, reported that ethanolic extract of the plant also lowered blood sugar. The powder of its roots is taken with water to stop vomiting and their paste is applied in headache.

Its leaves are applied externally in eruptions of skins and the juice of its leaves and roots is given in diabetes. The leaves are boiled in gingelly oil and are applied externally to ring worm and psoriasis. Its fruits are also advised to be eaten in diabetes. Besides, various plant parts are used in slow pulse, convulsions, scrofulosa, colli, sores, syphilis and gravel.

Folk Uses

Its green fruits are chewed to cure sores of the tongue and plant juice is dropped into ear in earache in different parts of the country .

Genus: *Trichosanthes* Linn.

The generic name has been derived from the Greek words Trichos = hair and Anthos: flower, referring to the hairy flowers.

The genus comprises of annual or perennial climbing herbs. Leaves are entire or 3-9-lobed and denticulate; the tendrils 2-5-fid. The flowers are white and usually dioecious; male peduncles are in exillary pairs, one 1-flowered, the other racemose; female flowers are solitary. The fruits are fusiform, ovoid or globose, smooth and acute or obtuse at the apex. Seeds are numerous, horizontal, compressed, ellipsoid and sometimes angular on the margins.

About 15 species occur in Indo-Malayan region and Australia. All have been reported from India.

Species: *Trichosanthes cucumerina* Linn. (*T. brevibracteata* Kunda; *T. pachyrrhochis* Kundu)

Description of Plant

It is annual herb with twining and more or less pubescent stem. Tendrils are 3-fid. Leaves are reniform-suborbicular, 3-7-lobed or angular, denticulate, obtuse, glandular hairy and 5.0-20.0 cm across. The flowers are white and dioecious; male flowers: bracts absent or rare, pedicels jointed at the top and 1.0-2.0 cm long, beaked, ovoid and conical. Seeds are compressed and corrugate and are embedded in red pulp.

Flowering and Fruiting

During July-December.

Distribution

It is found distributed throughout India.

Plant Parts Uses

The fruits, leaves, stems, roots, seeds and whole plant are used.

Uses

The fruits of the plant are bitter and laxative and ripe fruits are cooked and eaten. The juice of its leaves is emetic and externally applied to the bald patches of alopecia. The roots are purgative and tonic and their juice is used as cathartic. A decoction of the leaves and stem is used for bilious disorders, skin diseases and as emmenagogue. Its seeds are antifebrile and anthelmintic and are given in the disorder of stomach.

Hakims consider the whole plant as general and cardiac tonic, alterative and antifebrile. In obstinate cases of fever, an infusion of the whole plant and coriander, infused for a night is recommended to be given in the morning and the evening with honey.

Folk Uses

In Mumbai, a decoction of the plant with ginger, chiretta and honey is taken as laxative, febrifuge and alternative by the people and over the whole body in remittent fevers. The roots of this plant and of *Cynoglossum glochidiatum* are taken in equal quantities and made into a powder which is taken for impotency by the local people in Orissa.

Preparation

Patoladi kwath.

Family: AIZOACEAE

Genus: *Tritanthems* Linn.

The generic name has been derived from Greek words, Tri = three and Anthemis = a flower; as flowers of this genus are usually clustered in threes.

The genus comprises of diffuse, prostrate and branched pubescent and glabrous herbs. Leaves are opposite, linear, ovate orobovate and entire. Flowers are axillary, sessile or peduncled, solitary, cymose or panicled. Calyx-tube is short or long and its lobes 5; petals absent; stamens 5-10 or more and ovary is free, sessile and 1-2 celled with 1 to many ovules. The capsules are membraneous below with a thick hard cap with one or more seeds. Seeds are reniform.

About 20 species are known to occur in tropical and subtropical Africa and Asia and Australia and 1 in tropical America. About 3 species have been recorded from India.

Species: *Trianthema portulacastrum* Linn.

Description of Plant

This is a diffuse, prostrate, glabrous and succulent herb. Leaves are subfleshy, opposite, ovate and unequally paired; Upper 1.8-2.5 cm and lower 0.5-1.2 cm. Flowers are pink or white and borne solitary. Calyx–lobes are 10-20 and ovary is truncated. The capsules are small, concealed in petiolar pouch and 0.3-0.5 cm long with 2 spreading teeth, carrying one seed and lower part 3-5 seeded. Seeds are reniform, muriculate and dull black.

Flowering and Fruiting

During rainy season.

Distribution

It is found almost throughout India in waste places, cultivated fields and along river beds.

Plant Parts Used

Whole plants, its roots and the leaves are used medicinally.

Uses

The plant in ayurveda, is described as alexiteric, alterative, analgesic, bitter, laxative and stomachic and it is used to cure *"Kapha"* and *'vata'* diseases, bronchitis, diseases of blood and heart, anaemia and inflammation. Its powdered roots are bitter and are used with ginger as cathartic and abortifacient. An infusion of its roots is given in constipation, jaundice, strangury and dropsy and decoction is used in rheumatism and is also said to be antidotal to alcoholic poison. Besides, leaves are used in oedema, dropsy and as cites.

Folk Uses

The leaves of the plant are cooked and are eaten to cure pustules of mouth throughout India as a house hold remedy.

Family: APLACEAE

Genus: *Angelica* Linn.

The generic name is derived from Latin word Anelus meaning divine messenger; referring to angelic qualities as the plant was used in medicine for curing contagious and for purifying blood.

The genus comprises of tall and aromatic herbs. Leaves are 1-2-3- pinnatifid; pinnae toothed and usually large. Flowers are white or lurid purple in compounded umbels. The fruits are ovoid or ellipsoid, dorsally compressed with much dorsally compressed seeds.

About 80 species are found distributed in Northern hemisphere and New Zealand. Only of them have been reported from India.

Species: *Angelica glauca* Edgw.

Description of Plant

It is a glabrous herb about 0.5-3.5 m high. Leaves are tri-partite or 2-pinnate; leaflets few, 2.6-7.5 cm, ovate, serrate and glaucous beneath. Flowers are white in compound and long stalked umbels. The fruits are subquardrate, glabrous and 1.2 x 0.6 cm with grooved seeds.

Flowering and Fruiting

During July-September.

Distribution

It is known to be distributed in Western Himalayas from Kashmir to Shimla between 2438-3038 m.

Plant Parts Used

The roots of the plant are used medicinally.

Uses

A roots of the plant are considered cordial and stimulant and are used in constipation and dyspepsia.

Folk Uses

In Garhwal Himalayas, its roots are taken for strength and vigour by the women after delivery.

Genus: *Centella* Linn.

The generic name has been derived from Latin word Centrum=hundred, referring to profusely branched prostrate herb.

This genus comprises of prostrate herbs. Leaves are cordate or hastate, orbicular or reniform, subentire or palmately lobed and long petioled; stipules small. The flowers are white and are borne in umbels. Carpels are laterally compressed or sub pentagonal; lateral primary ridges concealed within the commissure or distant there from and prominent and vittae absent or obscure and carpophore absent. The fruits are laterally compressed with narrow commissure. Seeds are laterally compressed.

About 40 species are known to occur in Africa, Australia, New Zealand and America; but only one species is reported from India which occurs throughout the country.

Species: *Centella asiatica* (Linn.) Urban (*Hydrocatyle asiatica* Linn.)

Description of Plant

It is a prostrate perennial herb. Stem is creeping with long stolons and nearly glabrous or hairy on young parts. Leaves are long-peioled, reniform or orbicular, crenate and lobed and glabrous. The flowers are 3-6 and are borne in umbel in the axils of small bracts and purple whites. The fruits are sub-cylindric, curved, slightly compressed, reticulate and 0.2 x 0.35 cm and each is with 9 curvilinear sub similar ridges within to commissure.

Flowering and Fruiting

Flowering: In July–August and Fruiting later on during October–December.

Distribution

The plant is found throughout India in marshy places upto 1828 m.

Plant Parts Used

The whole plant is used.

Uses

An Ethanol extract of the plant is antiprotozoal and spasmolytic. The juice of its leaves is given in fever and is also useful in cataract and other eye problems. The leaves are taken as tonic for improving memory and both externally as well as internally are also useful in syphilitic skin diseases. Its leaves are stated to contain an alkaloid which produces toxic symptoms similar to those of strychnine and besides, they have diuretic action.

The plant is also useful in treatment of swollen painful joints as external application. The plant is also used as alternative and tonic in diseases.

Species: *Centella asiatica* (Linn.) Urban (*Hydrocatyle asiatica* Linn.)

Description of Plant

It is a prostrate perennial herb. Stem is creeping with long stolons and nearly glabrous or hairy on young parts. Leaves are long-peioled, reniform or orbicular, crenate and lobed and glabrous. The flowers are 3-6 and are borne in umbel in the axils of small bracts and purple whites. The fruits are sub-cylindric, curved, slightly compressed, reticulate and 0.2 x 0.35 cm and each is with 9 curvilinear subsimilar ridges within to commissure.

Flowering and Fruiting

Flowering: In July–August and Fruiting: Later on during October–December.

Distribution

The plant is found throughout India in marshy places upto 1828 m.

Plant Parts Used

The whole plant is used.

Uses

Ethanol extract of the plant is antiprotozoal and spasmolytic. The juice of its leaves is given in fever and is also useful in cataract and other eye problems. The leaves are taken as tonic for improving memory and both externally as well as internally are also useful in syphilitic skin diseases (Chopra *et al.*, 1956). Its leaves are stated to contain an alkaloid which produces toxic symptoms similar to those of strychnine and besides, they have diuretic action.

The plant is also useful in treatment of swollen painful joints as external application. The plant is also used as alternative and tonic in diseases of skin, nerves and blood.

Folk Uses

A syrup is prepared from the herb which is given to children for increasing memory throughout the greater parts of the country, particularly to the children. Besides, the entire plant is boiled in water and its decoction is given in the treatment of leprosy.

Preparation

Brahmi rasayana, Brhami taila, Brahmi sarbat, Brhami ghirt, Brahmi pak, Brahmi panak, Saraswatarisht, Saraswat ghrit, Brahmi halwa, Brahmi malhar and Majun brahmi.

Genus: *Ferula* Linn.

The generic name is derived from Latin word Ferio = to strike; as the gaint stems of the plants were used for beating slaves for minor offences.

The genus comprises of perennial herbs. Leaves are 2-4 pinnated. The flowers are often polygamous and yellow in compound umbels. The fruits are ellipsoid or orbicular and compressed laterally. Seeds are compressed .

About 133 species are known to be distributed from Mediterranean to Central Asia. Only 4 of them are reported for India.

Species: *Ferula narthex* Boiss

Description of Plant

This is a perennial herb about 1.5-2.5m high. Leaves are pubescent; lower ones 30-60 cm long, ovate; secondary and tertiary pinnae decurrent, entire or irregularly crenate-serrate. Flowers are in compound and large umbles; ovary glabrous. The fruits are 0.8 x 0.5 cm with compressed seeds.

Flowering and Fruiting

Flowering: During April-May and Fruiting later.

Distribution

It is known to grow abundantly in Kashmir.

Plant Parts Used

Gum resin and the leaves of the plant a re used.

Uses

Its gum resin is anthelmintic, antiseptic and expectorant and is used in asthma, whooping cough, flatulent colic and in pneumonia and bronchitis. Besides, its leaves are also known to possess diaphoretic and carminative properties and are given to the persons suffering from rheumatism.

Folk Uses

A paste of gum resin is applied locally in ringworm diseases in Kashmir.

Family: ALNGIACEAE

Genus: *Alngium* Lamk.

The generic name has been derived from its Malayalam name Angolam.

They are shrubs or small trees. Leaves are alternate, petioled, oblong, entire, 3-nerved at base and persistent. The flowers are white and bisexual and are borne in axillary fascicles or cymes. Calyx-tube is adnate to the very, limb is 5.-10 toothed or truncate. Petals are 5-10, linear-oblong and valvate. Stamens are as many as petals or more and anthers very long. Ovary is inferior, 1-celled and surrounded by a disk. Fruit is a berry which is crowned with persistent calyx limb. Seeds are oblong with crumpled cotyledons.

About 17 species occur in tropical Africa, Madagascar, Comoro island, China, South east Asia, Indomalayan region and East Australia. 3 of them have been reported from India.

Species: *Alangium salvifolium* (Linn.f.) Wangerin ssp. *Salvifolium* Mukerjee (A lamarckii Thw.; *A. salvifolium* ssp. *decapetalum* (Lamk.) Wangerin; *A. decapetalum* Lamk; *A tomentosuum* Lamk.; *Grewia salvifolia* Linn. f.)

Description of Plant

It is a small tree or shurb upto 10 m in height and almost evergreen often with spinescent branchlets. Leaves are alternate, simple, varying in shape, usually oblong or elliptic oblong and pubescent or tomentose when young and glabrous or pubescent below when full grown. The inflorescences a r woolly. Flowers are white, fragrant and solitary of fascicled. The drupes are ellipsoid, black, crowned with enlarged calyx-limb and edible. Seeds are oblong, large, enclosed in red mucilaginous and sweet but rather with a stringent pulp.

Flowering and Fruiting

Flowering is in March–May and fruiting in June–July.

Distribution

It occurs throughout the drier parts of India; also grows vigorously in the forests of South India.

Plant Parts Used

The fruits, seeds, leaves and root-bark are used.

Uses

Its fruits are antiphelgmatic, laxative, nutritive, refrigerant and tonic, and are used in emaciation, curing for eye ailments and haemorrhage, The seeds are cooling and tonic and are used in the treatment of haemorrhage and as a cure for boil. As a poultice, its leaves are used in rheumatic pains and their ethanol extract is considered as spasmolytic, antiprotozoal and hypoglycaemic. The root-bark is purgative and anthelmintic and sntipyretic, astringent, bitter, diaphoretic diuretic, emetic, laxative, pungent and alterative and is used in leprosy and other skin diseases and in syphilis. A powder of its root bark is used in poisonous bites. An oil derived from its roots is applied externally in rheumatic pain and the powder of root–bark with black pepper is used in piles.

Folk Uses

A pasts of its bark is applied locally in snake-bites, cat-bites and mad dog-bites and about 10 gm is given orally by the people of Gandhamardan in Orissa. Kondh tribes in Phulbani, Orissa, apply a paste of its roots of paralytic parts twice daily for two months to cure the disease.

Preparation

Atisarnashak vati and Ankol taila.

Family: RUBIACEAE

Genus: *Catunaregam* Wolf

The generic name is derived from Malayam name, Katu = forest and Naregam = Citrus, reffering to its general appearance to citrus fruits.

The genus comprises of unarmed or spinous trees and shrubs. Leaves are opposite or with one arrested pair; stipules short. Flowers are white or yellowish, solitary and terminal or in axillary or leaf-opposed cymes. Calyx tube is covoid orturbinate and corolla tube is short or long; lobes 5. Stamens are 5 or more and ovary 2-celled with numerous ovules. The berries are globose, ovoid or ellipsoid, 2-celled and many seeded.

About 200-300 species occur in tropical parts of the world; 10 of which have been reported from India.

Species: *Catunregam spinosa* (Thunb.) Tiruv. (*Gardenia spinosa* Thunb.; *Randia dumetorum* (Retz.) Poir; *R. brandisii* Gamble; *Xeromphis spinosa* (Thunb) Keay).

Description of Plant

It is a small tree or rigid shrub armed with strong straight nearly opposite decussate, 1.3-3.0 cm long spines. Leaves or 3.2-5.5 x 2.0-3.2 cm. obovate, obtuse, wrinkled, shining above with cuneate base; stipules ovate and acuminate. Flowers are solitary or rarely 2-3 on short peduncles and are greenish yellow or white. Calyx rube is campanulate; teeth variable, sometimes spathulate. Corolla tube is 0.5-0.6 cm long; lobes obovate-oblong rounded at apex, pubescent outside and spreading. The berries are yellowish, globose or broadly ovoid, smooth or obscurely longitudinally ribbed and crowned by large calyx-limb. Seeds are numerous, compressed and are embedded in pulp.

Flowering and Fruiting

Flowering: During March-June and Fruiting: Later.

Distribution

It is known to occur throughout India; especially in Sub-Himalayan tract eastwards ascending in Sikkim upto 1219 m and southwards extends to Chittagong and Penninsula India.

Plant Parts Used

Fruits and bark are used.

Uses

Its fruits in ayurveda are described as aphrodisiac, bitter, carminative, emetic, alexiteric and antipyretic and are used to cure abscess, inflammations, tumours, skin diseases and ulcers. While in Unani system of medicine, these are used in chronic bronchitis, pain of muscles, paralysis, leprosy, boils, eruptions of skin and in the diseases of brain, asthma and leucoderma. The pulp of its fruits is applied on tongue and incidental ailments of children during teething and is also used as anthelminitc and abortifacient and in dysentery. Besides, its bark is known to possess astringent property and is externally applied in bone ache during fever and as anodyne in rheumatism. It is also used as a substitute of Ipecacuanha.

Folk Uses

In colic, its fruits are rubbed to make a paste with rice water and are applied over the navel region in different parts of the country (Kirtikar and Basu, 1935). In Baster (Bihar), roots of this plant and *Allium sativum* are made into a paste with salt which is given internally in cough with phlegm (Hemadri amd Rao, 1989).

Preparation

Pancha kashaya and Madanadi lap.

Species: *Catunaregam uligionosa* (Retz.) Sivarajan. (*Gardenia ulgigosa* Retz; *Randia ulilginosa* (Retz.) Poir; *Xeromaphis uliginosa* (Retz.) Mahes.)

Description of Plant

It is a small tree with about 6 m height and reddish brownish scaly bark. Leaves are 6.2-12.5 x 3.5-5.0 cm, obovate or obvate-oblong, obtuse, glabrous above and more or less pubescent and reticulately veined beneath. Flowers are white, fragrant and 35-5.0 cm in diameter and are borne solitary at the ends of suppressed branches. Calyx is about 1.32 cm long and corolla lobes 5-7, orbicular and much overlapping. Stamens are 5-7 and ovary is 2- celled with many ovules. The fruits are yellow, 5.0-6.0 cm long, ovoid and smooth and are crowned with persistent calyx-limb. Seeds are many, compressed and smooth.

Flowering and Fruiting

Flowering: During April-June, and Fruiting: Later.

Distribution

It is known to be distributed throughout India; chiefly in Eastern, Central, Western and Southern India but not so common in the North.

Parts Uses

Roots and the fruits of the plant are used medicinally.

Medicinal Uses

The roots of the plant are known to posses cooling and diuretic properties in ayurvedic and the unani system, and are used in thirst, heart problems, biliousness, dysuria and strangury. These after boiling in ghee are diarrhea and dysentery.

Its unripe fruits are roasted in wood ashes and are used as a remedy in diarrhea and dysentery; the central portion consisting of the stone and seeds being rejected as it a astringent.

Folk Uses

Sometimes, people take roots of the plant after boiling it in ghee for dysentery and diarrhea in certain parts of the country.

Genus: *Hedyotis* Linn.

The generic name has been taken from the Greek words Hedu = sweet and Otos– an ear, referring to the fragrant and sometimes ear-shaped leaves of the plant.

The genus comprises of annual or perennial herbs with slender and erect or di- to trichotomously branches. Leaves are small and narrow; stipules are acute or bristly., The flowers are small and pink or white in dichotomous axillary and terminal panicled cymes; rarely solitary. The capsules are small and usually membranous terete didymous or angled and rarely indehiscent. Seeds are angled or globose.

About 150 species are distributed in tropical and subtropical parts, chiefly in Asia; of which 20 species are found in India.

Species: *Hedyotis auricularia* Linn. (*Oldenlandia auricularia* (Linn.) K. Schum.; *Exallage auricularia* (Linn.)

Description of Plant

It is a suberect or diffuse, branched and glabrous or hirsute annual herb. Leaves are sessile or sub-sessile of petioled, ovate or lanceolate, acuminate and smooth or scrabid above of often pubescent beneath. Flowers are small and white in axillary, sessile or shortly stalked cymes. The capsules are ovoid or globose, 2-celled with numerous, minute seeds.

Flowering and Fruiting

During July-October.

Distribution

It is found almost throughout India.

Plant Parts Used

Leaves are usable.

Uses

The leaves of the plant are considered emollient and are externally applied to abscesses and as a salve for wounds; also in deafness.

Folk Uses

In South Kanra, its leaves are boiled with rice and are used as a remedy in bowel complaints including diarrhea and dysentery.

Species: *Hedyotis corymbosa* (Linn.) Lamk. (*Oldenlandia corymbosa* Linn.; *O. burmanniana* R. Br.)

Description of Plant

It is an erector decumbent ascending or prostrate herb with woody base. Leaves are sessile, 1.5–2.5 cm long, linear orlinear-lanceolate, acute, glabrous and some what pale or whitish beneath and bristly on the revolute margin. Flowers are exillary in 0.3-1.5 cm long peduncled and 2-5 flowered cymes. The capsules are usually broad, didymus, glabrous with truncate mouth and 0.2 cm across. Seeds are 0.02 cm long.

Flowering and Fruiting

Flowering: During August–September and Fruiting: Later in November–December.

Distribution

It is common throughout India upto 1828 m on the Himalayas.

Plant Parts Used

The whole plant is used as panchang.

Uses

The entire plant is considered anthelmintic and in used in jaundice and diseases of liver. A decoction of the plant is given in remittent fever and gastric irritation and in nervous depression. The juice of the plant is applied in burnings of the palms of hands and soles of the feet in fever. The decoction of the plant with that the *Gurbel, Nagarmotha, Papra* and *Chandan* constitutes '*Panchbhadra kwath*' which is given in all types of fever.

The plant because of the similar action, is used as a substitute for *Fumaria indica*.

Folk Uses

The plant is ground in to a powder and mixed with honey and *mulethi* is given in asthma and also smoked through chilam in different parts of country.

Preparation

Parpatadi kwath, Parpatadhyarisht and Shandgapaniya.

Genus: *Rubia* Linn.

The generic name is derived from Latin word Ruber = red; in allusion to the reddish dye obtained from the roots.

The genus comprises of scabrid, hispid or prickly erect, diffuse or climbing herbs with square and slender stems. Leaves are 4-8 in a whorl, rarly opposite and stipulate. The flowers are small or minute in axillary and terminal cymes. Calyx–tube is ovoid or globose and corolla is rotate or shortly bell or funnel shaped. Stamens are 4-5 and a re inserted on the corolla–tube and ovary is 2-celled with single erect ovule in each cell. Seeds are sub-erect with membranous testa and thin cotyledons.

There are 60 species known to be distributed in Western and Central Europe. Mediterranean region, East tropical and South Africa, temperate Asia, Himalayas and from Mexico to South America. Only 10 species of them are reported from India.

Species: *Rubia cordifolia* Linn.

Description of Plant

It is a climbing perennial herb with stout and smooth branches. Leaves are in whorls of 3-4, ovate o-oblong to lanceolate, with a cordate or rounded base, actue or acuminate with short–bristly margins.=, glabrous except prickly nerves beneath and 2.0–1-.0 x 1.5–5.0 cm. Flowers are greenish yellow in cymes arranged in eaxillary inflorescences which are combined into a many-flowered and widely branched panicle. The fruits are bluish black and glabrous with single seed and 0.4–0.6 cm in diameter.

Flowering and Fruiting

Flowering: During July–September and Fruiting: In November–December.

Distribution

It is found throughout the hilly regions in India.

Plant Parts Used

Roots are usable.

Uses

Its roots are alterative, astringent and tonic and are used in eye sores, paralysis, liver complaints, enlargement of spleen, rheumatism, leucorrhoea, leucoderma and dysentery. Hakims prescribe it much in paralytic affections, jaundice, obstruction of the urinary tracts and in amenorrhoea. An infusion of its roos is prescribed as a drink to women after delivery to procure copious flow of lochia. Besides, a paste made of its roots with honey is applied externally on freckles and discolouration of the skin; also in external inflammation, ulcers and various skin diseases such as pityriasis and vesicular etc.

Folk Uses

Local women of Bhagirathi valley in Uttarkashi take a decoction of its roots in menorrhagia. It is also taken for irregular menstruation in Darjeeling.

Preparation

Manjisthadi kwath, Manjisthadyarak, Karanjadi vati, Laghu-vis-garva taila and Pnda taila.

Family: NAUCLEACEAE

Genus: *Haldinia* Ridsale

The generic name is probably taken from the Haldi, *Curcuma longa* because of its yellow woods and antiseptic properties .

According to Ridsdale (1978), this monotypic genus is found distributed in Sri Lanka, India Eastwards to South China and Vietnam and Southwards to Peninsular Thailand (Surat Thani).

Species: *Haldinia cordifolia* (Roxb.) Ridsdale (*Abina cordifolia* (Roxb.) Hook. F. ex Brandis *Nauclea cordifolia* Roxb.)

Description of Plant

A large deciduous tree which often is buttressed with grey or brownish-grey and rough bark. Leaves are opposite, petioled, deciduous, 10.0-22.5 cm across, orbicular-cordate, abruptly acuminate, pubescent beneath and coriaceous. Flowering heads are lateral, 2-4 (-10) per node and flowering axis solitary and unbranched. The stipules are at the nod4es, bract-like and not surrounding the young flowering heads. Flowers are pentamerous and subsessile on the receptacle which is hairy. Calyx is short and its lobes are oblong, persistent without a deciduous apical portion. Corolla is hypocrateriform and its lobes are valvate but strongly imbricate at apex. Stamens are inserted in the upper part of tube and filaments short, labrous with basifixed anthers. Ovary is 2-locular and ovules are many and pendulous. The fruits is a head of loose dehiscent fruitlets with hard endocarps. Seeds a re ovoid, slightly bilaterally compressed and shortly winged.

Flowering and Fruiting

Flowering: During June-July and Fruiting: In October-November.

Distribution

It is found distributed from the Himalayas to Central and South India, especially common in Eastern ghate,. Mysore and part of Konkan.

Plant Parts Used

Bark and leaf-juice and used.

Uses

The bark is considered febrifuge and antiseptic and its decoction is given internally to check yellow colour of the urine. Besides, its leaf- juice is used to kill the worms in sores.

Family: VALERIANACEAE

Genus: *Nardostachys* DC.

The generic name is derived from Greek words, Nardos, Nard = the Indian spike nard and Stachys = spike.

The genus comprises of erect and perennial herbs. Leaves are entire, radically elongated and spathulate. Flowers are in cymose heads. Calyx-limb is 5-lobed; corolla tubular-campanulate, lobes 5; stamens 4 and ovary is 3-celled and 1-ovuled. The fruits are obovate, compressed, 3-celled and 1-seeded. Seeds are obovate and compressed .

Only 2 species are known to occure in West China and Himalayas including India.

Species: *Nardostachys gradiflora* DC.
(*N. jatamansi* DC.; *Valeriana jatamansi* auct non. Jones)

Description of Plant

It is a perennial herb with woody rhizomatous root stocks covered with fibres from the petioles of withered leaves. Stem is about 10-60 cm long, more or less pubescent upwards and often glabrate blow. Radical leaves are 15.0–20.0 x 2.0 02.5 cm, glabrous or slightly pubescent narrowed into the petiole and cauline leaves 11 or 2 paired, 2.5-7.5 cm long, sessile and oblong or obovate. Lower heads are usually 1-3 or 5; corolla-tube about 06 cm long sometimes hairy within. The fruits are about 0.4 cm long, covered with white hairs and crowned by ovate, acute and often dentate calyx teeth.

Flowering and Fruiting

Flowering: During February-May and Fruiting: In May-June.

Distribution

It is distributed in alpine Himalayas between 3352-4572m, extending eastwards from Kumaon to Sikkim upto 5181 m.

Plant Parts Used

Roots of the plant are usable.

Uses

In ayruvedic system, the roots of the plant have been using since a long time and these are said to cure '*tridosha*', '*kapha*' and biulousness of the blood, burning sensation, erysipelas and various skin diseases and in Unani system, theses are used to promote the growth and blackness of the hair and also to increase the appetite.

Its root are known as aromatic, astiseptic, bitter, stimulant and tonic and are used for the treatment of epilepsy, hysteria and convulsive affections. An infusion of its fresh roots in the doses of 25-50 gm three times daily is recommended in spasmodic, hysterical affections, palpitation of heart, nervious headache, chorea and flatulence (Nadkarni, 1954), Besides, a powder of its root is also said to be useful in menopause disturbances, hystero-epilepsy and similar nervious and convulsive ailments.

The roots and aromatic rhizomes of the plant are used as a good substitute of Valerian drug, obtained from *Valeriana officinalis* and adulterant with rhizomes of *Cymbopogon shoemanthus*.

Folk Uses

In Garwhal Himalayas, local people use its roots for asthma, spasmodic pains and hysteria.

Preparation

Mansyadi kwath, Raktshoghna ghrit, Sarvousadhi snan, Laghu-vis-garva taila, Pippalayasav, Jimadsumbhuttib and Rogan nardin.

Genus: *Valeriana* Linn.

The generic name has been derived from Latin word, Valere = to be heating; in allusion to the medicinal uses of the plants.

The genus comprises of perennial herbs. Leaves are pinnatifid or pinnatge, radical often long-petioled and undivided. Flowers are borne in corymbose panicles. Calyx-linmbs in flowers are obscure and unrolling in fruits into 5-15 plumose bristles, united at base into a short wide funnel shaped tube. Corolla tube is funnel-shaped, lobes 5 and pink or white coloured; stamens are 3 and ovary is 3-celled and 1-ovuled. The fruits are oblong-lanceolate, compressed and crowned with persistent calyx.

About 200 species are known to be distributed in Eurasia, South Africa, temperate North America and Andes. 12 species of them have been reported from India; chiefly from the Himalayas between 1320-5610 m.

Species: *Valeriana hardwicikii* Wall

Description of Plant

It is an erect, perennial herb with a stout rootstock. Stem is simple and pubescent below, corymbosely branched and glabrous upwards, often with bearded nodes, Basal leaves are long petioled and entire; cauline pinnatipartite with 1-5 lanceolate segments. Flowers are white and are borne in corymbose, cymose panicles. The fruits are oblong lanceolate and compressed and are crowned with pappus-like persistent calyx.

Flowering and Fruiting

Flowering: During July-August and Fruiting: In September-October.

Distribution

It is found in temperate Himalayas, from Kashmir to Sikkim between 1219-3657 m and in Khasia hills between 1219-1828 m.

Plant Parts Uses

Its roots are used medicinally.

Uses

Considered its roots antiseptic, carminative and stimulant and these are used in epilepsy, hysteria, chorea, shell shock and neurosis.

It is a good substitute of *Voleriana jaeschkei* and *V. jatamansii.*

Folk Uses

Khasi and Jaintia tribals in Meghalaya apply a leaf-paste of the plant on scaly skin diseases. Besides, in Garhwal, whole plant is pounded and is applied in various skin-diseasew (Purohit *et al.*, 1985).

Species: *Valeriana jaeschkei* Cl. Var. *kaschmiriensis* (Gurbov) Nasir (*V. kaschmiriensis* Keryer ex. Grubov., *V. officinalis* auct. non Linn.)

Description of Plant

This is a perennial herb with short rootstock and erect, furrowed, smooth, hairy below and foetid stem reaching a height of 0.3-0.9 m. Leaves are alternate; leaflets in 6-10 pairs, narrowly oblong or linear often entire, much or sparingly toothed. Flowers are white or flesh coloured; petals united to one another so as to form a tube ending above in 5 loges. The fruits are hairless.

Flowering and Fruiting

Flowering: During June-July and Fruiting: In August-September.

Distribution

This plant is confined only to Kashmir between 2438-2443 m.

Plant Parts Used

Its tuberous roots are used medicinally.

Uses

The roots of the plant are considered antiseptic, scarminative and stimulant., As an antispasmodic, these are used in hysteria, epilepsy, chorea and allied affections and as a stimulant, in advanced stages of fever and asthemic inflammations.

It is used as a substitute of *Valeriana jatamansii.*

Species: *Valeriana jatamansii* Jones (*V. wallichi* DC.)

Description of Plant

It is a slightly hairy perennial herb with thick and horizontal root stock. Basal leaves are 2.5-7.5 cm in diameter, long stalked, deeplycordate-obovate, usually toothed or sinuate and sharp pointed. Stem leaves are few, much smaller and entire or pinnate. Flowers are white or tinged with pink in teminal corymb; often unisexual. The fruits are hairy or nearly hairless.

Flowering and Fruiting

Flowering: In June-July and Fruiting: During September-October.

Distribution

It is found distributed in temperature Himalayas, from Kashmir to Sikkim upto 3048 m and in Khasia hills between 1219-1828 m.

Plant Parts Uses

The roots are used.

Uses

Valerian roots are the very old drug and in ayurveda, these are used for head problems, eye diseases, blood diseases and also in suppression of urine, and as laxative, astringent, carminative, emmenagogue and hypnotic in Unani system of medicine.

Its roots as antiseptic, carminative and stimulant and according to them, these are successfully used for the treatment of chorea, epilepsy and hysteria.

The rootstocks of the plant are used as a good substitute for other species *viz.*, *V. jaeschkei* and *V. hardwickii*.

Folk Uses

Gandwal tribals in Garhwal Himalayas, use its roots in sprains and contusions and in Pauri Garwhal, people crush entire plant with roots of *Hedychium spicatum* and *Cynodon dactylon* to make a paste which is applied on body before bath as a fragrant and refreshing agent.

Preparation

The drug is much used in making perfumed powder and cardiac preparations. Some important yogas include Sudarshan chruna, Pippalayasav and Dasanga lep.

Family: ASTERACEAE

Genus: *Achillea* Linn.

The generic name has been assigned in the name of Achilles, the hero of the Trojan wars, who learned the healing properties of the plant *Achillea millefolium* from Chiron, the centraur.

The genus comprises of perennial, pubescent and villous herbs. Leaves are alternate, narrow, serrulate or pinnatisect. Flowers are white, pink or yellow in corymbose heads. The achenes are oblong or obovoid, dorsally compressed and glabrous with 2 cartilaginous wings.

About 200 species are known to occur in North temperate region of the world. Only 2 of them have been reported from India.

Species: *Achillea millefolium* Linn.

Description of Plant

It is an erect and pubescent, perennial herb about 15-60cm in height. Leaves are alternate, oblong-lanceolate, 5.0-10.0 x 6.0-1.8cm, 3 pinnatiscet; segments linera and acute; radical leaves stalked, upper sessile, The flowers are white or pink coloured in radiate heads crowded in compound corymbs. The achenes are oblong, flattened and shining.

Flowering and Fruiting

During August-December.

Distribution

It is found in Western Himalayas from Kashmir to Kumaon between 1066-3657 m, especially around Shimla.

Plant Parts Used

Whole herb is used medicinally.

Uses

The herb is known to possess diaphoretic, emmenagogue, stimulant and tonic properties and is useful in colds, obstructed perspiration and commencement of fevers. Nadkarni (1954), stated that the powder of its leaves and that of flower heads is useful as carminative and tonic in the doses of 0.3-2.3 gm and a hot infusion of leaves along as emmenagogue.

Genus: *Ageratum* Linn.

The generic name has been taken from Greek words A = without and Geras = old age, referring to the plants as they continue in flower for a long time.

The genus comprises of erect and weedy herbs or shrubs. Leaves are opposite or the upper ones alternate. The heads are corymbose or panicled and homogamous. Involucre is campanulate. Bracts are bi- to triseriate, linear and subequal. Receptacle is flat, naked, or with caduceus scales. Fruits are 5-angled achenes.

About 60 species occur in tropical America; 2 species have been naturalized and occur throughout India ascending upto 1650 m in the Hiamalayas.

Species: *Ageratum conyzoides* Linn.

Description of Plant

It is an annual and soft hairy herb about 30-45 cm high. Stem is often decumbent and rooting at the base. Leaves are broadly ovate or rhomboid-ovate to triangular, subcordate, rounded or truncate at the base, obtuse or acute, often acuminate at the apex, serrate-dentate, glabrate or thinly long hairy and 2.0-10.0 x 1.5-5.0 cm. Heads are violet or white and are arranged in dense terminal corymbs. Involucral bracts are linear, acute and ribbed on their backs. Flowers are many. Corolla is in fundibuliform. Fruits are glabrous or thinly hairy achenes and about 0.16-0.18 cm long.

Flowering and Fruiting

During cold season.

Distribution

It is a native of America, now got naturalized and found throughout India ascending upto 1650 m in the Himalayas.

Plant Parts Used

The roots, leaves, flowers and panchang are used.

Uses

The juice of the roots is used a antilithic. Its leaves are styptic and anti-tetanic and are applied to cuts and sores and externally in ague. Its leaves are given with water as an emetic. The flower-extract is useful in Asiatic form of Schwartz leukemia and is known to prolonge the life-span of mice.

The whole plant is also said to be used a s nervine tonic. An infusion or the decoction of entire plant is given in diarrhea and colic with flatulence and other gastro-intestinal ailment. The extract of the whole plant possesses insecticidal properties too.

Folk Uses

Its leaves are crushed in week long stored water and their extract is given on Sunday only to control bed wetting by the children during sleep. It may be given on subsequent Sundays till the disease is cured.

Genus: *Arctium* Linn.

The generic name is based on Greek name *Arction* of the plant.

The genus comprises of tall, coarse, erect and branched herbs. Leaves are alternate or radical, broad and undivided. Floral heads are terminal, sessile, solitary or fascicled; involucre sub-globose. Involucre bracts become hooked and woody after the flowers wither and by clinging to fur etc. aid in dispersing the fruits. The achenes are oblong, sub-triquetrous, ribbed, glabrous, shining and truncate with subequal base.

Only 5 species are known to be distributed in Paleotemperate region of the world; 1 of them has been reported from India.

Species: *Arctium lappa* Linn.

Description of Plant

It is and erect, coarse and rough herb with about 0.6-1.2m height. Leaves are radical and altgernate, stalked, broadly ovate, 7.5-30.0 cm across, cordate and sinuate-toothed; lower surface is hoary or white-cottony. Flower heads are globose, discoid and 1.8-7.5 cm across in terminal clusters. Involucre bracts are many, upper half spreading; tips rigid and hooked. The flowers are purple coloured; corolla tube long and lobes 5. The achenes are large, glabrous, angled and finely ribbed.

Flowering and Fruiting

During September-October.

Distribution

It is known to be distributed in Western Himalayas from Kashmir to Shimla between an altitude of 1828-2743 m.

Plant Parts Used

Roots of the plant are used.

Uses

The roots of the plant are used as alterative, antiphologistic, diuretic and diaphoretic. Besides, plant extract is reported to cause a sharp long lasting reduction of blood sugar with an increase in carbohydrate tolerance and less toxicity *in vitro*.

Genus: *Artemisia* Linn.

The generic name has been named after Artemis, daughter of Jupiter and Latona of Grteek mythoilogy; Artemis was also virgin Goddess of Moon and of hunting, supposed to kill without pain, in allusion to the soothing properties.

This genus comprises of aromatic herbs or shrubs. Leaves are alternate, entire and serrate or 1-3-pinnatisect. Heads are small, solitary or fascicled, racemose or panicled, hoimogamous or heterogamous and disciform; outer ones are female, 1-series and fertile. Disk flowers are bisexual and fertile or sterile. Involucre is ovoid and subglobose or hemispheric. The achenes are minute, ellipsoid and oblong or subovoid.

About 400 species occur in North temperate region, South Africa and South America. 34 species have been recorded from India, chiefly between 1650-5940 m in North Western Himalayas.

Species: *Artemisia brevifolia* Wall. Ex DC.

Description of Plant

It is a hairy or tomentose shruby herb with erect or ascending much branched stems. Leaves are ovate, 2 pinnatisect; segments linear and obtuse; upper simple linear. Flower heads are 3-8 fid; involucre bracts linear-oblong, outer herbaceous tomentose and inner scarious, acute and glabrous. The achenes are minute.

Flowering and Fruiting

During August-October.

Distribution

It is known to be found in Kashmir, Himachla Pradesh and Uttarakhand between 1219-3657 m.

Plant Parts Used

Flower heads and leaves are used medicinally.

Uses

Its flowers heads are considered anthelmintic and they are principally use as a avermifuge against ascarides. Besides, a decoction or infusion of leaves of the plant is used in ague and in intermittent and remittent fevers.

Species: *Artemisia capillaries* Thunb.

Description of Plant

It is a glabrous and pubescent annual or perennial herb with 90-180cm height. Stems are slender, branched. Grove and usually tinged with purple. Leaves are stalked or nearly sessile, 2.5-7.5cm long and 1.3 pinnatisect; segments thread like. The floral head are yellow, minute and sessile or on short capillary pedicels arranged in second panicled racemes. Outer female flowers are fertile, inner are hermaphrodite or sterile and with larger corollas. Achenes are very minute.

Flowering and Fruiting

During September-December.

Distribution

It is found distributed throughout India ascending upto3657 m in the Himalayas.

Plant Parts Used

whole plant is usable.

Uses

The plant is known to possess purgative property and smoke of the herb is used for burns. Besides, P-OH-acetophenone, an active constituent of the plant is said to promote the secretion of the bile and bile salts.

Folk Uses

In Punjab, an infusion of the plant is give as a purgative and smoke of the plant is considered goof for burns.

Preparation

Artemisol, which is used against urolithiasis.

Species: *Artemisia indica* Willd.
(*A. vulgaris acut.* non Linn.)

Description of Plant

It is a hairy-tomentose shrubby herb with much branched stem. Lower leaves are 1-3 pinnately lobed or 1-3 pinnatifid,2.5-7.5 cm long; segments narrow or rather broad, entire or nearly so, acute or obtuse, upper surface pubescent or hairy and lower tomentose or densely hairy. Floral leaves are 3-lobed near to base or lanceolate.

Flowering and Fruiting

During August-October.

Distribution

This is found distributed throughout the mountainous regions of India, ascending upto1524-3048m in the Western Himalayas and upto 1524-2438 m in Sikkim and Khasia hills and in Mount Abu and Western ghats.

Plant Parts Used

Whole plant is usable.

Uses

The herb possesses anthelmintic, antiseptic, emmenagogue and stomachic properties and its roots are used as antiseptic and as tonic. Besides, an infusion of its leaves and of flowering tops is administered in nervous and spasmodic infections, in asthma and brain diseases.

Folk Uses

Its leaf juice is applied on skin allergy and in itching by the people in Nagaland. Besides, in North east India, 10 ml of leaf infusion of the plant is taken twice daily for cough and fever.

Genus: *Blumea* DC.

The generic name has been assigned in the honour of Dr. C.L. Blume, Dutch botanist of the 19th century.

The genus comprises of annual or perennial woolly or glandular-pubescent herbs. Leaves are alternate and lobed. The floral heads are corymbose panicled or fascicled, rarely racemed, heterogamous and purple or yellow. Outer flower is many seriate and female fertile and 2-3 toothed. Disk flower is many seriate and female fertile and 2-3-toothed. Disk flowers are hermaphaorodite, few, fertile, tubular and slender and limb 5-toothed. Involucre is ovoid or campanulate. The achenes are small, angled or sub-terete and ribbed or not ribbed.

Its 50 species are known to occur in tropical and South Africa, Madagascae, East Asia to Australia and Pacific. Of these, 30 species are found in India, mostly in Southern and Western India and tropical Eastern Himalayas upto 1650m.

Species: *Blumea lacera* (Burm. F.) DC. (*B. Subcapiatata* DC.; *Conyza lacera* Burm.f.)

Description of Plant

It is an erect, aromatic and somewhat viscid annual-biennial herb with a stout taproot. Stem is simple or branched and very leafy. Leaves are oblong-obovate or elliptic-oblong, obtuse or acute, narrowed into short petiole at base, dentate or serrate, silky on both surfaces and 2.5-5.0 x 1.0-1.5cm. Heads are 0.5-0.6 cm across on 0.5-1.5 cm long peduncles and are crowded at the tops of branches in terminal speciform panicles and rarely corymbose, Involucral bracts are linear, acute, glandular hairy and 0.2-0.7cm long. Corolla-lobes of disk flowers are hairy with colleters and unicellular or multicellular hairs. The achenes are hairy and 0.05-0.06 cm long.

Flowering and Fruiting

During winter season.

Distribution

It is found distributed throughout the plains of India ascending upto 609m.

Plant Parts Used

Whole plant and its roots are used.

Uses

The plant is considered bitter and antipyretic and its alcoholic extract is said to exhibit marked antinflammatory activity against carrageein and bradykinin induced inflammation in rats.The juice of its leaves is anthelmintic, astringent, febrifuge, stimulant and diuretic. This is mixed with black pepper and is given in piles. The essential oil obtained from its leaves is considered antimicrobial, antifungal and insect repellent. Besides, its roots are astringent and febrifuge and are used in cholera.

Preparation

Arshkuthar ras.

Genus: *Centratherum* Cass.

The generic name is derived from Greek words Kentron = spur and Antheros = spine; referring tot eh presence of spines.

The genus comprises of herbs. Leaves are alternate. The floral heads are terminal or axillary, solitary cymose or panicled and homogamous. Involucre is ovoid, globose or hemispheric and equal or shorter than flowers. The bracts are in many series. Receptacle is naked or pitted, sometimes shortly hairy. The achenes are striate and ribbed or angled or rarely terete.

About 20 species are found distributed in the tropical parts of the world; 11 species occur in India, chiefly in Western ghats.

Species: *Centratherum anthelminticum* (Linn.) Kuntze [*Conyza anthelmintica* Linn.; *Vernonia anthelmintica* (Linn.) Willd.]

Description of Plant

These are annual, tall, robust and glandular-pubescent herbs with terete stems. Leaves are elliptic-obovate or ovate, acute or acuminate, 7.5-20.0 cm long, coarsely serrate and glabrescent or thinly hispidly hairy. Heads are 1.5cm across, subsolitary, leaf-opposed on long peduncles and combined intoa terminal and leafy corymb.Involucral bracts are in 4-5 series. Achenes are terete,10-ribbed, thinly hairy and pubescent.

Flowering and Fruiting

Flowering: During the rainy season and Fruiting: In the spring season.

Distribution

The plant occurs throughout India upto 1676 m in the Himalayas; often cultivated near villages.

Plant Parts Used

Seeds are used.

Uses

Its seeds as highly anthelmintic, alteative, astringent, febrifuge and diuretic and are given in cough, fever dropsy and flatulence. An infusion of its powdered seeds is used as tonic, diuretic and stomachic. Crushed seeds with honey are given in two equal doses, followed by an aperient for the treatment of round worms. These are also used in white leprosy and other chronic skin diseases as psoriasis etc. For the purposes, the seeds with black pepper or black sesame in equal parts are powdered and are given in doses of 3.8gm daily in the morning with tepid water. It should be taken after perspiration has been induced by exercise or exposure to Sun. It is recommended as daily dose for one year, for a complete cure to be effected.

Externally, the seeds are used in skin diseases in variety of forms such as oil and paste etc. Their juice is used locally to destroy pediculi of the head and body. The seeds with that of *Cassia sophera, Cassia tora,* turmeric and common sea-salt in equal parts are rubbed together into a paste with whey and fermented paddy water and are applied over eruptions attended with itching.

Folk Uses

In Andhra Pradesh, the seeds boiled with pepper and garlic are used to expel worms from the intestine and seed-paste is applied externally in skin-diseases.

Preparation

Somraji taila and Somraji ghirt.

Genus: *Cichorium* Linn.

The generic name is taken from its Arabic name, Chikourych which has been adopted by the Greeks from the Egyptians as Cichora.

The genus comprises of erect and glabrous or hispid herbs with divaricate, sometimes spinescent branches. Upper leaves are subentire and lower ones pinnatifid. Flower heads are sessile on the branches or on thickened peduncles; involucre bracts; inner 1-seriate and outer few shorteor. The achenes are glabrous and sub-penta-angled with contracted base.

Its 9 species are known to occur in Europe, Mediterranean region and in Abyssinia. Only 3 of them have been recorded from India.

Species: *Cichorium intybus* Linn.

Description of Plant

It is an annual or perennial herb with thick tap roots. Radical and lower leaves are short petioled, lyrate, pinnatifid, obtuse or acute and remotely dentate or lobed; segments hairy or glabrescent, upper ones sessile and 4-10 together in axillary and terminal clusters combined into a widely branched, leafy panicle. Involucre bracts are 2-seriate, lanceolate-oblong, sub-acute, 1.0cm long, outer ones with a gland-hairy margin. The achenes are 5-angular, turbinate and about 0.2-0.3 cm long.

Flowering and Fruiting

During August-September.

Distribution

It is known to occur wild in Kashmir, Punjab, Himachal Pradesh; also cultivated else where.

Plant Parts Used

Whole plant is used medicinally.

Uses

Wild form of the plant is considered as alexiteric, emmenagogue and tonic and cultivated one as tonic and both form are used in fevers, diarrhea, vomiting and in spleen enlargement, stated that a strong infusion of its powdered seeds is used successfully in obstructions or torpor of liver and to check bilious enlargement of the spleen with general dropsy.

Its roasted and powdered seeds are used as a substitute of coffee.

Folk Uses

A sharbat of its flowers is taken by the inhabitants of Punjab and Kashmir. Besides, a paste of the herb is also applied in inflammation over the skin externally.

Genus: *Echinopos* Linn.

The generic name has been taken from Greek words Echinos = a hedgehog and Opsis=appaarance, referring to the spinous involucral bracts.

The genus comprises of thristle-like herbs. Leaves are alternate, pinnatifid and spinous. The floral heads are in globose involucrate balls, blue or white and sessile or shortly stiptate on a common receptacle and 1-flowered. The flowers are bisexual; all are fertile, tube slender and limb with 5 slender segments. Involucre is oblong and bracts are many-seriate, rigid and pungent or spinescent; of which outer is shorter, inner spathulate and innermost is linear lanceolate or sometimes all connate to form a tube with one long rigid spine on outer side. Achenes are elongate and usually villous.

About 100 species occur in Eastern Europe, Africa and Asia; 4 species of them have been reported from India.

Species: *Echinops echinatus* Rox.

Description of Plant

It is a thristle-like much-branched, spreading and rigid annual herb. Stems are branched from the base and clothed with white cottony pubescence. Leaves are alternate, pinnatifid, spinous and thickly white beneath with cottony wool. The heads are 1-flowered and are crowded in a globose white ball which is about 5.0 cm across. Achenes are obconic and densely silky.

Flowering and Fruiting

During July-December.

Distribution

It is found more or less throughout India ascending upto 1524 m in the hills.

Plant Parts Used

The whole plant and its roots are used.

Uses

The entire plant is a nervine tonic, alterative and diuretic and is used in hoarse cough, hysteria, dyspepsia, scrofula and opthalmia. An infusion of the plant is given in seminal debility and impotence.

Its powdered roots are applied to cattle wounds to destroy maggots and to kill lice.

Folk Uses

A piece of its roots is tied on forehead or hairdo of the expectant mother tofacilitate andhasten the childbirth in case of difficulty.

Genus: *Eclipta* Linn.

The generic name has been taken from the Greek word Ecleipo means to omit or leave out, referring to the absence of the pappus.

The genus comprises of annual or perennial herbs. Leaves are opposite. The floral heads are small, axillary or terminal, peduncled and heterogamous rayed. Raya flowers are female, sub-2-seriate and fertile or sterile. Disk flowers are bisexuial, fertile, tubular and limb 4-5 fid. The fruits are achenes of ray, and are narrow, triquetrous, often empty and laterally compressed with entire, toothed or 2-aristate top.

About 3-4 species occur in the warmer America, Africa, Asia and Australia. Only 1 species has been reported to occur throughout India.

Species: *Elipta prostrate* (Linn.) Linn. [*E. alba* (Linn.) Hassk.; *Verbesina prostrate* Linn.]

Description of Plant

A prostrate decumbent-ascending or erect, annual or perennial herb. Stems are creeping and rooting at the base. Leaves are sessile, ovate-lanceolate, elliptic-oblong or oblong, acute or obtuse, narrowed to the base and entire or serrate. The floral heads are axillary and terminal and 0.6-1.0 cm across on 5.0-7.0 cm long peduncles. Involucral bracts are 2-seriate, ovate-lanceolate, acute, appressed-pubescent and 0.3-0.6 cm long. Marginal flowers are white and with 2-dentate and 0.25 cm long ligules. The achenes are oblong–turbinate and tuberculate with thickened margins and about 0.2 cm long.

Flowering and Fruiting

During the greater parts of the year.

Distribution

The plant is found distributed throughout India upto 1980 m in the Himalayas.

Plant Parts Used

Whole plant is useful.

Uses

The entire plant is used as tonic and deobstruent in hepatic and spleen enlargement and emetic. Its juice in combination with aromatics is administrated for catarrhal jaundice. The roots are known as emetic and purgative and are applied externally as antiseptic to ulcers and wounds of cattle. Its leaves are said to be used in scorption stings also. Their juice along with honey is used as a remedy to catarrh in infants. Crude extract of gum resin obtained from herb possesses anticancerous activity against Ehrlich ascites carcinoma. The pulp of the bruised herb roasted in ghee is applied to wounds and burns and the powder of the herb cured patients suffering from infective hepatitis. The aqueous extract of the herb is a good myocardial depressant and hypotensive and the juice is used to cure shoulder pain caused by carrying heavy loads.

The fresh plant mixed with sesame oil is externally applied in elephantiasis. The juice of the plant is also applied in anus to remove anal worms.

Folk Uses

In Mumbari, the natives use its juice in combination with 'ajwain' seeds as tonic anddeobstruent and give two drops of it with eight drops of honey to newly born child suffering from catarrh. The leaf-juice with buttermilk is taken as antidote for snake bite by the Yanadi tribals in Andhra Pradesh.

Preparation

Bhringraj taila, Shadbindu taila, Bhringraj ghrit, Bhringrajadi churna,Tafroli, Hab miskeen nawaz, Roghan amala khas and Majun murrawah-ularwah.

Genus: *Inula* Linn.

The genetic name *Inula* is the old Latin Classical name of the plant used by the Romans and was derived from the Greek name *Helenion*, which was derived from Helenium of Helena, wife of Menelaus. As pr the fable, Helen of Troy had *Inula* flowers in her hand when Paris abucted Helen. According to Greek mythology, this plant arose from the tears of the fair Helen of Troy.

The genus comprises of herbs rarely of shrubs. Leaves are radical and alternate, Floral heads are solitary, corymbose or panicled, heterogamous, radiate and rarely disciform. Involucre is broad or rather narrow; bracts seriate, inner usually rigid and narrow, outer herbaceous and outer most often foliaceous. The achenes are subterete and usually ribbed.

About 200 species are found distributed in Europe, Asia and Africa. Out of which only20 species have been recorded from India; mostly confined between 1320-4620 m in North Western Himalayas.

Species: *Inula racemosa* Hook, f.

Description of Plant

it is a tall and stout herb reaching a height upto 0.3-1.5 m. Leaves are leathery, rough above, densely hairy beneath and crenate; basal leaves are 20.0-45.0 x 12.5-20.0 cm, long stalked and elliptic-lanceshaped and stem leaves oblong and often lobed at the base. Flower heads are many, large and 3.5-5.0 cm across in racemes. The fruits are 0.4 cm long, slender and hairless achenes.

Flowering and Fruiting

During September-October.

Distribution

It occurs in temperate and alpine western Himalaya between 1524-4267 m and in Kashmir between 1524-2133 m.

Plant Parts Used

Roots of the plant are used medicinally.

Uses

The roots are used to cure pains of the heart, spleen, liver and joints and in hemicrania, eruptions, inflammations, ear pain, cough and boils in Unani system of medicine. Roots of the plant are also used as expectorant and as resolvent in indurations. Its seeds also possess aphrodisiac and bitter properties and are useful in strengthening the hair and in preventing them to fall.

The roots of this plant are adulterated with Kuth (Saussurea lappa) and Digitalis.

Folk Uses

Its seeds also possess aphrodisiac and bitter properties and are useful in strengthening the hair and in preventing them to fall in certain parts of the country.

Preparation

Puskarmuladi churna, Puskaradi churna, Devdaradi kwath, Sudarshan churna, Punarnava mandur, Sringyadi chruna and Kukuvadi churna.

Genus: *Sphaeranthus* Linn.

The generic name is derived form Greek words, Sphaira = a ball, sphere and Anthos = a flower; referring to the globular shape of inflorescence.

The genus comprises of divaricately branched herbs. Leaves are alternate, toothed and decurrent along the stem. Heads are small, numerous and sessile and are crowded on a large common receptacle into more or less globose terminal compound heads with or without a general involucre of empty bracts at the base; outer flowers female, few or many fertile, disk flowers hermaphrodite, solitary or few, fertile or sterile. The achenes are oblong and compressed.

About 40 species are known to be found in Africa excluding North-West Africa, Madagascar, Iraq to Persia, Southeast Asia, West Malayasia, Celebes and in North-East Australia. Only 3 of them have been recorded from India; chiefly from the plains with 1 extending to 1650 m in the Himalayas.

Species: *Sphaeratnthus senegalensis* DC.
(*S. indicus auct.* non Linn.)

Description of Plant

It is a prostrate or decumbent–ascending, aromatic and viscid annual herb. Stem is much branched with coarsely dendate wings of decurrent leaf-base and glandular pubescent. Leaves are obovateoblong, with narrow base, obtuse, mucronate coarsely dendate and glandular villous. Heads are compound, globose-ovoid, ebracteate, about 1.0-1.5 cm across on solitary glandular peduncles with toothed wings. The achenes are 0.1 cm long, glabrous and stalked.

Flowering and Fruiting

During December-March.

Distribution

It is known to be distributed throughout India ascending upto 1524 m; commonly seed in dry rice fields.

Plant Parts Used

Whole plant, its flowers, roots and seeds are used medicinally.

Uses

In ayurveda, the herb is known to possess to alterative, anthelmintic, alexipharmic, bitter, laxative and tonic properties and is used in insanity, tuberculous glands, indigestion, bronchitis spleen diseases, elepohantiasis, anaemia, pain in uterus and vagina, piles, stangury, biliousness, epilepsy, asthma, leucoderma, dysentery and in hemicrania. Its roots and seeds are also used as anthelmintic. An oil, obtained from its roots by steeping them in water and then boiling in sesame oil until the all water is expelled and this is taken on empty stomach daily in the morning for 41 days in the doses of about 60 gm as a valuable aphrodisiac. Besides, leaves of the plant dried in shades, in powder form are used in the doses of 1.5 gm 2 times daily in skin diseases as antisyphilitic and nervine tonic and also in urethral discharges and jaundice. Its flowers are highly esteemed as alterative, depurative, refrigerant and tonics and are also useful as blood purifier in skin diseases.

Folk Uses

In Chota Nagpur, the entire plant is pounded with a little water and its expressed juice is used as a styptic and juice of the leaves along is boiled with milk and sugar candy is drunk in cough. Its flowers are also chewed or swallowed for curing conjunctivitis in different parts of the country. In Baharich, people apply a paste of whole plant along with seed paste of *Psoralea coryfolia* and *Azadirachta aindica* oil for two months in leucorrhoea. In Tamil Nadu, adecoction of its leaves and flowers is taken as a cooling remedy. Besides, Sauria Paharis in Santhal Paragana (Bihar), give one teaspoonful infusion of the plant for 5 days a anthelmintic and blood purifier.

Preparation

Atriphal mundi, Mundi ark, Majunmundi, Sharbat mundi, Rogan mundi and Choa mundi.

Genus: *Taraxacum* F.H. Wigg.

The generic name Taraxacum is derived from the Persian name Tarashqum of the plant.

The genus comprises of scapigerous milky herbs. Leaves are radical, entire, sinuate-pinnatified or runcinate-pinnatifid. Flowers are solitary on leafless scapes, yellow and homogamous. Involucre is campanulate or oblong; bracts herbaceous, innermost 1-seriate, erect, subequal and sometimes connate below, outer shorter, many-seriate and usually recurved. The achenes are oblong, obovoid or narrow, 4-5-angled, or the outer dorsally compressed, beaked, glabrous and 10-ribbed.

About 60 species are known, mostly found in North temperate region, 2 in temperate South America. About 25 species are reported from India, usually confined between 330-5940 m in the Himalayas.

Species: *Taraxacum officinale* Webber

Description of Plant

It is a perennial milky herb with a stout rootstock. Leaves are basal, sessile, oblanceolate or linear, entire to lyrate pinnatifid; lobes acute and denticulate. Flowers heads are solitary on scapes. Outer involucre bracts are ovate and inner ones linear. The achenes are obovoid and ribbed; beak as along as or longer than the body.

Flowering and Fruiting

During March-November.

Distribution

It is known to occur throughout the Himalayas from 304 to 5486 m and in Mishmi hills.

Plant Parts Used

Roots and leaves of the plant are used.

Uses

About 25-50 gm decoction of its roots and that of Podophyllum is used in jaundice, hepatitis and in indigestion effectively. Further its root as aperient, diuretic and tonic and these are used as a remedy for chronic disorders of kidney and liver. Besides, its leaves are also used for fomentation.

Its leaves are used as a substitute and adulterant of Henebane (Dried leaves and flower tops of *Hyoscyamus niger*) and roots as a substitute or adulterant of coffee.

Folk Uses

In Almora District of Uttarakhand, 5-10 gm crushed roots or its root paste is given orally with one glass of water in case of urinary problems, liver disorders and in constipation. Besides, latex of its leaves and stalkes is applied externally on the affected parts of the body for a month in corns and warts.

Genus: *Tussilago* Linn.

The generic name is derived from Latin word, Tussia = a cough; as the leaves and flowers of the plant are used for curing coughs.

This monotypic genus is known to be distributed in North Africa and Eurasia including India and excluding China.

Species: *Tussilago faffara* Linn.

Description of Plant

This is a white, woolly and scapigerous herb with a perennial stoloniferous root stocks. Leaves are radical, 7.5-15.0 cm long, coming after the flowers, orbicular-cordate and toothed. Flowers heads are solitary, heterogamous, radiate and yellow. Involucre is campanulate or cylindric; bracts 1-series and equal with a few very small outer ones. The achenes are 5-10 ribbed with slender rough pappus hairs.

Flowering and Fruiting

Flowering: During spring and fruiting later.

Distribution

It is found distributed in West Himalayas, from Kashmir to Kumaon between 1828-3352m.

Plant Parts Used

Roots and leaves of the plant are used.

Uses

Roots and leaves of the plant are used in chest problems, chronic bronchitis and in asthma. Besides, leaves are demulcent and are smoked in pulmonary complaints. Juice of its fresh leaves is also taken in small quantities every day to heal scrophulous ulcers.

Folk Uses

In hilly parts of India, its leaves and rots are smoked like tobacco as a remedy for asthma, obstinate cold, cough and other chest problems.

Genus: *Xanthium* Linn.

The generic name has been derived from Greek word Xanthos means yellow, as the ancients were extracted a dye from the plant.

The genus comprises of annual coarse rough herbs or under shrubs which may be unarmed or with trifid spines. Leaves are alternate and toothed or lobed. Heads are monoecious and axillary. Bisexual heads are in the upper axils and are globose and many flowered; flowers are sterile, tubular and 5-toothed. Female heads are 2-flowered, fertile, apetalous, involucre of bisexual heads is short; bracts few and 1-2 seriate and receptacle is cylindric, Involucre of female heads is with 2-beaked, ovoid and herbaceous 2-celled utricle of bracts with one flower in each cell and clothed with hooked bristles. The filaments are monoadelphous and anthers free. Style of bisexual flowers is slender and undivided and arms of female flowers are free. The achenes are enclosed in hardened involucral cells and are obovoid and thick.

About 80 species are known to be distributed cosmopolitically throughout the world; of these 2 species have been reported to found in India upto 1650 m in the Himalayas.

Species: *Xanthium strumarium* Linn.

Description of Plant

A coarse, erect, simple or branched and aromatic annual herb, about 0.5-2.0 m high. Leaves are petioled, ovate-triangular, palmately 3- to 5-lobed or angled, cuneate base, acute or acuminate, irregularly coarsely denate, glandular and upto 1-x12 cm. The heads are in terminal and axillary racemes. Male heads are short peduncled and involucral bracts are 2-3 seriate, linear, acute, patent, ciliate and 0.2-0.25 cm long. The involucres of female heads are fine, 2.0-2.5 cm long and patent hairy with incurved beaks and 0.3-0.5 cm long prickles.

Flowering and Fruiting

During rainy season.

Distribution

It is a native of America, now got naturalized and found throughout the hotter parts of India ascending upto 1828 m in the Himalayas.

Plant Parts Used

The whole plant, its roots, leaves, buds, fruits and the seeds are used.

Uses

The entire plant is known as diaphoretic, sedative, sudorific and sialogogue and its extract is given in malaria and leucorrhoea as styptic and in urinary problems. An ethanol extract of the entire plant in flowering and fruiting conditions, when applied in the form of a paste externally for 1-10 days on breast cancer, shows growth regression. Its roots are bitter and tonic and are locally applied instrumous diseases and cancer.

Its leaves as astringent, alterative, diuretic and antisyphilitic which are given inscorfula and herpes I the doses of 6.0 gm. Its leaves to possess antirpanosomal activity also.

The buds are sued as tonic, diuretic and sedative. The fruits are cooling, tonic, diuretic, diaphoretic, sedative and demulcent. And their powder is given for regulation of hormonal activity and in urinogenital disorders. Its seeds are used for resolving inflammatory swellings and their oil for bladder affections, herpes and erysipelas.

Folk Uses

Its seed-paste is applied on forehead to cure headache by Bhils. Besides, two teaspoonfuls of plant decoction twice a day are taken for week by the tribal women for leucorrhoea in Central India.

Family: CAMPANULACEAE

Genus: *Lobelia* Linn.

The generic name has been assigned in the honour of Mathias de l' obel (1538-1616), born at Lille, France and later become Physician and Botanist to James I of England. He traveled widely in Europe in search of medicinal plants and published important botanical treatises.

The genus comprises of herbs or undershrubs.Leaves are alternate. Peduncles are 1-flowered, solitary in the axills of leaves or bracts, sometimes in terminal racemes. Calyx is superior limb 5-partite, corolla oblique, more or less distinctly 2-lipped; staminal tube free from corolla and ovary is inferior and 2-celled. The capsules are loculicidally 2-valved with calyx lobes. Seeds are many, minute, ellipsoid and compressed.

About 200-300 species are known to be distributed cosmopolitically; mostly in tropics and substropics, especiallyin America, 22 of them have been reported from India; chiefly from Eastern Himalayas, with a few descending to the plains of Eastern India.

Species: *Lobelia nicotianaefolia* Roth ex R. and S. (*L. beddomeana* Wimmer; *L.colorata* Wall; *L. aerecta* Hook. F. and Thomas; *L. pyramidalis* Wall; *L. rosea* Wall).

Description of Plant

it is large biennial or perennial herbs reaching height upto 1.0-2.3 m. Leaves are alternate, sessile or nearly so, lower sometimes reaching 40.0-45.0 x 5.0-7.5 cm and uppermost 7.5x2.0 cm and then passing into floral leaves or bracts, all oblong-lanceolate, acute, serrulate, usually glabrous above and pubescent or glabrous beneath. Flowers are white in terminal racemes which may be 30.0cm long. Clayx tube is pubescent or glabrous, lobes 1.3cm long; corolla lobes linera and 3 usually connate throughout. The capsules are 0.8 cm in diameter, globose and open by 2 valves. Seeds are very small, ellipsoid, compressed and yellowish brown.

Flowering and Fruiting

During August-October.

Distribution

This is found in Western ghats from Mumbai to Travancore between 914-2133 m; Chennai, Kerala and In Malabar and Mysore.

Plant Parts Used

Leaves and seds of the plant are used medicinally.

Uses

The leaves in ayurveda system of medicine are considered acrid, aphrodisiac, bitter, diuretic, sweet and stomachic and are used ot cure 'kapha' blood diseases and to cure burning sensation, biliousness and erysipelas. Besides, the seeds are also used as acrid and for poisonous purposes.

Folk Uses

Todas people in Nilgiris hills (Tamil Nadu), use its stem bark for toothache and place a piece of stem bark over the aching tooth to get relieve in pain. A paste of its leaves is applied locally for hydrocele by the inhabitants of Madhya Pradesh. Besides, Katkare tribals in Maharashtra, inhale the smoke of its dried leaves 2-3 times a day to cure asthma.

Family: ERICACEAE

Genus: *Gaultheria* Linn.

The generic name has been given in the honour of JeanaFrancois Gaulthier (C. 1708-1756), a French Physician and Botanist of Quebee, Canada.

The genus comprises of erect or procumbent shrubs. Leaves are persistent, alternate and seerulate, Flowers are greenish-white or reddish and small in racemose or axillary solitary. Calyx is 5-fid and much enlarged in fruits; corolla ovoid tubular. Stamens are 10 and ovary is 5-celled with many ovules in each cell. The capsules are 5-celled with numerous, minute and subglobose seeds.

About 200 species occur in Circumpacific (West to Western Himalayas and South India); 2 in East north America and 8 in Brazil. 12 of them have been recorded from India.

Species: *Gaultheria fragrantissima* Wall

Description of Plant

This is a small, low and much branched shrub with orange-brown bark and pink twigs. Leaves are persistent, 3.5-6.0 cm long or short stout petioles, oblong oval or oblong-lanceolate, obtuse, apiculate, serrate and glabrous. Flowers are small, many on short drooping pedicels with a pair of bractlets below the flowers and a bract at base and are closely placed in dense pubescent axillary racemes. The capsules are small, pubescent and are completely enclosed in fleshy ovoid and enlarged calyx.

Flowering and Fruiting

Flowering: During June-July and Fruiting: Later.

Distribution

It is distributed from Nepal to Bhutan between 1828-2438 m and in Assma and Nilgiris.

Plant Parts Used

Oil, obtained from its leaves is used medicinally.

Uses

Oil of Gaultheria, the genuine winter green oil is obtained from its leaves by distillation. Which is aromatic, carminative and stimulant and is used in neuralgia and rhemumatism and also as antiseptic.

Besides 'oil of Gaultheria' in the doses of 6.0 ml, gradually increased, preferably in capsules is given successfully in acute reheumatism and sciatica and is also applied externally. It can also be used as a substitute for the true oil of wintergreen.

Family: MYRSINACAE

Genus: *Embelia Burm*, F.

The generic name is taken from Sri Lankan name 'Acumbelia' of the plant.

The genus comprises of large and usually climbing shrubs. Leaves are alternate and entire or toothed. Flowers are whitish, small and hermaphrodite or polygamo-dioecious in axillary or terminal simple or compound racemes or panicles. Calyx is 4-5–fid or 4-5–partite and persistent; petals 4-5, free or slightly cohering a the base, erect-patent or reflexed, elliptic and imbricate, stamens 4-5 and ovary is avoid or globose or rarely beaked. The fruits are small and globose with 1 or 2 seeds.

About 130 species are distributed in tropical and sub-tropical parts, Africa, Madagascar, East Asia, Indomalayan region and in Pacific. Only 16 of theme are known to found in India.

Species: *Embelia ribes Burm.* F.

Description of Plant

It is a large scandent shrub with long, slender, flexible and terete stem and long internoded branches. Leaves are 5.0-9.0 x 2.0-3.5cm, coriaceous, elliptic or elliptic-lanceolate, acuminate, entire, glabrous on both sides and shining above; petiole 0.6-1.6cm long, more or less margined and glabrous. Flowers are yellow, pentamerous and small in lax panicled racemes. The fruits are globose, smooth, succulent, black when ripe and 0.3-0.4 cm in diameter.

Flowering and Fruiting

Flowering: During July-August and Fruiting: In September-October.

Distribution

It is known to occur throughout India upto 1524 m.

Plant Parts Used

Fruits, roots and leaves of the plant are usable.

Uses

The fruits of the plant in ayurveda are known to possess anthelmintic, alexiteric, alterative, appetizing, carminative and laxative properties and these are used to cure ascites, bronchitis, mental diseases, dyspnea, urinary discharge, jaundice, hemicrania and the heart diseases. *Sushurta* recommends their use along with liquorice root for strengthening the body and preventing the effects of age.

A decoction of its dried fruits is useful in fevers and the diseases of chest and skin. Besides, an infusion of roots is also given in coughs and diarrhea. Its young leaves cominbed with ginger are used as a gargle in sore throat, aphathe and indolent ulcers of the mouth.

The fruits of the plant resemble somewhat black pepper drupes and are used as an adulterant or substitute for them and also for the powder of *Piper cubeba*.

Folk Uses

In Bihar, people take 500 gm roots of the plant and 51 black pepper and ground them and then fried in50 gm thee; from this 21 pills are prepared and one pill is taken daily with milk in swelling of the body.

Preparation

Vidangadichurna, Vidang louha, Vidaangarista and Vidanga taila.

Species: *Embelia tsjeriam-cottam* DC. (*E. robusta* auct. non Linn.)

Description of Plant

It is a rambling small tree or shrub with glabrous branches. Leaves are 6.3-11.5 x 3.8-5.5 cm, elliptic, acuminate and gland-dotted; margins entire or sometimes irregularly toothed, glabrous above and often reddish beneath; petiole 1.0-2.0 cm long and rusty-pubescent. Flowers are pale-greenish yellow and pentmerous in axillary rusty-puberulous racemes which are shorter thanthe leaves. Calyx is 0.1 cm long and petals 0.3 cm long, with few glands outside. The fruits are globose and red when ripe.

Flowering and Fruiting

Flowering: During April-July and Fruiting: In September-November.

Distribution

It is known to be distributed throughout the greater parts of India upto1524 m.

Plant Parts Used

Fruits and root bark of the plant are used medicinally.

Uses

Its fruits are anthelmintic, antiseptic and carminative and are internally used for piles and sometimes in tuberculous glands of the neck. Besides, dried root bark is used for toothache and indolent ulcers of the mouth and gums.

Its fruits are used as an adulterant to Vidanga (*Embelia ribes*) and also to adulterate mustard seeds and as a substitute for Kamela (*Mallotus phillippensis*).

Genus: *Mrysine* Linn.

The generic name is the Greek name of the plant used by Dioscorides to Myrrh.

The genus comprises of usually glabrous shrubs or trees. Leaves are coriaceous and usually entire. Flowers are white, small, sessile or shortly pedicellate and polygamous or often dioecious in fascicles. Calyx are small, 4-5-fid and persistent and corolla 4-5-partite. Stamens are 4-5 and are inserted at the base of corolla tube and ovary is globose or ovoid and 1-celled.The fruits are small, globose, subfleshy or dry and red with single seed.

About its 7 species are known to occur in Azores and from Africa to China; of which only 2 have been reported from India.

Species: *Mrysine Africana* Linn.

Description of Plant

An evergreen shrub reaching a height upto 0.6-1.2 m with pale brown or grey bark. Leaves are 1.2-2.5 cm long, obovate or lanceolate, sharply toothed and dotted with resinous glands when young and minutely puberfious on the mid rib above. Flowers are minute and nearly sessile in axillaryclusters of 3-8.Calyx and corolla are 4-lobed. Stamens are 4 and stigma capaitate and covered with minute protuberances. The berries are 0.5-0.6 cm in diameter, globose and dark purple coloured with single seed.

Flowering and Fruiting

Flowering: During March-April and Fruiting: In May-June.

Distribution

It is found in Himachal Pradesh, Uttar Pradesh, Uttarakhand and Kashmir between 609-2743 m.

Plant Parts Used

Its fruits and gum, obtained from the plants are used.

Uses

About its fruits are used as an anthelmintic, especially for tapeworms. Besides a gum is also obtained from the plant which is used in dysmenorrhoea and also as laxative, in colic and dropsy.

Its fruits are used as a good substitute for anthelmintic drug Embelia ribes.

Folk Uses

A decoction of its leaves is taken as a blood purifier in certain parts of the country.

Family: SAPOTACEAE

Genus: *Madhuca* Buch.-Ham, ex J.F. Gmel.

Madhuca is a Latinized term of its Sanskrit name, Madhuka.

The genus comprises of trees with milky juice. Leaves are coriaceous and crowded towards the ends of the branchlets. The flowers are on axillary,l generally fasciculate pedicels. Calyx is 4-8 lobed which are 2-seriate, outer valvate and inner sub-imbricate in bud. Corolla is campanulate and 8-10 lobed. Stamens are twice of the corolla lobes and anthers are cordate, acute, connective and often produced. Ovary is villous and 6-8 celled. The betties are globose or ovoid and fleshy and are 1-3 seeded.

About its 80 species are known to occur in Indo-china and Indo-Malayan regions, especially in West Malaysia and Australia. Out of which 4 species have been reported from India.

Species: *Madhuca butyracea* (Roxb.) Macbirde (*Bassia butyracea* (Roxb.)

Description of Plant

It is a medium sized deceiduous tree upto 1.8 m in girth and 12 m in height. Leaves are 20.0-35.0 x 9.0-15.0 cm, obovate or acute and entire with acute base. Flowers are white with a sickly fragrance and about 2.0-2.5 cm across and are croweded in fasacicles, usually just below the leaves and sometimes a few between the leaves. The berries are ellipsoid, 2.0-4.5 cm long, green shining with 1-2 seeds.

Flowering and Fruiting

Flowering: During January-February and Fruiting: Later in summer.

Distribution

It is found in Sub-Himalayan tract from Kumaon to Sikkim between 304-1524 m.

Plant Parts Used

The fat, obtained from the plant is used for medicinal purposes.

Uses

A fat or butter is obtained from the plant which is used as an ointment in cases of rheumatism. Besides, it forms an excellent emollient for chapped hands etc. during the winter season.

Folk Uses

In certain parts of the country, butter obtained from the plant is applied externally over chapped hands in cold season.

Species: *Madhuca longifolia* (Koen.) Macrbride (*M. Indica* Gmel; *M. latifolia* (Roxb.) Mac Bride; *Bassia longifolia* Koen; *B. latgefolia* Roxb.)

Description of Plant

It is a large deciduous tree with about 17 m height, The bark is grey or blackish with shallow wrinkles and vertical cracks. Young branches, leaves and petioles are pubescent or tomentose. Leaves are clustered at the ends of branches, 12.5-17.5 x 7.5-10.0 cm, elliptic or oblong-elliptic and shortly acuminate with cuneate base. The flowers are many, at the ends of the branches and are borne on drooping pedicels. Calyx is coriaceous and densely rusty-tomentose. Corolla is yellowish-white and early caduceus. Stamens are 20-30 and anthers are hispid at the back with stiff hairs. The fruits are ovoid, fleshy and green betties and about 2.5-5.0 cm long. Seeds are 1-4.

Flowering and Fruiting

Flowering: During April-May and Fruiting: In June-July.

Distribution

It occurs in the plains and lower hills of India upto 1200m. Common throughout West Bengal, Bihar, Madhya Pradesh, Orissa, Punjab and Uttar Pradesh including the plains of Uttarakhand.

Plant Parts Used

Seeds, bark, flowers, roots and the fruits are used.

Uses

The seeds are laxative and are used in piles and constipation. An gummy juice of its seeds is applied in rheumatism and skin-affections and an oil derived from the seeds is also used in skin-diseases. The residual cake 'mowrah meal', obtained after the extraction of the oil from the seeds is said to be used as fish poison and as a worm-killer for lawns. The decoction of its bark is useful in itching and ulceration. The bark is also used as astringent, emollient, hypoglycaemic and tonic and its powder is in bleeding gums, stomachache, tonsillitis and diabetes. Besides, the stem-bark also possesses some insecticidal properties.

The flowers are considered demulcent, tonic and cooling and are used incoughs, colds and bronchitis. The formentation with its dried flowers produces relief in orchitis and with milk, they are administered in impotence due to general debility. The roots of the plant are also said to be applied on ulcers. Besides, the fruits of the plant are eaten raw or cooked.

Folk Uses

Its oil cake is burnt inside the room to keep away snakes by the Bhil tribals in Madhya Pradesh. Its seeds are ground with water and are used as a collyrium in eye in snake-bitten fainting. In Santhal Pargana of Bihar, Sauria Paharias use two teaspoonfuls of pickle prepared form the flowers of the plant, twice daily for six months as a supplementary treatment in tuberculosis. A powder of its stem bark mixed with the latex of '*Udumbar*' (Ficus racemosa) in 5.0 gm doses is administered in '*Swas roga*' with warm water twice a day by the tribal people in Rajasthan.

Preparation

Madhukasav, Kutajarishta, Kanakasav, Parthadyarishta and Madhukadi him.

Family: EBENACEAE

Genus: *Diospyros* Linn.

The generic name has been taken from Greek words Dios =divine and Puros = wheat *i.e.*, Celestial food, as the fruits are said to be celestial.

The genus comprises of shrubs or trees. Leaves are alternate and rarely sub-opposite. The flowers are dioecious, very rarely polygamous and axillary on short pedicelled or in small cymes; sometimes, often the females are solitary and 4-5 merous. Calyx is loboed, rarely truncate and usually accrescent in the female flowers. Corolla is tubular, salver-shaped or campanulate and shortly or deeply lobed. Male flowers contain 4-many, distinct stamens, paired or united filaments and linear or lanceolate anthers. Female flowers show 0-16 staminodes and globose or conical ovary with solitary ovule. The fruits are globose, ellipsoid or ovoid conic and are supported by the enlarged or woody calyx. Seeds are oblong and usually compressed.

About 500 species are found in warmer parts of the world. Of them, only 44 species have been reported from the India.

Species: *Diospyros malabarica* (Desr.) Kostel (*D. embryopteris* Pers.; *D. peregrima sensu* Gurke; *Embryopteris peregrima* Gaertn.; *Garcinia malabarica* Desr.)

Description of Plant

It is a medium-sized tree with many spreading branches forming a compact head and almost glabrous except the young parts and infloroescences. The bark is dark grey or greenish-black. Leaves are oblong-obtuse at the base, coriaceous, 10.0-25.0 x 3.0-7.5 cm and shining. The flowers are white or cream coloured and scented; the male flowers are in small axillary, drooping and pendunculate cymes of 3-6 flowers and female flowers are large, solitary, axillary and dropping on short pedicles. The fruits are globose, usually solitary and are supported by the enlarged calyx-lobes and covered with rusty-coloured powdery tomentum and yellow when ripe. Seeds are 4-8, compressed, oblong, smooth, reddish-brown and shining and are embedded in viscid glutinous pulp.

Flowering and Fruiting

Flowering: During March-April and Fruiting: In May-June.

Distribution

It grows wild practically throughout India. Common in West Bengal, South India and the Sub-Himalayan tract; also cultivated along road-sides and in villages.

Plant Parts Used

The fruits, flower, seeds, bark and the leaves are used.

Uses

its fruits are considered astringent and the juice of its unripe fruits is used as styptic and the infusion as a gargle in aphthae or stomatitis and sore throat. Stated that its ripe fruits are edible and are useful in gonorrhoea and teprosy. A decoction of the fruit-rind is used in chronic diarrhea and dysentery. Its flowers are diuretic and aphrodisiac and are useful in leucorrhoea, aurethrorrhoea, splenomegaly, anaemia, scabies and nyctalopia. Its seeds and seed-oil are used as antidiarrhoeal and astringent.

Its bark is known as astringent, acrid, cooling, anti-inflammatory, constipating, depurative and febrifuge and is useful in intermittent fever, burning sensation, inflammations, diarrhea, dysentery, leprosy, skin-diseass, pruritus, dyspepsia, piles, diabetes, spermatorrhoea and vaginal disorders. The paste of is bark is applied to boils and tumours. Besides, the leaves are diuretic, carminative, laxative and styptic and are useful in strangury, dyspepsia, flatulence, scotoma, nyctalopia, opothalmia, epistaxis, haemotysis, burns, tubercular glands, scabies and wounds.

Folk Uses

About 10 gm of its fruit-pulp is made into a paste in goat's milk which is given to cure leucorrhoea in Orissa.

Family: SYMPLOCACEAE

Genus: *Symplocos* Linn.

The generic name is based on Greek word, Symploke meaning combination, connection; referring to the basally united stamens of the plants.

The genus usually comprises of glabrous trees or shrubs. Leaves are alternate, coriaceous or membraneous, toothed or entire and turning yellow when dry. Flowers are usually white and are borne in axillary spikes or racemes, sometimes, reduced to few-flowered fascicles or to a single flowers; bracts usually solitary at the base of each pedicels and are caduceus. The drupes are subglobose or ellipsoid with woody stone and 1-3-seeded. Sees are oblong.

There are 350 species known to occur in tropical and sub-tropical Asia, Australia, Polynesia and America. 40 species of them have been reported from India.

Species: *Symplocos paniculata* (Thunb.) Miq. (*S. crategiodes* Buch.-Ham. Ex D. Don; *Prunus paniculatus* Thunb.)

Description of Plant

This is a large shrub or medium sized tree with pilose young branches. Leaves are 5.0-10.0 cm long, membranous, elliptic, acute or acuminate, rounded or cuneate at the base and deeply serrate towards apex. Flowers are pediciellate, white or yellow and fragrant arranged in cymose corymbs on elongated, terminal and axillary panicles.

The drupes are ovoid or globose and about 0.3-0.8 cm across and are crowded with the remains of the calyx-limb.

Distribution

This is found in the Himalayas upto 2743 m from the Indus to Assam and in Khasia hills.

Plant Parts Used

The bark is used medicinally.

Uses

In ayurveda, its bark is described as acrid, cooling, digestive and astringent and is used in eye diseases, in spongy gums and bleeding and to cure 'kapha' diseases, especially in biliousness, blood diseases, dysentery, inflammations and leprosy. The bark is also used as tonic and in opthalmia.

Its bark is used as a better substitute of *Symplocos racemosa* and is considered better of two species in ayurveda.

Species: *Symplocos racemosa* Roxb.

Description of Plant

It is a small glabarous and evergreen tree with stem upto 6.0 m high and 15.0 cm indiameter, Leaves are coriaceous, elliptic-oblong, serrulate and 9.0-18.0 x 3.2-5.0 cm with glabrous or pilose midrib. Flowers are yellow, fragrant, nearly sessile and are borne on short axillary and compound spikes., The drupes are oblong or cylindric, 1.0-1.2 cm long, more or less distinctly ribbed, often slightly curved and purple coloured when ripe.

Flowering and Fruiting

Flowering: During October-January and Fruiting: In December-May.

Distribution

This is very common throughout Chota Nagpur, North-east India including Assam and Terai of Kumaon upto 762 m.

Plant Parts Used

The bark and wood of the plant are used medicinally.

Uses

The bark in ayurveda is known for its acrid, astringent, cooling and digestible properties and is used in eye diseases, spongy gums, bleeding and to sue 'kapha' diseases and in Unai system of medicine. Considered its bark as astringent and cooling and according to them, it is useful in bowel complaints, eye diseases, menorrhagia and in ulcers. A decoction of its bark is used as a gargle for giving firmness to spongy and bleeding gums. *Chakradutta* recommends a paste of its bark, rasot, tubers of *Cyperus rotundus* and honey to be applied in bleeding of gums, Besides writes that a paste prepared from its bark with liquorice roots, burnt alum and rasot in equal parts with water is applied successfully round the eyes in opthalmia.

A paste of its wood is also applied externally to boils to promote suppuration and discharge of pus.

Folk Uses

Tribals of Sundergargh in Orissa, sieve the bark leachate and apply it in conjuctivits and besides this, they take a pinch of powder of its bark in soft drinks for dysentery and other bowel complaints. A pinch of dried bark of the plant is used in vaginal diseases with a cup of milk for a week by the tribal people in Western Maharasthtra. Sauria Paharias of Santhal Paragana in Bihar, take a decoction of its bark to check the bleeding.

Preparation

Lodhrasav, Lodhradi kwath and Pippalayasav.

Family: OLEACEAE

Genus: *Jasminium* Linn.

The origin of the generic name is uncertain but it is said to be a form of the Greek la=a violet and Osme=smell; or derived from the Arabic name Ysmym.

The genus comprises of erect or scandent shrubs. Leaves are opposite of alternate and simply trifoliate or unequally pinnate. The flowers are in tri-or dichotomous cymes or rarely solitary. The berries are didymous or often by suppression simple. Seeds are 1 or rarely 2 in each carpel and are exalbuminous.

According to Airy Shaw (1973), its 300 species are found in tropical and subtropical parts of the old world; 40 species have been reported from India.

Species: *Jasminium grandiflorum* Linn.

Description of Plant

It is a large scandent and glabrous shrub with striate branches. Leaves are opposite and imparipinnate; leaflets 7-11, rhomboid-oblong, the lowest larger than uppermost pair. The flowers are fragrant, white and tinged with purple outside and are borne in divaricate cymes. Corolla is 5-lobed, elliptic and obtuse or acute. The fruits are didymous and ovoid berries. Seeds are erect and exalbuminous.

Flowering and Fruiting

During March-August.

Distribution

It is found wild on the Western Himalayas upto 2133 m; also cultivated in gardens throughout India.

Plant Parts Used

The flowers, roots and the leaves are used.

Uses

The flowers are known as cooling and are used externally in headache and skin-diseases and in weak eyes. Mohammedan writers mentioned that its flowers are used as a plaster to loins, genitals and pubes as an aphrodisiac. Oil of is flowers is applied as cooling agent externally and the roots are used in ringworm. Its leaves are astringent and they are chewed as a treatment for ulcerations or eruptions in the mouth. The fresh juice of the leaves is applied to corns between the toes. Besides, a bland oil prepared with the juice of its leaves is poured into the ear in otorrhoea .

The leaves also possess antibacterial activity against *Staphylococcus aureus*.

Folk Uses

The fresh juice of this leaves is dropped into the ear in earache and is also applied to remove the corns of the feet throughout the country.

Preparation

Jatyadi taila, Jatyadi ghrit, Jatyadi varti and Jatipatradi kwath.

Genus: *Nyctanthes* Linn.,

The generic name has been taken from the Greek words Nyctos=night and Anthos=flowers, referring to tis flowers which open in the evening and fall off the following morning.

The genus comprises of shrubs or small trees. Leaves are opposite, ovate and entire or toothed. The flowers are sessile in small peduncled and bracteate heads disposed in terminal trichotomous cymes., Calyx is ovoid-cylindric. Corolla is salver-shaped and orange coloured. Anthers are 2 and sub-sessile near the top of the corolla-tube. Ovary is 2-celled with one ovule in each cell. The capsules are orbicular, compressed and separating when ripe into 2 sub-discoid 1-seeded carpels. Seeds are erect, orbicular and flattened.

About 2 species have been recorded from Siam, Sumatra and Java. One of which is reported from India.

Species: *Nyctanthes arbor-tristis* Linn.

Description of Plant

It is shrub or small tree with rought and 4-angular branches. Leaves are petioled, opposite, ovate, entire or with distant teeth and acute with rounded or cuneate base. The flowers are very fragrant and are borne in small sessile bracteate heads disposed in terminal trichotomous cymes. Calyx is ovoid-cylindric and subtruncate. Corolla is salver-shaped with cylindric tube and orange-red with white and spreading; lobes 4-8 which are imbricated in bud and patent. Anthers are 2, sub-sessile and inserted near the mouth of corolla-tube. Ovary is 2-celled with one ovule in each cell. The fruits are orbicular and compressed capsules and about 0.7-1.2 cm long. Seeds are erect, orbicular and flattened.

Flowering and Fruiting

Flowering: During August-November and Fruiting: In cold season.

Distribution

The plant is wild in outer Himalayan range from Kashmir to Sikkim, Assam, Bengal, Madhya Bharat and southwards to Godavari, also cultivated throughout India.

Plant Parts Used

Leaves, bark and the seeds are used.

Uses

The plant is said to be used is dysentery, menorrhagia, sores and ulcers in tribal medicines. Its leaves are useful in fever and rheumatism and their fresh juice is given with honey in chronic fever. The decoction of its leaves prepared over a gentle fire is recommended for obstinate sciatica and the expressed juice of the leaves is considered cholagogue, laxative and mild bitter tonic and is given with a little sugar to children as a remedy for intestinal worms.

Besides, the extract of its leaves possesses antileishmanial and antiamaoebic properties. Its flower-extract can be used as antiallergic and immunostimulant agent. The bark of the plant is used as expectorant and the seed-paste is applied in dandruff and seed powder in scurfy affections of the scalp.

Folk Uses

The leaves of this plant are used as a poultice in sciatica pain and in chronic and bilious fever and as safe purgative for children. In Northern India, its one fresh leaf is ground with one teaspoonful of water and is given to infants suffering from diarrhea. The tribals of Gorakhpur use, a decoction of its leaves with black pepper, salt and ginger thrice daily for three days to cure malarial fever.

Family: APOCYNACEAE

Genus: *Alstonia* R. Br.

The genetic name has been assigned in honour of Dr. Charles Alston (1685-1760), Professor of medicine at Edinburgh University during 1716-1760.

The genus comprises of trees or erect shrubs. Leaves are usually whorled. Flowers are in sub-terminal corymbose cymes. Calyx is short and 5-lobed or partite. Corolla is salver-shappedm, closed by a ring of reflexed hairs and its lobes overlap to the right or left. Stamens are included and anthers free from stigma. Disk is anular and lobed or absent. Carpels are 2 and distinct. Ovules are membranous. The follicles are two, linear and slender. Seeds are oblong or linear, flattered and often ciliate; cotyledons oblong and flat.

About its 50 species occur in Indomalyan region and Polynesia. Only 6 of them have been reported from India.

Species: *Alstonia scholaris* (Linn.) R. Br.

Description of Plant

It is a medium-sized evergreen tree with grey bark, whorled branches and bitter milky juice. Leaves are in whorls of 5-7, coriaceous, 10.0-20.0 x 2.5-3.7 cm, obovate-oblong or oblong and obtuse; main lateral nerves parallel. Flowers are greenish-white or greenish-yellow and fragrant and are borne in compact umbellate-corymbose cymes. Fruits are slender follicles, 10.0-12.0 cm long and 1.25 cm across and hanging in pendulous clusters. Seeds are 0.7 cm long, slender, flattened, peltately attached and densely ciliate with long brown hairs all around.

Flowering and Fruiting

Flowering: During October-December and Fruiting in January-March.

Distribution

It is found wild in Sub-Himalyan tract, West Bengal, Bihar Peninsular India and in the Andamanns; also cultivated.

Plant Parts Used

Leaves, bark and wood are used.

Uses

The decoction of its leaves is sued in beriberi disease and in congestion of liver and the bruised leaves boiled in oil are given internally in dropsy. Their juice mixed with ginger juice is prescribed after confinement and poultice is applied to ulcers. Its bark is alterative, anthelmintic, antidysenteric, antimalarial, astringent and galactagogue and is used to cure gastro-intestinal troubles and as sedative and tonic. Its wood paste with water is applied in rheumatism and wounds. The bark-latex is used in tuberculosis, also as antimicrobial, antipyretic and antimalarial. The milky juice is applied in rheumatic pains, sores, toothache, tumours and ulcers and mixed with oil is used as ear drops. The milky juice of the plant is also useful in dental diseases. Besides, the stem-bark of the plant with the leaves of *Vitex negundo* is said to be used as poultice in body-ache and joints pains.

Ditamine, the uncrystallizable substance obtained from its bark is as efficacious as quinine in malarial fever without disagreeable secondary symptoms of that drug.

Folk Uses

In Assam, aqueous extract of its bark-latex is used in tuberculosis and milky, viscous and white sap of the plant mixed with water is taken in cough and gonorrhoea.

Preparation

Saptparnasatvadi vati, Saptchhadadi kwath and Malariasamhar vati. Besides, the plant is also used in homoeopathy for its bitter, tonic and astringent properties.

Genus: *Carissa* Linn.

The generic name has been derived from its Sanskrit name.

The genus comprises of shrubs with spinous branches. Leaves are opposite, small and coriaceous. The flowers are nearly sessile, white and tinged with pink and are arranged in lax terminal umbel-like or corymbose and trichotomous cymes. Fruits are ellipsoid or globose berries. The seeds are usually 2 and pelately attached at the septum with fleshy albumen or horny.

About 35 species occur in warmer Africa and Australia. Out of which, 12 species have been reported from India.

Species: *Carissa congesta* W. (*C. carandas* var. *congesta* (W.) Bedd)

Description of Plant

It is a small and evergreen shrub having a short stem with 2-3 chotomously branches and glabrous with pairs of divaricate simple or branched thorns at the nodes. Leaves are opposite, 2.5-3.5 x 1.7-2.5 cm, elliptic-ovate or obovate, coriaceous, glabrous and shining on both the surfaces. The flowers are white and are borne in terminal lax cymes and inodorous. The fruits are ovoid or globose berries which are black, 4-8 seeded, 0.7-0.5 cm long and shining. Seeds are concavo-convex and hairy.

Flowering and Fruiting

Flowering: During April-May and Fruiting in June-July.

Distribution

This is found wild throughout India; sometimes cultivated in gardens for its fruits.

Plant Parts Used

The fruits, leaves and roots are used.

Uses

An unripe fruits are used as astringent and ripe fruits as cooling and acidic. A decoction of its leaves is used in remittent fever. The roots are anthelmintic, bitter, purgative and stomachic and are used locally in scabies. A paste of its root is used as insect-repellent and internally in diabetic ulcers.

Besides, various plant parts are said to be used in dropsy, anasarca, madness, rheumatism, hemiplegia, epilepsy, convulsions, postnatal complaints, sores and bite of rabbit, jackal or dog.

Folk Uses

In Konkan, its leaves are pounded with horsepiss, lime juice and camphor and are used as a remedy for itches. The oil of its seeds is used in chilblain. The cow's ghee is applied on its leaves which are warmed gently and tied around the affected site in abscess by the tribal of Purvanchal zone in Uttar Pradesh.

Genus: *Hotarrhena* R.Br.

The generic name has been derived from Greek words Holos means the entire and Arren means male, referring to the anthers being from the stigma.

The genus comprises of tree or shrubs. Leaves are opposite and membranous. Flowers are white and are borne in terminal sub-axillary corymbose cymes. The follicles are elongated, terete, spreading and incurved. Seeds are linear or oblong, concave, compressed and tipped with a deciduous coma.

Its 20 species are found in tropical Africa, Madagascar, Southeast Asia, Philippines and Malaysia. Only one of them is known to occur in India.

Species: *Holarrhena antidysenterica* (Linn.) Wall.

Description of Plant

It is a small deciduous tree with rough brown bark which exfoliates in irregular flakes. Leaves are elliptic, oblong and ovate or ovate-oblong and their nerves are 10-14 paired and strong arched. The flowers are white and slightly scented. The fruits are of 2 distinct and divaricate follicles and are slender, parallel, terete, coriaceous and usually with long white spots. Seeds are numerous, 0.8 cm long, linear and glabrous.

Flowering and Fruiting

Flowering: During April-July and Fruiting in August-October.

Distribution

It is found in tropical parts of India; abundant in the Sub-Himalayan tracts. Also planted on road-sides.

Plant Parts Used

The seeds, leaves, bark, roots and fruits are used.

Uses

The seeds, and bark are considered anthelmintic, antidysenteric, astringent, bitter, febrifuge and stomachic. Its bark is used in dysentery and it is dried and ground and is rubbed over the body in dropsy. The seeds are antiperiodic and carminative and a decoction of them is used inchronic dysentery and bleeding piles. Its seeds are also used in pessaries for promoting the conception. The powdered seeds are mixed with honey and are given inchronic chest affection, asthma and in colic pain. Its leaves are used in chronic bronchitis; also applied to boils and ulcers. The roots are given to cattles in disease in which tongue ejects out and gets swollen. An oil made by mixing the decoction of its bark, a number of astringent and aromatic substances in small quantities and sesame oil which is used as external application for various purposes. Besides, its fruits are said to be applied in snake-bite to allay swelling and irrigation.

An ethanol extract of the stem bark is hypotensive and of fruits possesses antiprotozoal, anticancer and hypoglycaemic properties.

Folk Uses

A combination of its seeds, old cotton and honey is used by the natives as a local application to the os uteri in cases of the inflammation of the uterus. '*Asurs*' of Jharkhand prepare, a paste of its roots by grinding them with watger, which is massaged at bed time to cure rheumatic pain, gout and paralysis for seven days. A decoction of its bark of stem mixed with black peeper is given three times a day for two days to cure malarial fever by the tribals in Varanasi district. Besides, about 5.0 gm of raw seeds are consumed as antitetanic by the people in Western Maharashtra.

Preparation

Kutajarista, Kutajavaleha, Kotari-ras, Bhuinimbadi churna, Brihat sudarshan churna, Pathadya, Pradoranilauha, Amritastak, kwath, Sudarshan churna, Sarvajwar hara louha, Atsayakadi kwath, Majunregmahi, Majunsalab, Majun bawasir and Sufuf habis.

Genus: *Rauvolfia* Linn.

The genetic name has been assigned in the honour of L. Rauvolf, a German physician, who traveled extensively and cultivated medicinal plants.

The genus comprises of herbs, shrubs or small trees. Leaves are whorled or rarely opposite. The flowers are small and are arranged in terminal or pseudo-axillary

2-3 chotomous umbel-like or corymbiform cymes. Peduncles are alternating the terminal leaves and finally become lateral. Calyx is 5-fid or partite. Corolla is salver-shaped, tube cylindrical. Stamens are included and attached at or above the middle of the tube and anthers are small, acute and free from the stigma. Disk is large, cup shaped or annular and entire or slightly lobed. The carpels are 2, distinct or connate; style is filiform; stigma is capitate and ovules are 2. The drupes are usually 1-seeded. Seeds are ovoid.

About its 100 species occur in the tropics of both hemispheres; 5 species have been reported from India, chiefly from the mountains of Southern and Western India with 2 extending to the Himalayas upto 1320 m.

Species: *Rauvolfia serpentina* Benth. ex Kurz

Description of Plant

It is an erect and glabrous perennial suffruticose herb or undershrub. Leaves are in whorls of 3 or 4, rarely opposite, shining, green above and pale beneath, lanceolate or oblanceolate, acute or acuminate and narrowed into a short petiole. The flowers are white or pinkish and are borne in corymbose cymes. Corolla is salver-shaped, tube cylindrical and white ringed with red; mouth constricted and throat usually hairy within. The drupes are obliquely ovoid and purple black when ripe. Seeds are ovoid.

Flowering and Fruiting

Throughout the year but chiefly during April-may.

Distribution

It grows in waste places and shady forest in different parts of India from Punjab eastwars to Sikkim and Assam; also in some parts of Central India and Western ghats.

Plant Parts Used

Its leaves and the roots are used.

Uses

The most common use of the plant mentioned in the literature is to cure snake-bite, but Mhaskar and Caius could not establish the efficacy to the drug in snake-bite. Actually, the roots of the plant are a powerful sedative and whenever these were administered to victims of snake-bite, who became nervous, shocked, excited and often got hysterical fits, they calmed them. The sedative effect of the drug calms the victim and when the sedative action was over the victim returned to normal condition. Even today, in general treatment of snake-bite, a sedative is administered as first aid. But, if the victim is bitten by a poisonous snake with a full fatal dose, the sedative action of roots is ineffective and the victim dies. Earlier this fact was not realized and it was believed that the victim was cured due to administration of the roots of this plant.

The roots are hypnotic and sedative. A decoction of its roots is employed to increase uterine contractions and for the expulsion of foetus in difficult cases and root extract is used in intestinal disorders. The roots are supposed to be anthelmintic, febvrifuge and bitter tonic. Its alkaloid extract induces bradycardia, hypotension and sedation and produces tranquillizing effect and is used in hydropohobia, neuropsychiatric disorders, psychosis and schizophrenia. Respirine, an indole alkaloid principle is sedative in actionand is widely used in fevers. Besides, the juice of the leaves is applied for removal of opacities of the cornea of the eyes.

Folk Uses

The working women smeared their breast with root-paste of sarpgandha and crying babies are put on smeared breast to such the milk to sleep in different parts of the country. Kondh tribals, take its dried fruits with that of *Piper longum* and *Zingiber officinale* in equal quantities and make them into a powder; three gm of which is given thrice a day for 3-4 days to regularize menstruation.

Preparation

Sarpgandhaadi churna, Sarpgandha yog, Sarpgandha vati and Hala fisher.

Family: ASCLEPIADACEAE

Genus: *Calotropis* R. Br.

The generic name has been derived from Greek words Kalos=beautiful and Tropis=the keel of aboat, referring to the shape of the coronal scales.

This genus comprises of erect and glabrous or hoary shrubs with milky juice. Leaves are usually sub-sessile, rather fleshy and opposite. The flowers are large and are borne in umbelliform cymes. The follicles are single, thick, straight on the ventral and convex on the dorsal side, oblique and sharply incurved near the base. Seeds are comose.

About 6 species are found in tropical Africa and Asia. Three species grows in India, of these one is found throughout India and one is confined to Western and Central India and one is to Eastern India.

Species: *Calotropis gigantia* (Linn.) R. Br. Ex Ait

Description of Plant

Usually it is a shrub, rarely small tree. Its young branches, inflorescence, and underside of leaves are covered with soft, white, adpressed and woolly tomentum. Leaves are 10.0-15.0 cm long, sessile or sub-sessile, obovate or obovate-oblong with cordate base. Flowers are purplish-lilac or white and are borne in axillary pedunculate-corymbs; corolla lobes spreading or reflexed. The follicles are 8.0-10.0 cm long, recurved and turgid. Seeds are many and are broadly ovate.

Flowering and Fruiting

Almost throughout the year; chiefly Flowering: During January-July and Fruiting: In February-August.

Distribution

It commonly occurs in waste lands, road sides and railway embankments, ascending upto 1000 m in the Himalayas from Punjab to Assam.

Plant Parts Used

Flowers, leaves, latex and root bark are used for medicinal purposes.

Uses

Its flowers are known to possess digestive, stomachic and tonic properties and in powder form, these are used in colds, coughs, asthma and indigestion. The leaves of the plant are roasted and are applied to painful joints or swellings; and in powder form boiled in sweet oil are applied in eczema, skin erupations, ulcers and on wounds and their tincture is given in intermittent fevers. The latex, obtained from the plant is considered a violent purgative and gastro-intestinal irritant and it is used for producing abortion and uterine contraction by simply by inserting a stick smeared with the latex into vagina of women. *Charaka* recommends its root bark to be taken for piles and leaves to cover boils and *Sushruta* mentions its use in ear ache, asthma and in dog bite. However, root bark is diaphoretic, emetic and expectorant and is applied in elepohantiasis and hydrocele in the form of a paste.

The floss of the plant is employed to adulterate Indian kapok (*Bombax ceiba*) and latex as an adulterant to Persian opium; also as an adulterant or substitute of Ipecac (*Cephaelis ipecacuanha*).

Folk Uses

Natives of Bhadrak district in Orissa, stich few of its leaves together so as to form a leafly cap which is put on the head of a patient sufferieng from pneumonia to alleviate fever.

Species: *Calotropis procera* (Ait.) Ait. F. ssp. *hamilatonill* (W.) Ali (*C. hamilatonii* W.; C. *procera* auct. non (Ait) f.)

Description of Plant

It is a middle-sized shrub. Its young parts are covered with adpressed white tomentum and young leaves are hoary and glabrous when full grown. Leaves are 10-18 cm long, ovate-obovate or obvate-oblong and acute. Inflorescence are covered with woolly tomentum. The flowers are purplish-red and silvery outside with odour; buds are hemispherical. The follicles are 6.5-9.5 x 2.5-5.0 cm and recurved. Seeds are numerous, ovoid and 0.6cm long with a bright silky white coma.

Flowering and Fruiting

During February-September.

Distribution

This is known to occur more or less throughout India in warm dry places from the Punjab to Western, central and Southern India. Abundant in the Sub-Himalayan tracts and the adjacent plains in the North-East.

Plant Parts Used

The flowers, leaves, latex and root-bark are used.

Uses

The flowers are known as digestive, stomachic and tonic and are useful in asthma, cold, cough and anorexia. Powdered flowers with black pepper are given with ash of barley seeds in cholear. Its leaves are roasted and are applied to painful joints or swellings. The powder of its leaves boiled in sweet oil is used in skin eruptions, eczema, toothache, ulcers and wounds and the tincture is given in intermittent fever. The latex from the plant is considered as irritant and purgative. It produces abortion when a stick smeared with it is applied locally to induce uterine contractions. It is given orally with *'gur'* in bite of rabid dog and is also used in piles along with powder of *'javakhar'* and *'batasha'*.

Besides, the juice of the aak is also said to be used to cure snake-bite. For the purpose, the white rometum from leaves and a paste with this and milky juice is made, from which pills about the size of a common marble are made. These are given in a betel leaf to chew it to the bitten person. This dose is given after half an hour till the numbness begins and afterwards by everyone hour. Only nine doses are enough. If the bitten person is unconscious and not able to swallow, then these pills are given after dissolving in water. This is considered a good remedy for snake-bite, but it should be used with great caution. The root bark is antidysenteric, antispasmodic, diaphoretic, emetic, expectorant and purgative and is used in piles and syphilis and its paste is applied in hydrocele and elephantiasis.

The powder of its roots mixed with milk of goat is dropped in nostrils in epilepsy; also used to treat elepohantiasis.

Folk Uses

A decoction of its roots is used by Santals in infantile convulsions and in wandering of the mind during fever. The triblas of Varanasi, use the latex to remove worms from the teeth and in toothache. The people in Southern Uttar Pradesh, take a decoction of its root-bark along with black pepper twice a day for three days to cure malarial fever. Besides a preparation is made by taking 400 ml cows milk and adding the sugar in required quantity in it for taste. Then, the milk is put on the fire and stired with the twig of this plant and allow to boil and latex is continuously added to it while stirring with fresh twig. The milk is boiled till the milk is reduced to its one-fourth. This preparation is used as an antimalarial dish by the tribal people in Shetrunjaya hill of Palitana in Gujarat.

Preparation

Nagbhasm and Shankh bhasm.

Genus: *Gymnema* R. Br.

The generic name is derived from Greek words, Gymnos = naked and Klados = a branch; referring to the appearance of branches during winter season.

The genus comprises of twining shrubs. Leaves are opposite. Flowers are small and are arranged in crowded lateral umbellate cymes. Calyx is 5-partite; corolla subrotate to the middle or beyond it, lobes thick and overlapped to right of bud. Staminal column is arose from the base of the corolla. Style-apex is often exerted beyond the anthers. The follicles are smooth with comose seeds.

About 25 species occur in Paleotropical parts, South Africa and Australia. Of which 10 species are known to be found in India.

Species: *Gymnema sylvestre R. Br.*

Description of Plant

It is a large, stout, much branched and woody climber with densely appressed hairy branchlets. Leaves are 3.2-5.0 x 1.3-3.0 cm, rarely pubescent above, thinly coriaceous and elliptic or obovate-acute. Flowers are yellow, small incroweded umbelliform cymes; corolla sub-rotate with thick lobes and fleshy coronal processes on the throat. The follicles are 6.0-7.5 x 0.7-0.8 cm, slender and glabrous; one follicle often suppressed. Seeds are about 1.2cm long, narrowly obovoid-oblong and flat with a broad thin wing and pale brown in colour.

Flowering and Fruiting

Flowering: During July-August and Fruiting: In October-December.

Distribution

It is commonly occurred in the hills of Bihar, Orissa, Madhya Pradesh and Deccan Peninsula.

Plant Parts Used

Whole plant, its seeds, leaves and roots are used.

Uses

This plant use as a destroyer of *madhumeha* (glycosuria) and other urinary disorders. On account of its abolishing property to sugar, it has been given the name '*Gurmar*' and it has been proved that it neutralizes the excess of sugar present in the body in diabetes mellitus.

In ayurveda, the plant considered as acrid, bitter, alterative, anthelmintic, alexiteric and tonic and is used in eye complaints, to cure corneal opacity, heart diseases, piles, leucorrhoea and biliousness and its fruits in bronchitis, heart disease and urinary discharges. The seeds of the plant are used as emetic and as a remedy for colds. Its leaves are hypoglycaemic and are used incough, fever and in the management of maturity onset ediabetes. Besides, these are mixed with castor oil and are applied externally to swollen glands and enlarged spleen and in powder form used as diuretic and stimulant. Its roots are used as astringent, emetic, expectorant, refrigerant, stomachic and tonic.

Folk Uses

The leaves either in powder form or fresh, are chewed by the natives to control sugar level of blood in diabetes mellitus in certain parts of the country; also taken with a cup of milk or honey. The tribals of Madhya Pradesh, mix five or six leaves of the plant with a little portion of kernel of Syzygium cumini (Jamun) and take with water in empty stomach early in the morning for regularizing the menstrual cycle. In asthma, people of Madhya Pradesh, take about 12 gm leaves of the plant and add 5-6 black pepper and prepare a powder. 30 gm of which is taken thrice a day for 15 days to cure the disease. Besides, leaf juice is also applied externally on eyes to cure iritis and about 60 gm shrub is ground with 25 gm ginger and rocksalt; 4.0 gm of this preparation is taken thrice daily for 15 days to cure parkinsonism.

Preparation

It is an important ingredient of various ayurvedic formulations for diabetes, important one includes Madhumehantaka churana.

Genus: *Hemidesmus* R. Br.

The generic name is derived form Greek words, Hemi = half and Desmos = a tie; alluding to the filaments of the plant.

This monotypic genus occurs in South-east Asia, Malaysia and South India.

Species: *Hemidesmus indicus* R. Br.

Description of Plant

It is a perennial, prostrate or twining shrub with terete stem and woody rootstocks. Leaves are obovate to oblong-elliptic, 5.0-10.0 x 0.8-3.5 cm, linear or linear-lanceolate and obtuse or apiculate. Flowers are small, greenish-purple and are borne in opposite, crowded subsessile cymes. The follicles are 10.0-15.0 x 0.6-0.7 cm, glabrous, often purplish coloured and divaricate. Seeds are ovate-oblong, thick, about 2.5 cm long and black coloured.

Flowering and Fruiting

Flowering: During July-August and Fruiting: In October-November.

Distribution

It is found commonly as hedges and in waste places throughout India; chiefly in South India.

Plant Parts Used

The roots and root bark are used medicinally.

Uses

The fragrant root bark of this plant is known as Indian sarsaparilla. Its roots and root bark are considered alterative, demulcent, diuretic, diaphoretic and tonic and are used in anorexia, dyspepsia, fever, leucorrhoea, chronic rheumatism, skin-diseases. A hot infusion of its roots or root-bark with sugar and milk is prescribed in chronic cough and diarrhea to children. It is also given with cow's milk inscanty and highly coloured urine, dysuria and urinary calculus. Besides, a liquid extract of root-bark or its roots is given as a tonic and in skin-disease and also as an alterative in chronic rheumatism. In vomiting and nausea, roots of the plant are boiled in water, strained off and then ground with a little asafetida and made into a thin paste and then mixed with ghee; this is given in morning to stop the vomiting. Besides, a decoction prepared of its roots with that of Colocynth and *Hedyotis biflora* with addition of powdered long pepper and bdellium is administered in chronic diseases, syphilis, elephantiasis, loss of sensation and in hemiplegia.

The roots of this plant are used as a good substitute for European sarsaparilla (*Smilax ornate*).

Folk Uses

In Ranchi (Jharkhand), people take root juice of this plant and that of *Cyperus rotundus* in equal quantities and mixed; 10 ml of this mixture is given orally a at the interval of 30 minutes each to patient till he regains consicuousness in snake-bite besides, root paste is also applied externally over bite point. Gonds of Madhya Pradesh, grind 20 gm roots of this plant and take it with milk in various skin-diseases. Bondo tribals in Orissa, take 10.0 gm powdered root of this plant and mixed with 3.0 gm stem bark of *Syzgium cumini* which is given with water on empty stomach within 3 days of delivery to increase lactation and 30.0 gm root powder in six divided doses, twice a day to cure diarrhea.

Preparation

Sarivadi kwath, Sarivadi vati, Sarivadaya leha, Sarivadyasava and Pinda taila.

Genus: *Pergualria* Linn.

The generic name is based on Latin word, Pergula = an arbor, trellis work; referring to the habit of the plants.

The genus comprises of twining pubescent or tomentose undershrub. Leaves are opposite and cordate. Flowers are greenish-white in axillary racemose or corymbose pedunculate cymes. Calyx is 5-partite and 5-glandular; lobes acute and corolla-tube is short, campanulate or funnel shapepd, lobes overlapping to the right in buds. Stamens are 5, adnate to the corolla-tube; ovary consists of 2 distinc carpels and style slender. The follicles are smooth or softly echinate and often recurved with ovate and minutely pubescent seeds.

Its 3-5 species occur in Africa and from Madagascar to India. Only 2 of them have been reported to be found in India.

Species: *Pergualria daemia* (Forsk.) Chiov.

Description of Plant

It is a slender climber with foetid smell and subhispidly hairy or glabrate stems. Leaves are membranous 5.0-12.0 x 3.5-9.0 cm, broadly ovate, acuminate, deeply cordate and pubescent beneath. Flowers are yellowish-green or dullwhite in long stalked corymbose drooping panicles. The follicles are reflexed, 5.0-7.5 x 1.3-1.4 cm, lanceolate and are clothed with long soft spines. Seeds are pubescent, broadly ovate, about 0.8 cm across with entire margins.

Flowering and Fruiting

Flowering: During August-September and Fruiting: In October-December.

Distribution

It is common throughout the hotter parts of India.

Plant Parts Used

Whole plant, its leaves and bark are used medicinally.

Uses

The whole plant is considered as anthelmintic, emetic and expectorant and its extract is given in uterine and menstrual problems to facilitate parturition. Leaves of the plant possess expectorant property and their juice is given in catarrhal affection, asthma and in infantile diarrhea and mixed with lime, these are applied to rheumatic swellings. Besides, a poultice of leaves is also applied to carbuncles. Its root bark mixed with cow's milk is used as a purgative in rheumatism .

Folk Uses

In Western parts of the country, a juice of its leaves and that of '*tulsi*' (*Ocimum spp.*) is expressed by rubbing between the palms of hands which is given to infants as stimulating emetic.

Genus: *Sarcostemma* R.Br.

The generic name is derived from Greek words, Sarkos =flesh and Stemma=a crown, referring to the fleshy flowers.

The genus comprises of usually leafless, trailing or twining and jointed shrubs with pendulous branches. Flowers are small in sessile umbels which are produced terminally or laterally from the nodes. Calyx is 5-partite and corolla rotate or sub-campanulate and deeply divided with 5 lobes. Staminal column is arose from the base of corolla; style apex shortly conical or oblong-fusiform and included or exerted. The follicles are smooth with comose seeds.

About 10 species are found distributed in tropics and subtropics of the old world. 3 of which are reported from India.

Species: *Sarcostemma acidum* (Roxb.) Voigt. (*S. brevistigma* Wt. and Arn.; *Asclepias acida* Roxb.)

Description of Plant

This is a leafless, jointed shrub with terete, green and straggling branches. Flowers are whitish and fragrant in terminal and sessile umbels; bracts minute and lanceolate. The follicles are 10.0-12.5 x 0.7-0.8cm, thinly coriaceous with fine and straight tip. Seeds are 0.5 x 0.3 cm, ovate and flattened.

Flowering and Fruiting

During August-November.

Distribution

It is occurred in Deccan Peninsula and other arid rocky places of India.

Plant Parts Used

Whole plant is used.

Uses

The whole plant in ayurveda, is considered as alterative, bitter and cooling and is much used to cure 'tridosha', biliousness and thirst. Besides, its dried stem is used as emetic.

Species: *Sarcostemma secasome* (Linn.) Bennet (*S. esculentum* (Linn.f.) Holm; *Oxystelma sedamone* (Linn.)

Karst.; O. esculentum R. Br. Ex Shult; *Periploca esculenta* Linn, f.; *P. seccamone* Linn.)

Description of Plant

It is a twining, glabrous and slender herb with milky juice. Leaves are 4.5-9.0 x 0.3-0.8 cm, opposite, lanceolate, linear-lanceolate and acuminate. Flowers are large and about 2.5 cm across, in loose umbelliform cymes or solitary. Calyx is glabrousand divided nearly to base; corolla broadly campanulate or terete and white or terete and white or rose coloured with purple veins. The follicles are 3.5-6.5 cm, long, thick, oblique or curved and smooth or narrowly 2-winged. Seeds are many, 0.3 cm long, broadly ovate, flat and with thin margins.

Flowering and Fruiting

During July-December.

Distribution

It occurs throughout the plains and lower hills of India including paddy fields and semi-marshy places.

Plant Parts Used

Whole plant, its latex and roots are used medicinally.

Uses

In ayurveda, the plant has been described as anthelmintic, aphrodisiac, bitter and hot and it is used in leucoderma and bronchitis. In Unani system of medicines, the juice of the plant is given in gleet, gonorrhoea, pain of the muscles, cough and in leucoderma. Besides, the decoction of while plant is also used as a gargle in apthous ulceration of the mouth and in sore throat.

The latex of the plant is used to hasten the healing and resh roots efficaciously in jaundice.

Folk Uses

In certain parts of the country, milky sap (latex) obtained from the plant, is used to wash ulcers by the natives.

Genus: *Tylophora* R. Br.

The generic name has been derived from the Greek words Tylos = a swelling and Phorein=to bear, referring to its pollen masses.

The genus comprises of perpennial twining or erect herbs or undershrubs. Leaves are opposite. The flowers are small and are borne in simple or branched umbelliform or racemose cymes. Calyx is 5-partite and its segments are ovate or lanceolate. Corolla is rotate and deeply 5-lobed; corona is of 5 fleshy tubercles and adnate to the short staminal coloumn. Anthers are small. Stigma is disciform and 5-angled. The follicles are acuminate, slender and smooth. Seeds are comose.

About 50 species occur in the Paleotropic and South Africa. About 19 species occur in India in Southern and Western parts and subtropical Himalayas upto 1980m.

Species: *Tylophora indica* (Burm, f.) Merr. (*T. asthmatica* (Linn.f.) W.&A.; *Cynanchum indicum* Burm.f.)

Description of Plant

It is a glabrous or pubescent twining herb. Leaves are ovate, ovate-oblong, acute or apiculate, glabrous and dark green above and pale beneath. The flowers are inumbellate cymes. The follicles are striate and attenuate towards the tip into fine point. Seeds are ovate.

Flowering and Fruiting

During July-October.

Distribution

It is found in North and East Bengal, Assam, Orissa, Konkan, Deccan and in Chennai and upto 914 m in the hilly parts.

Plant Parts Used

The leaves, roots and whole plant are used.

Uses

The leaves of the plant are3 diaphoretic, emetic and expectorant and their powder is used in diarrhoea and dysentery and decoction in asthma, bronchitis, diarrhoea and dysentery. The roots are alterative, aromatic, bitter, antirheumatic, expectorant, cathartic and stimulant and are used in chronic bronchitis and in early stages of whooping coughs and to reduce lochia and also to relieve pain in gout. The powder of the roots is used in intermittent malarial fever. Stated that half a teaspoonful of an infusion of its roots is used to induce vomiting in children who cannot bring up phlegm after coughing. The whole plant as general and cardiac tonic, alterative, antipyretic and febrifuge and it is useful in intestinal worms and for boils.

The powder of its dried leaves is considered as one of the best indigenous substitute of Ipecac.

Folk Uses

A decoction of its leaves and that of roots is taken in asthma and bronchitis by the natives in India.

Family: STRYCHNOACEAE

Genus: *Strychnos* Linn.

The generic name *Strychnos* is the Greek name for poisonous plants; especially for nightshade plants.

The genus comprises of trees and climbing shrubs with short axillary, often hooked tendrils, Leaves are opposite, usually coriaceous. Flowers are white in terminal of axillary cymes; with small bracts. Calyx is 4 or 5-partite and corolla campanulate or hypocrateiform with 4 or 5 lobes. Stamens are usually 5 or 4 and are inserted in the throat of corolla. Ovary is 2-celled throughout or 1-celled in upper part many ovules in each cell. The berries are globose with hard rind. Seeds are 1 or 2 or numerous embedded in a fleshy pulp.

About 200 species are distributed in the tropics of the world. Of them, 22 are known to occur in India.

Species: S*tryhnos nux-vomica* Linn.

Description of Plant

It is an evergreen and glabrous tree; often with short sharp strong axillary spines and thin grey, smooth or rough bark and reaching a height upto 30 m. Leaves are ovate, 5-nerved, glabrous, shining with obtuse base and 7.5-15.0 x 4.5-7.5 cm. Flowers are greenish-white in terminal pubescent, pedunculate and corymbose cymes. Calyx is 0.2 cm long and pubescent outside and corolla- tube cylindric and 4-5 times longer than calyx. Style is filiform and stigma undivided. The berries are globose, about 2.5-5.0 cm in diameter, slightly rough bur shining and yellow when ripe. Seeds are usually many, discoid and about 1.2 cm across.

Flowering: During March-April and Fruiting: In winter.

Distribution

It is found throughout tropical parts of India, including Bihar, Orissa, West Bengal and Uttar Pradesh ascending upto 1350 m.

Plant Parts Used

Root-bark, fruits, leaves, wood and seeds are the parts, used medicinally.

Uses

In ayurveda, the fruits of the plant, are said to acrid, appetizer, bitter, pungent and tonic and are used to cure leucoderma, 'vata' and 'kapha' diseases, blood diseases, anaemia, piles, jaundice and urinary discharge. In Unani system, these are said to be aphrodisiac, bitter, diuretic, emmenagogue and tonic and are used to cure pain and weakness of joints, lumbago, piles and paralysis. Root bark of the plant is ground into a fine paste with lime juice and made into pills which are given efficaciously in cholera.

The leaves as a paste are applied to sloughing ulcers and wounds; especially in cases when maggots are formed. Its wood is also used in cases of dyspepsia. The seeds of the plant as tonic, antidiarrhoeal, antidysenteric, antispasmodic, emetic, febrifuge, stimulant and tonic and these are used in cholera, diabetes, emotional disorders, hysteria, epilepsy, intermittent fever, gout, rheumatism, hydrophobia, impotence, insomia, paralytic and neuralgic affections, prolapse of rectum, retention or nocturnal incontinence of urine and spermatorrhoea. *Llajul-gurba* prescribed, a paste made of the equal parts of seeds of nux-vomica, seeds of *Momordica charantia, red ochre, siah jira* fand roots of *Bismari-ki-jhad,* to be applied in tympanitis. Besides, oil obtained by heating its fresh seeds is also applied externally in chronic rheumatism, palsy and relaxation of the muscles and tendrons. It is also used as antidote to alcoholism and opium poisoning.

Folk Uses

in Konkan, seeds are given in small doses with aromatics in colic and juice of fresh wood, which is obtained by applying heat to the middle of a straight stick to both ends of which a small pot has been tied, is given in few drops in cholera and acute dysentery. In South Orissa, people macerate its bark with lemon juice and made pills from it which are taken orally in cases of acute diarrhoea.

Preparation

Agnitundi vati, Bismastli, Navjivan, Lakshmivilas ras, Krimi mudgar and sul hara yaga.

Species: *Strchynos potatorum* Willd.

Description of Plant

A moderate-sized tree with black and cracked bark, reaching a height of 12 m. Leaves are 5.0-7.5 x 2.5-4.5cm, sessile, subcoriaceous, ovate of elliptic, acute or subacuminate and glabrous and shining with rounded or acute base. Flowers are white and fragrant in axillary cymes. Calyx is glabrous with 5 segments and corolla with 5 oblong lobes. Stamens are 5 and ovary is 2- celled, ovoid and tapering into a long glabrous style. The berries are block and about 1.5-1.7 cm across with hard rind. Seeds are 1 or 2, circular, Ienticular and yellow.

Flowering and Fruiting

Flowering: During April-May and Fruiting: In cold season.

Distribution

It is know to occur wild in Central India, Konkan, Deccan Peninsula and from Karnataka to South Travancore.

Plant Parts Used

Seeds of the plant are used for medicinal purposes.

Uses

The seeds of the plant in ayurveda, are considered acrid, alexipharmic and lithontiptic and are used inj eye diseases, strangury and diseases of urinary tracts and of heads while in Unani system, these are used as astringent, aphrodisiac, diuretic and as a tonic; also in kidney problems, gonorrhoea and to improve eye sight. A powder of its seeds is given internally with milk in irritation of the urinary tracts and in gonorrhoea. In long standintg diarrhoea, a fine paste of its one half ofr full seed is made with butter milk and is given internally for a week. Besides, its seeds are rubbed up with honey and little camphor into a paste which is applied much to the eyes in lachyrmation or copius watering; also as emetic in dysentery, diabetes and gonorrhoea.

Seeds of this plant species are used to adulterate the seeds of Nux vomica (*Strychnos nux-vomica*).

Folk Uses

In Melghat, seeds of the plant are used on eyes sores. In Keonjhar forest area, in Ranch (Jharkhand), people put in fruits in water in the ration of 1:20 and take out these after 10 days; then seeds are removed and are rubbed with water to make a paste which is applied in conjunctivitis of eye. This is also applied with honey in eyes once daily for 5 days in eye cataract. In Prakassam district of Andhra Pradesh, people rub its seeds on stone and apply the paste thus obtained externally to relieve pain of scorpion stings. Besides, 3 drops of its seed extract are applied in eyes once daily for 3 days in jaundice.

Family: GENTIANACEAE

Genus: *Gentiana* Linn.

The generic name has been assigned in the honour of Gentius, the king of Illyria, 500 BC. who according to legend forest discovered the medicinal properties of the plant.

The genus comprises of annual of perennial herbs, which are sometimes, woody below. Leaves are opposite and often connate at the base. Flowers are blue, yellow or white and are borne axillary or terminal, usually sessile. Calyx is tubular, terete or keeled, lobes 5 or 4 and corolla tubular, campanulate, funnel shaped or subrotate with 5-4 lobes. Stamens are 5 or 4, which are attached with middle of the tube and ovary is 1- celled with numerous ovules. The capsules are stalked or sessile and oblong or ellipsoid. Seeds are numerous and small.

About 400 species are distributed cosmopolitically excluding Africa; chiefly in alpine region. 60 species of them have been reported from India.

Species: *Gentiana kurroo* Royle

Description of Plant

It is a perennial herb with thick root stock and decumbent and tufted stem reaching a height of 12.5-30.0 cm. Leaves are oblong; radical rosulate, 7.5-12.5 x 6.0-12.0 cm. and stem leaves are narrow and about 2.5 cm long. Flowers are blue dotted with white and are borne solitary or racemose; calyx half of corolla length with 5 linear lobes and corolla is 5-lobed. The capsules are oblong.

Flowering: During August-September and Fruiting: Later.

Distribution

It occurs in Kashmir and North-west Himalayas between 1524-3352 m.

Plant Parts Used

Entire plant and roots of the plant are used medicinally.

Uses

The plant in Unani system of medicine, is known for its bitter taste and for emmenagogue and tonic properties and is used successfully to treat leucoderma and syphilis. Its roots are used as febrifuge, stomachic and tonic; also in urinary affections. A decoction of its roots with equal amount of 'sunthi' and 'dikamali' or an infusion of its roots in cold water with addition of 0.2 gm camphor and shilajit and 5.0 gm honey is used as a remedy in debility after fevers, indigestion and loss of appetite; also in catarrh, syphilis, leprosy and other skin diseases. Besides, an infusion of entire plant mixed with a little powder of long pepper is used to relieve in fevers accompanied with coughs and difficulty of breathing and a powder of roots mixed with honey is given in hiccup and to stop the vomiting.

Folk Uses

An infusion of about 50 gm of its roots and 1.05gm cloves and cinnamon is boiled for about six hours and then strain which is taken as a tonic before meals twice daily. 10 gm of smashed roots and coriander seeds are boiled in 400 ml watger till reduced 50 gm which is taken with a few drops of honey in torpid liver as a house hold remedy in different parts of the country.

Preparation

It is much used in combination with other drug as *'Chirayatta'* and *'Katuki'*; as infusion or in pill forms for different ayurvedic preparation. Important yogas include Bhumnibadi churna and Sudarshan churna.

Genus: *Swertia* Linn.

The generic name has been assigned in the honour of Emmanuel Swert of Haarlem, a famous Cultivator of bulbs and flowers in Holland of the 17th century and author of *Florilegium.*

The genus comprises of perennial herbs. Leaves are usually opposite. Fiowers are blue, lurid or white, rarely with yellow glands and are borne in terminal corymbose or paniculate cymes. Calyx is deeply 4-5-partite; tube either absent or very short. Corolla is subrotate, lobes 4-5 and acute. Stamens are 4-5, inserted at the base of the corolla and ovary 1-celled with numerous ovules. The capsules are ovoid or oblong and 2- valved with numerous minute and compressed seeds.

About 100 species are distributed in North America, Eurasia, Africa and Madagascar. 34 of them, have been reported form India.

Species: *Swertia chirayita* (Roxb.ex flemming) Karsten

Description of Plant

This is an annual herb with subterete stem reaching a height of 5.0-12.0 cm. Cauline leaves are subsessile, elliptic, acute and 5-nerved; the lower are often much larger and sometimes petioled. Flowers are green-yellow in many-flowered, leafly and large panicles; often fascicled. Calyx and corolla are 4-lobed; corolla is with 2 glands on each lobe and terminated by long hairs. The capsules are sessile and ovate with polyhedral and smooth seeds.

Flowering and Fruiting

Flowering: During July-August and Fruiting: In September-October.

Distribution

It is found in temperate Himalayas ascending upto 1350-3350 m from Kashmir to Sikkim and in Khasi hills between 1200-1500 m.

Plant Parts Used

Whole plant is usable.

Uses

In ayurveda, the plant is known for its antihelmintic, antipyretic, antiperiodic, bitter, cooling, galactagogue and laxative properties and is used to cure asthma, bronchitis, biliousness, burning sensation, inflammations, leucoderma, pain of the body, piles, urinary discharges and ulcers; while in Unani system of medicine, it is considered astringent, bitter, stomachic and tonic and is used to improve aye sight, as a sedative to pregnant women, pain of joints, leucoderma and skin diseases.

An infusion of entire plant is generally used as antidiarrhoeal, febrifuge, stomachic and tonic. Besides, it is also taken daily in the morning on an empty stomach to keep fit and healthy. It is used as a substitute of Quinine.

Folk Uses

The panchang of the plant or plant infusion is given to children, daily early in the morning to deep them fit and healthy for sometime and again resumed until the desired action produced in different parts of the country.

Preparation

Kiratadi kwath, Panchbhadra kwath, sudarshan churna, Satyadi kwath, Sarva-jwar-har louha, Phalatrikadi kwath and Chandraparbha vati.

Family: BORAGINACEAE

Genus: *Onosma* Linn.

The generic name is derived from Greek words, Onos = an ass and Osme = smell; referring to the bad smell of plant.

The genus comprises of hispid herbs or undershrubs. Leaves are alternate. Flowers are yellow or purple and sessile or shortly pedicelled in simple or cymose racemes. Calyx is 5-fid or 5-partite; segments in fruits, sometimes enlarged. Corolla is tubular or ventricose; lobes 5 which are very short. Stamens are 5 and ovary is deeply 4-lobed. The nutlets are useally 4, ovoid, acute, erect or incurved and smooth or tuberculate.

There are 150 species found distributed from Mediterranean to Himalaya and China. Only 14 of which have been reported from India.

Species: *Onosma bracteatum* Wall

Description of Plant

It is a small hirsuate herb with erect and stout stem reaching a height of 35-40 cm. Radical leaves are 15.0 x 2.5 cm; cauline 5.0 x 1.5 cm, acuminate and lower leaves lanceolate and silky white beneath. Flowers are bluish and are borne in denses silky heads which is 5.0-7.5 cm across. The nutlets are ovoid, 0.4 cm in diameter and acute with rough surface.

Flowering and Fruiting

Flowering: During July-August and Fruiting: Later.

Distribution

It occurs wild in Kashmir and Kumaon upto 3352 m.

Plant Parts Used

Whole plant is usable.

Uses

The entire plant is considered alterative and tonic. A decoction of the plant is used very much in rheumatism, syphilis and leprosy and as a refrigerant and demulcent in thirst and restlessness, in febrile excitement; also to relieve functional palpitation of the heart, irritation of the bladder and stomach and in strangury.

It is also used as a good substitute of Sarsaprilla (Nadkarni, 1954).

Preparation

Banafshadi kwath and Gaozoban ark.

Species: *Onosma hispidium* Wall ex D.Don (*O. echioides sensu* Cl.; *O. kashmiricum* Johnston)

Description of Plant

It is a biennial hispid herb upto 20-50 cm in height. Cauline leaves are 6.0 x 1.5 cm and oblong. Flowers are yellow or yellowish in elongated racemes. Calyx lobes are 0.8 cm long while in fruits sometimes may attain a length of 2.5 cm. Corolla is 0.2 cm long, cylindric and glabrous. The nutlets are 0.5-0.6 cm long, stony, white and smooth.

Flowering and Fruiting

During July-September.

Distribution

It is found wild in Kashmir between 1524-2743 m and in Kumaon.

Plant Parts Used

Whole plant is usable.

Uses

The plant in ayurveda is known to possess antihelmintic, alexipharmic, bitter, cooling and laxative properties and is useful in eye diseases, bronchitis, abdominal pain, leucoderma, fevers, piles, strangury and in urinary calculi. A powder of its leaves is given to children as purgative and bruised roots are used as local application on skin eruptions. Besides, according to flowers or the plant are used as cordial and stimulant in rheumatism and palpitation of the heart.

Family: EHRETIACEAE

Genus: *Cordia* Linn.

The generic name has been assigned in the honour of E. Cordus (1473-1543), a German physician and writer on medicinal plants in the 16th century.

The genus comprises of trees or shrubs. Leaves are alternate, rarely subopposite and petioled and usually coriaceous. The flowers are often polygamous in terminal or leaf-opposed cymes. Calyx is 4-5 toothed; teeth accrescent. Corolla is funnel-shaped; lobes 4-8. Stamens are 4-8. Ovary is entire and 4- celled with one ovule in each cell. Fruit is an ovoid or ellipsoid drupe with hard endocarp and usually 1-celled by abortion. Seeds are exalbuminous with plicate cotyledons.

There are about 250 species known to occur in the warmer parts of the world; of which 16 species have been reported from India.

Species: *Cordia dichotoma* Forest. (*C. myxa* sensu Cl.; *C. lowricana* Brandis)

Description of Plant

It is a crooke tree about 14 m high with grey or brown bark which possesses shallow, longitudinal wrinkles. Leaves are alternate, variable in form and size, ovate or oblong, repand-crenate or bub-lobate and 7.5-15.0 x 5.0-10.0 cm. The flowers are white, polygamous and hermaphrodite and are borne in many-flowered lax corymbs. The fruits are ovoid, subacute, yellow and glossy drupes which are usually one-seeded with a viscid sweetish pulp.

Flowering and Fruiting

Flowering: During March-April and Fruiting: In May-June.

Distribution

It commonly occurs wild throughout India ascending upto 1000 m; often planted.

Plant Parts Used

Fruits, seeds, leaves and the bark are of medicinal value.

Uses

Its fruits are antihelmintic, astringent, demulcent, diuretic and expectorant and are beneficially used in the diseases of lungs, spleen and urinary passage. The powder of its fruit-kernels mixed with oil is applied to ringworm. Its bark-juice mixed with coconut milk is given in gripes. Besides, a decoction of its bark is useful in dyspepsia and fever. The bark is also useful in calculus affections, strangury and catarrh. Its leaves are applied to ulcers and in headache and are also said to be used in snakebites and the decoction of its leaves is given in cough and cold. Besides, ethanol (50 per cent) extract of its leaves and stem is considered antimicrobial.

Folk Uses

The tribal people in Gujarat apply, a paste of its fruits externally in headache.

Preparation

Shaleshmatak-panak, Banafshadikwath, Laukspista, itriphal-i-zamani, Dayaquaza and Sufuf habis.

Family: CUSCUTACEAE

Genus: *Cuscuta* Linn.

The generic name is based on its Arabic anme Kechont and Cuscuta, medieval Latin name for parasitic dodders.

These are leafless, twining and yellowish or reddish herbs. The flowers are in fascicles and are small, white or rose-coloured and sessile or shortly pedicelled; bracts none or small. Sepals are 5 or 4, subequal and distinct or shortly connate at the base. The capsules are globose or ovoid and dry or succulent with fleshy albumen; cotyledons are none or obscure.

About 170 species occur in tropical and temperate regions of the world; 12 species are found in India, mostly in western Himalayas.

Species: *Cordia reflexa* Roxb.

Description of Plant

It is a leafless, climbing, yellowish-green and thread-like twining herb which germinates in soil but becomes parasitic on the plant on which it meets with. The stem is 0.1-0.3 cm across. Flowers are subracemose; bracts small, fleshy and subquardate. Corolla is whitish and campanulate and scales are remoted from the filaments. The fruit is a globose capsule which is acute with black warts. Seeds are black.

Flowering and Fruiting

Flowering: During October–December and Fruiting December-February.

Distribution

It is common as parasitic climber throughout the plains of India ascending the hills upto 2438 m.

Plant Parts Used

The whole plant, its fruits, seeds and stem are used.

Uses

The plant is purgative and is used internally in protected fevers. An infusion of entire plant is used as a wash for sores. Its fruits are used in cough and fever and the seeds as demulcent, diaphoretic and tonic. A cold infusion of its seeds is used as carminative and depurative. Its stems are purgative and their decoction is useful in bilious infections. Their paste is also applied over skin diseases.

In Unani system of medicine, its seeds are considered germinative and anodyne and their powder is used antifertility drugs. Besides, the juice of the plant is useful in hook worm and in diphtheria and paste of the herb is used to cure dishocated joints, to check falling of hairs and also in the treatment of swellings of testicles and headache.

The seeds of the plant are uses as a substitute for that of *Cassytha filiformais.*

Folk Uses

As folk medicine, the juice of whole plant is given to woman once only after menses to make the woman barren for even in various parts of the country. Its seeds are boiled and tied over stomach in belching and in pain of stomach due to digestive problem and gastric troubles. Besdies, in Tripura, tribal people use, a liquid obtained after prolonged boiling of the plant in water for bathing once a day in skin diseases

Preparation

Itrifal ustakhudus and Sharbat dinar.

Family: CONVOLVULACEAE

Genus: *Evolvulus* Linn.

The generic name is derived from Latin word, Evolvo meaning to unravel; as these plant do not twine like other groups of Convolvulaceous plants.

The genus comprises of erect or prostrate, small herbs or undershrubs. Leaves are small, entire and often distichous. Flowers are white or bluish, small and borne axillary, solitary or in few-flowered pedunculate cymes. Sepals are 5 and subequal and corolla infundibuliform or subrotate. Stamens are 5, included or excerted and ovary is usually 2-celled or rarely 1-celled with 4 ovules. The capsules are 2-4 valved usually with 2 or 4 seeds or rarely with single seed.

About 100 species are known to distribute in tropics and subtropics of the world; chiefly in America. Only 2 of them have been reported from India.

Species: *Evolvulus alsinoides* Linn.

Description of Plant

It is an erect-ascending or decumbent-ascending or prostrate herb with woody base. Leaves are linear-oblong or wide or wide-elliptic with acute or rounded base and 0.5-2.0 x 0.3-1.0 cm. Flowers are blue or rarely white, 1-3 together and peduncled. Sepals are lanceolate and hairy and corolla shallowly lobed. The capsules are globose and 4-valved with usually 4-seeds. Seeds are glabrous and about 0.1 cm long.

Flowering and Fruiting

During July-December.

Distribution

It is very common throughout the drier parts of India, ascending upto 1828 m.

Plant Parts Used

Whole plant and its leaves are used.

Uses

In ayurveda, the plant is known for its alexiteric, alterative, antihelmintic, bitter, pungent and tonic properties and is used in bronchitis, biliousness, epilepsy and leucoderma. A decoction of the herb with *Ocimum sanctum* is prescribed in the doses of 50-100 gm in fever associated with diarrhoea or indigestion and with oil is to promote hair growth. Besides, leaves of the plant are smoked in asthma and chronic bronchitis.

Folk Uses

Mahali tribal people in Bihar, take juice of the herb for 2 days to cure hysteria. Inhabitants of Khadbrahma in Gujarat, apply its leaf juice locally in ulcers. A leaf-paste of the plant is applied on fingers affected with whitlow disease in many parts of the country. Besides, in Western Maharashtra, tribals take half teaspoonful powder of the herb as a tonic for a month.

Genus: *Ipomoea* Linn.

The generic name has been taken from the Greek words Ips=binding weed and Omoios=similar, referring to the climbing habit of most of the species.

The genus comprises of twining or prostrate herbs but rarely may be shrubby or erect. Leaves are entire and lobed or divided. The flowers are axillary and solitary or in cymes. Corolla is campanulate or funnel-shaped. The fruits are 4-6 valved capsules which are rarely indehiscent. Seeds are 4 or 6, glabrous and beared or uniformly velvety or woolly.

A about 500 species are found distributed in tropical and warm temperate regions of wold. Of them, 60 species are met in India, chiefly in Southern, Western and Eastern India.

Species: *Ipomoea carnea* ssp. *fistulosa* (Mart. ex Choisy) Austin (*I. Fistulosa* Mart. ex Choisy; *I. crassicaulis* (Bentyh.) Robinson; *Batatas crassicaulis* Benth.)

Description of Plant

It is an erect-ascending or scandent and perennial undershrub about 1.2-1.5 m high with milky juice. Leaves are ovate-triangular with a cordate or hastate base and glabrous; petiole 1.5-15 cm long with a pair of glands at the apex on the lower side. Flowers are purple coloured in many-flowered cymes. The capsule are globose, glabrous and 1.5-2.0 cm long. Seeds are brown-villous and 0.7-0.8 cm long.

Flowering and Fruiting

Almost throughout the year.

Distribution

It is a native of tropical America, now naturalized and occurs throughout India; also grown as a hedge plant.

Plant Parts Used

Seeds and flowers are used.

Uses

An ethanol (50 per cent) extract of aerial parts of the plant is said to be central nervous system depressant and the extract of seeds and petals reported to posses antifungal properties. Though the plant in low doses is mild purgative but the plant ingestion produces haemolysis of blood in animals and causes death.

Folk Uses

The leaves of this plant are warmed and tied over the affected area in sprains and swellings throughout India.

Species: *Ipomoea mauritiana* Jacq. (*I. digitata* non Linn.; *I. paniculata* (Linn.) R. Br. Non Burm. f.; *Convolvulus paniculatus* Linn.

Description of Plant

It is a large, glabrous and scandent perennial with long, thick, twining tough and glabrous atem. Leaves are 10.0-15.0 cm long, usually broader than long and palmately 5-7-lobed; lobes lanceolate or elliptic. Flowers are borne in many-flowered corymbosely paniculate cymes. Corolla widely campanulate, glabrous and pink-purple and ovary is 4- celled. The capsules are ovoid, 0.8-1.2 cm long, 4-celled and are surrounded by enlarged sepals. Seeds are 4, woolly, 0.4-0.7 cm long and light brown coloured.

Flowering and Fruiting

During July-December.

Distribution

It is found throughout tropical parts in India in moist places.

Plant Parts Used

Whole plant and its roots are used.

Uses

The whole plant is used as diuretic and expectorant; also in fevers and bronchitis. Besides, roots of the plant are considered alterative, aphrodisiac, cholagogue, demulcent, mucilaginous, lactagogue, purgative and tonic and in powder form, these are given for emaciation to children and in liver and spleen diseases and menorrhagia. The juice of it fresh roots mixed with sugar and cumin is given in spermatorrhoea and with coriander and fenugreek seeds as lactagogue. Besides, as stated by powder of its roots macerated in its own juice is given with honey and ghee as aphrodisiac.

Its tubers are used as a substitute for *Pueraria tuberosa* tubers.

Folk Uses

In Western Maharashtra, tribal people take a teaspoonful of dried powder of the plant with a cup of milk daily as a tonic for 15 days.

Family: SOLANACEAE Genus: *Atropa* Linn.

The generic name is based on 'Atropos' one of the three fates of Greek mythology, who cut the thread of human life as this is a genus of poisonous herbs.

The genus comprises of poisonous, lurid and glabrous herbs. Leaves are entire and elliptic–lanceolate. Flowers are large and dirty purple or lurid yellow. Calyx is large and deeply 5-lobed, corolla widely tubular-campanulate. Stamens are attached near the base of corolla and ovary is 2-celled. The fruits are globose berries with numerous seeds and are subtended by enlarged calyx.

Species: *Atropa belladonna* Linn.

Description of Plant

This is and erect, glandular-pubescent or nearly glabrous herb reaching a height of 60-90 cm. Leaves are entire, elliptic-lanceolate and 10.0-15.0 cm long. Flowers are solitary, pale-purple tinged with yellow or green and are borne axillary. Calyx is deeply 5-lobed. Corolla is widely tubular-campanulate; lobes 5, triangular and imbricated in buds. Stamens are attached near the base of corolla; bases of filaments cover the ovary. The ovary is 2-celled with obscurely 2-lobed stigma. The berries are purple-black, globose and 2.0 cm across and are surrounded by enlarged calyx. Seeds are compressed and numerous.

Flowering and Fruiting

Flowering: During autumn and Fruiting: In winter season.

Distribution

It is found in Western Himalayas, from Kashmir to Shimla between 2000-4000 m.

Plant Parts Used

Whole plant, its fruits, leaves and the roots are used.

Uses

An extract or tincture of whole plant is applied as liniment and plaster in neuralgia, intercostals pains, pleurisy, rheumatism and sciatica as counter-irritants and anodyne; as suppositories is used to relieve pain of the piles colic and with alkalies in cystitis and dysuria. The fruits of the plant are highly poisonous but in case of poisoning by opium and muscarine etc., these are given as emetic with milk or honey and water. Besides, its leaves and fruits are used as diuretic, mydriatic, narctic and sedative. And an extract of leaves is externally applied to relieve pain and internally is given to suppress excessive perspiration.

Preparation

It is used as proprietary drug in various pharmaceutical preparations prescribed for bronchial asthma, spastic constipation and dysmenorrhoea, nocturnal enuresis, gastro-intestinal hypermotility, hypersecretion, peptic ulcer and whooping cough and as sedative and tranquiliser.

Genus: *Datura* Linn.

The generic name is taken from Dhattura, the Sanskrit name which Arabians adopted it as 'tatorah'.

The genus comprises of glabrous or minutely pubescent shrub-like herbs. Leaves are large, entire and sinuate or deeply toothed. The flowers are large, solitary, pedicelled and white or purple. Calyx is long, tubular, berbaceous, 5-toothed and circumscissile above the base in fruit. Fruit is an ellipsoid, 4-celled and usually spinous capsule, which is 4-valved or irregularly burst near the apex. Seeds are many, compressed and rugose.

About 10 species occur in tropical and temperate regions of the world, chiefly in America; All are reported from India too.

Species: *Dutura metal* Linn. (*D. fastuosa* Linn.; *D. fastuosa* Linn. Var. alba (Nees) Cl.; *D. alba* Nees)

Description of Plant

It is an erect, perennial, coarse and minitely pubescent herb about 60-120 cm high. Stem is woody below and purplish towards lips. Leaves are large, entire, sinuate and 7.5-15.0 x 5.0-10.0 cm with unequal base. The flowers are erect and whitish purple. Calyx is long-tubular and 5- toothed at the apex. Corolla is long and tubular to funnel-shaped with wide mouth. Fruits are globose or ellipsoid capsules which are hnodding, spinous, 1.2-2.5 cm across and 4-valved or irregularly breaking up near the apex. Seeds are compressed, rugose and brown.

Flowering and Fruiting

During April-June; sometimes during July-December.

Distribution

It is a native of Mexico, now naturalized and very common in waste places, along road-sides, railway lines and in scrub jungles throughout the tropical parts of India.

Plant Parts Used

The parts used are seeds, leaves and the roots.

Uses

Its seeds, leaves and roots are used in insanity, fever with catarrh, diarrhoea, cerebral complication, skin diseases and as antiseptic. Sambamurty and Subrahmanyam (1989), stated that an extract of its bruised leaves or seeds in oil is often very effective in allaying the pain in rheumatic swellings, boils and tumuours. Besides, various plant parts are used in headache, otitis, media suppurative, sores, mumps, pain, dropsy, anasarca, madness, rheumatism, rigid thigh muscles, haemiplegia, epilepsy, convulsions, cramps, delirium, febris, pimples, small pox, venereal sores, syphilis, orchitis, epididymitis and hydrocele.

The juice of the leaves is considered as a good substitute of belladonna.

Folk Uses

In Assam, to tighten flabby breast to women, its four or five leaves are heated over open fire after smearing them with mustard oil and tied over breast. This practice should be done once a day for 8-10 consecutive days. Its seeds are pounded and their juice mixed with mustard oil and certain other ingredients is applied locally in leprosy by the people of Purulia district in West Bengal. Naga tribal people in Nagaland, warm up its leaves on fire and rub on breasts for reducing or stopping milk when child dies after delivery.

Preparation

Kanakasava, Dhaturdhum, Martyuanjaya ras, Dattuphal vasma, Dhatturadi pralep, Laghu-vis-garva taila, Dugdha vati, Jawrankush, Haba shafa and Roghan dhatura.

Species: *Datura stramonium* Linn. (*D. stramonium* var. *tatula* (Linn.) Cl.)

Description of Plant

It is an erect, coarse and glabrous or farinose-puberulous annual berb about 60-120 cm high. Leaves are stalked, 15.0-17.0 cm long, ovate, deeply toothed or sinuate and pale-green. Flowers are large and purplish or violet coloured. Calyx is long, tubular and herbaceous. Corolla is long and funnel-shaped with wide mouth. Stamens are attached near the base of tube. Ovary is 2- or spuriously four-celled. The fruits are ellipsoid, 4.0-7.0 cm long and spinous capsules. Seeds are many, compressed and rugose.

Flowering and Fruiting

During July-December.

Distribution

It is found in the Himalayas from Kashmir to Sikkim upto 2743 m.

Plant Parts Used

The fruits seeds, flowers and leaves are used.

Uses

The fruits of plant are sedative and intoxicating and their juice is applied to scalp for curing dandruff and loss of hairs. The leaves and seeds are considered antiseptic, anodyne and narcotic. The leaves are applied to sores and fish-bites and are smoked like cigarettes to give narcotic effect. An ointment of its powdered leaves is used for the treatment of haemorrhoids. Besides, the juice of the flowers is used for earache.

Folk Uses

The seeds in unripe fruits are move on plant itself and clove buds are placed inside and are sealed. After 24 hours, cloves are taken out and given in asthmatic attacks in Western Maharashtra by the tribal people.

Preparation

Kankasav, Dhaturdhum, Martyuanjaya ras and Jawarankush.

Genus: *Solanum* Linn.

The generic name is derived from Latin word Solor=to soothe, in allusion to the soothing properties of the plant.

The gunes comprises of herbs or shrubs, sometimes subscandent or rarely small and unarmed or porckly trees. Leaves are alternate or subopposite, entire and lobed or pinnatifid. The flowers are in dichotomous racemose lateral or terminal cymes. Calyx is 5-10 lobed and rarely subentire. Corolla is rotate and rarely campanulate, tube short and limb usually 5-lobed. Stamens are usually 5, reraly 4 or 6 and are attached to corolla throat, filament short and anthers are oblong, often narrowed upwards, connivent in a cone and open by terminal pores or short slits. Ovary is 2-celled or rarely 3-4-celled and style is columnar. The fruits are globose or elongated berries. Seeds are many and discoid.

About its 1700 species occur in tropical and temperate regions of the world; out of which 40 species are found in India, mostly in the mountains of Southern and Western India and in the temperate Himalayas.

Species: *Soloanum dulcamara* Linn.

Description of Plant

It is a climbing or trailing pubescent herb with long branches. Leaves are 2.5-7.5 cm long, ovate or oblong and entire or sometimes lobed at the base. Flowers and purple and 1.2-2.0 cm across in loose–drooping cymes. Calyx–teeth are obtuse and corolla–lobes recurved. The berries are red and 0.6 cm in diameter.

Flowering and Fruiting

During July–September.

Distribution

It occurs wild in Kashmir, Himachal Pradesh and Uttar Pradesh between 1219-2438 m.

Plant Parts Used

Fruits and twigs are used medicinally.

Uses

The fruits of the herb are considered alterative, diruretic and diaphoretic and are used in shin-diseases, psoriasis, lepra and syphilitic affections; also in chronic rheumatism and enlargement of liver. Besides, a decoction of its dried twigs is also used as diuretic, resolvent and narcotic and to promote all secretions; also in rheumatism, obstinate cutaneous eruptions and scrofula.

It berries are considered a good substitute for those of *S. villosum* ssp. *Villosum*.

Species: *Solanum ferox* Linn. (*S. indicum* Linn.)

Description of Plant

It is an erect, stellate-tomentose and armed herb or undershrub. Stems are acuteate with hooked prickles. Leaves are ovate-oblong, sinuate-pinnatifid with subcordate or ruounded or truncate base, subacute or obtuse, stellate-pubescent and 6.0-12.0 x 5.0-8.0 cm. the flowers are in lateral cymose corymbs. Calyx is short and stellate-pubescent. Corolla is 2.0-2.5 cm, blue and stellate-pubescent; lobes elliptic-oblong and obtuse. Ovary is glabrous. The fruits are orange-yellow, fulvous, 0.8 cm across and spreading. Seeds are 0.2 cm across and smooth.

Flowering and Fruiting

Almost throughout the year.

Distribution

It occurs throughout India ascending upto 1524 m on the Himalayas.

Plant Parts Used

Fruits, roots and seeds are used.

Uses

Its fruits are used in asthma, chest pain, cough, dropsy, rheumatism and sore-throat.

Its roots as carminative and expectorant and these are used in difficult parturition, toothache, fevers, worm complaints, colic, dysuria and inchuria. These are also used to relieve labour pains. Besides, the seeds are used as appetizer.

Folk Uses

The crushed unripe fruits of this plant are eaten to reduce hyperacidity by the tribal people in Western Maharashtra.

Preparation

Brihatyadi kwath, Brihatyadi gan, Sudarshan churna, Dashmularishta, Devdarvadi kwath, Dashmula taila, Chavyanprash and Satayadi kwath.

Species: *Solanum surattense* Burm.f. (*S. xanthocarpum* Sch. and Wendl.; *S. mccanni* Sant.).

Description of Plant

It is an annual–perennial prickly diffuse herb. Stems are aculeate with 0.5-2.5 cm long straight prickles and glabrous except stellate- hairy young parts. Leaves are ovate or elliptic, sinuate or subpinnatified and glabrescent with many straight spines. The flowers are in 2-6-clowered cymose inflorescence. Calyx is 0.6 cm long, aculeate and stellate hairy. Corolla is blue and lobes shallow. The fruits are globose and glabrous berries and are about 1.5 cm long and yellow when ripe. Seeds are 0.2 x 0.1 cm and glabrous.

Flowering and Fruiting

During March-June.

Distribution

It is found commonly in waste places and along road-sides throughout India.

Plant Parts Used

Whole plant, its fruits, flowers, seeds, leaves, stem and roots are used.

Uses

The whole plant is alterative, anti-asthmatic, aperient, astringent, digestive, febrifuge, bitter and pungent which is used in bhronchitis, couth, dropsy and constipation. A decoction of entire plant is given in gonorrhoea and to promote conception. Its fruits, flowers and stems are bitter and carminative and are prescribed in ignipetiditis associated with a vesicular and watery eruptions. The bud and flower juice is used in watery eyes. The fine powder of its fruits with honey is used for chronic coughs of the children and also useful in sore throat.

The vapors of its burning seeds are used as an expectorant in asthma, coughs and in toothache. Its leaves are considered anodyne and their juice with black pepper is prescribed in rheumatism. The roots of the plant are considered expectorant, diuretic, anti-asthmatic and antiemetic and are used for cough and pain of chest. Its roots are beaten up and mixed with wine and are given to check the vomiting. Besides, the decoction of roots with that of Giloe (*Tinospora cordifolia*) is given in fever and cough.

Folk Uses

Its flowers are eaten to promote the digestion in Jammu and Kashmir (Kaul *et al.*, 1991). About 15-20 gm whole dried plant is boiled in 150 ml water for 15 minites to make its decoction; 25 ml of which is taken orally thrice for 2 days in fever and 5.0–10.0 gm ripe fruits are eaten orally in indigestion and stomach problems by the tribals in Kerala.

Preparation

Dashmula ghrit, Dashmularishta, Dashmula taila, Dashmula kwath, Nidigdhikadi kwath, Bayaghri haritaki, Bayaghri taila, kantkari ghrit, Kantkari avleha, Sudarshan churna, Sarva hara louh, Kanakasav, Devadarvadi kwath and Sringayadi churna.

Species: *Solanam torvum Swartz.*

Description of Plant

It is an erect and aculeate or rarely unarmed shrub about 2.5-3.0 m high. The stems are often tinged with dark purple, stellatetomentose and aculeate with short hooked prickles. Leaves are ovatesinuate or lobed, stellate hairy on the upper surface and stellate tomentose on the lower surface. The flows are white and are borne in many flowered, sessile or short peduncled cymose corymbs. Fruits are globose betties and are about 1.0 cm in diameter. Seeds are smooth.

Flowering and Fruiting

During December-June.

Distribution

It occurs wild throughout the tropical regions in India.

Plant Parts Used

Whole plant, its fruits, roots and leaves are usable.

Uses

The entire plant to be used as digestive, diuretic and sedative. Its fruits are very much useful in enlargement of spleen and their decoction is given inspleen and liver problems. The roots of the plant are used as a poultice over cracked feet. Besides, the leaves of the plant are used as haemostatic.

Folk Uses

Its roots are used on chapped hands and feet during winters by Pathari communities. Besides, a decoction of its fruit is also given in the enlargement of the spleem in doses of one tea cup daily in different parts of the country.

Preparation

Brihatayadi kwath, Brihatayadi gan and Dashmularisht.

Species: *Solanum trilobatum* Linn

Description of Plant

It is a scandent undershrub reaching a height of 2-4 m. Leaves are 2.5-5.0 x 2.0-3.5 cm, ovate or elliptic, irregularly obtuse sinuate or lobed and prickly. Flowers are large and showy, violet-purple in axillary racemose cymes. Calyx is cyathiform, 0.3-0.4 cm long and corolla lobes are triangular-lanceolate. The berries are 0.6-0.8 cm is diameter and scarlet when ripe with smooth seeds.

Flowering and Fruiting

During April-July.

Distribution

It occurs in waste places and swampy areas throughout the tropical parts of India; chiefly in southern India.

Plant Parts Used

Whole plants, its roots, leaves, fruits and flowers are used medicinally.

Uses

A decoction of the whole plant is used an asthma, bronchitis, cough, febrile affection and difficult parturition. Its roots and root decoction are prescribed in tuberculosis and lung diseases. Besides, fruits and flowers of the plant are given in cough.

Species: *Solanum villosum* Mill. Ssp. *villosum* Edmonds (*S. nigrum* Linn.; *S. nigrum* sensu Cl.; *S. pupureilineatum* Sabins and Bhatt)

Description of Plant

It is an annual glabrous or sparingly pubescent and erect herb. The stems are terete or angular, verrucose or smooth and glabrous or hairy. Leaves are ovate or oblong with a cuneate, often decurrent base and sinuate toothed or lobed. The flowers are nodding in umbelliform extra-axillary inflorescences. Peduncles are appressed-hairy; pedicels subumbelled. Calyx teeth are small and obtuse, Corolla is white and nearly glabrous. Anthers are 0.1-0.2 cm long. Ovary is glabrous and style base is hairy. The fruits are black or balckish purple betties, sometimes yellow and 0.5cm across. Seeds are smooth.

Flowering and Fruiting

During the whole year; chiefly during February-June.

Distribution

It is common in the waste places, road-sides and railway tracks throughout India; upto 2743 m in the Western Himalayas.

Plant Parts Used

Whole plant, its flowers, fruits and leaves are used.

Uses

The whole plant as anodyne, antidysenteric, antiseptic, diaphoaretic, expectorant, sedative and hydragogue and its is useful in cardalgia and gripe. An infusion of entire plant is given in dysentery, other stomachic problems and in fevers. The fresh extract of the plant is given in gonorrhoea, piles and enlargement of liver and spleen. Its fruits are antidiarrhoeal, antipyretic and tonic and are used in anasarca and in heart and eye diseases. The green fruits of the plant are pounded and are applied on ringworm and the decoction of fruits and flowers is given in cold and coughs.

The young shoots of the plant are prescribed in skin diseases and in psoriasis. A decoction of its leaves is used as diuretic and laxative and the hot leaves are applied to painful and swollen testicles and their paste as a poultice is used in gouts and rheumatic joints. Sometimes, fesh juice of the plant is used, produces dilatation of pupils in impaired vision. The plant is also said to be used in scorpion-stings.

It leaves are used as an adulterant of belladonna.

Folk Uses

A native people take a decoction of its roots mixed with gur to induce sleepness and with turmeric in jaundice. Sometimes, the juice of the plant is also taken as purgative by the people in various parts of the country.

Preparation

Ark-makoi.

Genus: *Withania* Pauquy

The generic name has been assigned in the honour of H. Witham, a Brithsh geologist and writer on fossil botany in the 19th century.

The genus comprises of shrubs or undershrubs. Leaves are entire. The flowers are axillary, sessile or shortly pedicelled and fascicled or solitary; sometimes polygamo-dioecious. Calyx is campanulate and 5-or 5 toothed. Cocolla is campanulate; lobes are 3-6 and short. Stamens are attached near the base of corolla and anthers oblong. Ovary is 2-celled, style is linera and stigma shortly bifid. The berries are globose, Seeds are numerous and dicoid.

About 10 species occur in South America, South Africa, Canaries and Mediterranean region. Out of these, 2 species grow in drier parts of India.

Species: *Withania somnifera* Dunal

Description of Plant

It is an erect and much-branched perennial undershrub about 0.5-2.0 m high with glutinous cell sap. Leaves are entire, ovate, subacute and 4.0-15.0 x 2.5-6.0 cm, Flowers are in 3-5 flowered axillary fascicles. Calyx is campanulate, acutely 5-6 toothed and stellate hairy. Corolla is greenish and lobes ovate, obtuse, stellate hairy outside and greenish or lurid yellow. Stamens are attached near the base of corolla. Ovary is glabrous and 2-celled and stigma shortly bifid. The berries are red and globose and are enclosed in the much enlarged, inflated, membranous, 5-angled and pubescent calyx and about 0.6 cm across. Seeds are numerous, discoid, 0.25 cm across and yellow.

Flowering and Fruiting

Flowering: In July-August and Fruiting: During January-February.

Distribution

Its occurs throughout the drier and subtropical parts of India.

Plant Parts Used

The fruits, seeds, leaves and roots are used.

Uses

The fruits as diuretic and these are also used in chest complaints. The seeds of the plant are hypnotic and diuretic and are used in chest complaints and coagulating milk. Its leaves are antipyretic, anthelmintic and bitter and bruised ones are applied in carbuncles, scabies, ulcers and painful swellings and their fomentation is used for sore eyes and swollen hands and feet. The roots are said to be adaptogenic, aphrodisiac, deobstruent, alterative, diuretic and tonic and are used in cough, dropsy, leucorrhoes, hiccough, menstrual problems, to restore loss of memory, in nervous exhaustion, spermatorrhoea and seminal debility. The powder of its root is given daily with sugar, honey and long pepper in the doses of 1.7 gm daily in spermatorrhoea and debility. A decoction of the roots is used with long pepper, ghee and honey in scrofula and consumption. Besides, these are also used for toning up the uterus of women who is habitual of miscarriage.

Folk Uses

Native people, use watery extract of its leaves for eye sores, boils and swollen parts of the body and apply over syphilitic regions to kill lice infecting the body. Besides, a decoction of its roots with black pepper and alligator pepper is an effective remedy for toning up the uterus of women who habitually miscarry.

Preparation

Ashwagandhadi churna, Ashwagandha rasayan, Ashwagandha ghirt, Ashwagandharishta, Narayana taila and Habb asgandh.

Family: SCROPHULARIACAEAE

Genus: *Bacopa* Aubl.

The generic name has been adopted from its South American Indian name Bacopa.

These are glabrous herbs. Leaves are punctuate and entire or toothed or multifid, when submerged. The flowers are yellow, blue or white and are borne inracemes; bracteoles are small or none. The fruits are globose or ovoid 2-grooved capsules. Seeds are many and minute.

About 100 species occur in the warmer parts of the world; 3 species are found in warmer parts of India.

Species: *Bacopa monnieri* (Linn.) Penell (*B. monnieria* (Linn.) Wettst; *Gratiola monnieria* Linn.; *Herpestis monnieria* (Linn.) Kunth; *Lysimachia monnieria* Linn.)

Description of Plant

It is a perennial, creeping-ascending and glabrous herb about 60-90 cm high. Leaves are subsessile or short petioled, oblong-cuneate to ovate, narrowed towards the base, 1.0-2.5 x 0.4-1.0 cm, entire, obtuse or rounded and gland dotted beneath. The flowers are white or tinged with purple and axillary or solitary. Capsules are 0.6 cm long and elliptical. Seeds are transversely rugose and 0.04-0.05 cm long.

Flowering and Fruiting

During winter season.

Distribution

It is found in marshy, damp and wet areas throughout India.

Plant Parts Used

Whole plant, its stem and leaves are used.

Uses

The whole plant is nervine tonic, diuretic and aperient and is used in insanity, epilepsy, asthma and hoarseness. About 5.0 gm of the fresh juice of its leaves is boiled with ghee and formed into a ghirt or mixed with 2.5 gm of the root of *Aplotaxis auriculata* and honey, which is used internally in insanity, epilepsy and bilious disorders. The plant is used externally in several cutaneous disorders and in the treatment of swollen painful joints. The juice of the plant mixed with petroleum is applied in rheumatism. Its leaves alone fried in clarified butter are given to relieve hoarseness and their powder is said to be very effective in asthenia. Besides, it stem and leaves are useful in the stoppage of urine which is accompanied by obstinate costiveness.

Folk Uses

A poultice of the boiled plant is placed on the chest tin acute bronchitis and other coughs of children in certain parts of the country.

Preparation

Brahmi ghirt is prepared in Bengal.

Genus: *Picrorhiza Royle* ex Benth.

The genetic name has been derived from Greek words, Pikros = bitter and Rhiza = a root; referring to the presence of bitter alkaloid picrorhizin in the root and root stocks.

The genus comprises of low more or less hairy perennial herbs. Leaves are subradical, spathulate and settate. Flowers are white or bluish and are borne on radical leafy flowering stems. Sepals are 5 and lanceolate and corolla lobes are ovate and acuminate. Stamens are 4 and ovary is 2-celled with numerous ovules. The fruits are ovoid, turgid, acute and septicidal and loculicidal capsules.

Only 2 species are known to occur in Western Himalayas including India.

Species: *Picrorhiza kurrooa* Benth.

Description of Plant

it is a trailing, perennial herb upto 15-25 cm in height and is clothed with withered leaf-bases. Leaves are 5.0-10.0 cm long coriaceous with rounded tip and narrowed base. Flowering stems or scapes are ascending, stout and longer that the leaves. Flowers are blue or violet coloured and are borne in 5.0-10.0 cm long, subcylindric, obtuse and many flowered spikes. Sepals are 0.6 cm long, ciliate and corolla of short stamened is 0.6-0.8 cm long, while that of longer filament is 0.8cm long. The capsules are about 1.2cm long.

Flowering and Fruiting

Flowering: During May-June and Fruiting: Later.

Distribution

It is known to occur wild in alpine Hiamalayas fromKashmir to Sikkim between 2743-4572 m.

Plant Parts Used

Roots of the plant is used.

Uses

The root of the plant in ayurveda are known to possess anthelmintic, antipyretic, bitter, cardiotinic, cooling and stomachic properties and are used to promote appetite, in biliouness, urinary discharge, asthma, hiccuough, blood-diseases, burning sensation, leucodrema and in jaundice. About 10 gm root powder with sugar and warm water is given as a mild purgative and about 1.2 gm powder with aromatics as pepper, asafetida, triphala and salts is used in constipation due to scanty intestinal secretions.

Its roots are used as a good substitute of Gentiana kurroo roots.

Preparation

Aryogyavardhini vati, Katukadya-lauh,Tiktadi kwath, Tiktadya ghrita, Amritastak kwath, Satyadi kwath, Aragbadhyadi kwath, Kirattikadi kwathm, Sudarshan churna, Sarvajwarhar louha, Amritarista, Phalatrikadi kwath, Punarnava mandur and Yograj guggulu.

Genus: *Veronic* Linn.

The generic name has been assigned in honour of St, Veronica.

The genus comprises of herbs or trees. Leaves are caulaine, opposite and rarely alternate. The flowers are blue or purple in solitary, axillary or in terminal or axillary racemes; bracts are conspicuous and bracteoles absent. The fruits are compressed or turgid 2-grooved capsules. Seeds are few or many and smooth or rugose, sometimes winged.

About 300 species occur in the world, mostly in North temperate zone, many in alpine and few in South temperate and tropical mountains. Of them, 32 have ben reported from India.

Species: *Vernoica anagallis-aquatica* Linn.

Description of Plant

It is annual erect or decumbent-ascending herb. Stems are glabrous downwards and patently gland-hairy towards the apex. Leaves are sessile, lanceolate-oblong or linera-oblong, entire or serrate usually with cordate base. The flowers are white or pale-purple and are borne in lax and axillary racemes. The pedicels are filiform and spreading longer than linera-lanceolate bracts. The capsules are compressed, gland-cliliat and 0.3 x 0.4cm.

Flowering and Fruiting

During cold season.

Distribution

It is found throughout the greater parts of India.

Plant Parts Used

Whole plant is used.

Uses

The herb is considered antiscorbutic and is used in scurvy, impurity of blood and in scrofulous affection, especially of the skin, The entire herb is bruised and is applied externally for healing burns, ulcersm, in whitlows and mitigation of swollen piles. Besides, its root-decoction is used for gargles.

Folk Uses

The entire herb is bruised and it applied for healing burns in different part of the country.

Family: BIGNONICEAE

Genus: *Oroxylum* Vent.

The generic name has been derived from the Greek words Orox = a mountain and Xylon=wood, referring to the supposed habitat.

The genus comprises of glabrous trees. Leaves are opposite, large and 2-3 pinnate; leaflets ovate and entire. Flowers are in long terminal arecmes. Calyx is large, campanulate and coriaceous or obscurely toothed. Corolla is white or purplish, large and campanulate; lobes 5, subequal, crisped and toothed. Stamensa re 5. Disk is wide, Fruit is a large linera and 2-valved capsule. Seeds are many and thinly discoid with broad wings.

About 2 species are found in China, Ceylon, Myanmar, Malaya and India.

Species: *Oroxylum indicum* Vent

Description of Plant

It is a small, deciduous and soft wooded tree with few branches and a small open crown. The bark is soft, light, brownish-grey and corky outside. Leaves are opposite, 3-pinnate near base, 2-pinnate about the midel and simply pinnate towards apex, very large upto 150 cm in length, rachis stout and cylindric; leaflets 2-4 aired, 6.0-12.0 cm long, ovate or elliptic, acuminate, entire and glabrous with a rounded or cordate base. The flowers are white or purplish and many in large and erect racemes. The peduncle is 60-90 cm long and pepdicels is 2.5-3.7 cm long, foetid and lurid. The fruits are 30.0-90.0 x 5.0-8.0cm, flat and straight capsules. Seeds are many, 2.5-3.7 cm long and are surrounded by a broad transparent white papery wing.

Flowering and Fruiting

Flowering: In June-July and Fruiting during cold season.

Distribution

It is found throughout India except in the Western drier are ascending upto 1219 m in the Himalayas.

Plant Parts Used

Root–bark, stem-bark, fruits, seeds and leaves are usable.

Uses

The root-bark of the plant is astringent and tonic and is used in diarrhea and dysentery. It is boiled in oil and is used in otorrhoea and its powder along with *'haldi'* is used to cure sore-backs of horses and in powdered form or infusion is used in acute rheumatism and as bitter tonic. Its tender fruits are used as refreshing and stomachic and the seeds as purgatie. Besides, the decoction of its leaves is said to be given in stomachache and rheumatism and externally used for spleen enlargement.

Folk Uses

The juice expressed from its bark after it has been enclosed within some leaves and layer of clay and roasted, is mixed with the gum of *'mocharasa'*, is given in dysentery and diarrhea. To cure dysentery and diarrhea, a powder of its stem bark along with the powder of grains of *Hordeum vulgare* and black pepper, is given for three to five days in various parts of the country.

Its seed-paste is sued externally on the lowr part of stomach in urinary complaints by the Bhoxaa tribals in Dehradun. The juice of its stem-bark mixed with cow's milk is given internally thrice a day for 10-15 days for the treatment of bone fractures. Besides, the bark is used a s gargle for sore throat and the powder of plant gum with milk is to relieve in cough.

Preparation

Dashmula ghrit, Dashmula taila, Dashmularisht, Chavyanprash and Shyonak putpal.

Genus: *Sterospermum* Chamaa.

The generic name has been derived from Greek words Stereos = tight and Sperma = a seed, referring to the sees tightly packed in the pods.

The genus comprises of trees. Leaves are large and 1- or 2-pinnate. The flowers are in large lax terminal panicles. Calyx is ovoid, truncate or shortly and unequally 2-5 lobed. Corolla is tubular-campanulae and limbs are spreading, 5-lobed and somewhat 2-lipped. Stamens are 4 with a rudimentary 5th. Disk is cushion-like and fleshy. Fruits is an elongate terete and sub-compressed 4-angled capsules. The seeds are many and trigonous in one or two series.

About 40 species are found in topical Africa and Asia. Only 4 species have been reported from India.

Species: *Stereospermum chelonoides* (Linn. f.) D.C. [*S. suaveolens* (Roxb.) DC.) *Bignonia cheloniodes* Linn. f.; *B. suaveolens* Roxb.]

Description of Plant

It is a large deciduous tree about 9-18 m high with dark grey bark which exfoliates in large flat scales. Leaves are opposite, about 30-50 cm long and paripinnate; leaflets are 5-9 or more commonly 7, 7.5-15.0 x 5.0-7.5 cm, elliptic, acute or acuminate, entire and coriaceous. The flowers are dark purple, pubescent and fragrant in viscid trichotomous panicles. The fruits are loculicidally 2-valved capsules, about 30.0-60.0 x 12.-2.7 cm, 4-ribbed and dark grey with white specks. Seeds are 3.0-3.7 x 0.8 cm with 2 oblong and lateral wings.

Flowering and Fruiting

Flowering: In May-June and Fruiting during cold season.

Distribution

The plant is found distributed in drier parts throughout India.

Plant Parts Used

Root-bark, seeds and flowers are used.

Uses

The root-bark of the plant is considered cooling, diuretic and tonic and its decoction is used to remove the calculi. A paste of its seeds which is applied on forehead in migraine. It's the flowers are rubbed up with honey and are given to check hiccough and in a form of confection as an aphrodisiac.

Besides, ethanolic extract of its flowers shows activity against 'Ranikhet disease' and anticancerous activity against human epidermoid carcinoma of the nasopharynx in tissue culture.

Folk Uses

A powder of its roots is given in the doses of 50 gm with water and paste of the roots is applied at the bite point in snake-bite by the tribal people in Ranchi in Jharkhand. Gond people in Uttar Pradesh, take 2 teaspoonful decoction of its roots along with that of Curcuma longa and jaggery for 8 days in fever accompanying child birth. Besides, root extract is also given in the doses of one teaspoonful to clear blood deposition of uterus after delivery.

Preparation

Dashmula ghrit, Dashmula taila, Dashmularisht, Dashmuladi kwath and Chavyanprash.

Family: MARTYNIACEAE

Genus: *Martynia* Linn.

The generic name has been assigned in the honour of John Martyn (1699-1768), Professor of Botany at Cambridge and author of *Historia Plantarum*.

This monotypic genus has been reported to be distributed in Mexico but got naturalized also in India.

Species: *Martynia annua* Linn. (*M. diandra* Gloxin)

Description of Plant

it is an erect, tall, viscid and densely glandular long hairy annual herb. Leaves are large, 20-40 cm across, cordate at base, subodtuse at apex and densely glandular hairy. The flowers are rose-colourd in axillary racemes. Fruits are green, ovoid and beaked by two strong curved spines. Pyrenes are black, very hard with serrate- denate ribs and 3.0 cm long.

Flowering and Fruiting

During rainy seasons.

Distribution

It is a native of Mexico, now got naturalized in India and springing up on rubbishheaps and in waste places.

Plant Parts Used

Leaves and fruits are usable.

Uses

Its leaves are given in epilepsy and are also applied to tuberculosis glands of the necks and the juice is used as a gargle for sore throat. Its fruits are alexeteric and are sued for inflammations. Besides, a paste of its nuts is used as a local sedative which is said to have a curative effect when applied to bites of venomous insects.

Folk Uses

Bhil people of Madhya Pradesh, apply its seed-oil in eczema. Besides, Korku tribal people in Melghat, tie a garland of its cleaned fruits in neck to cure 'galgan'.

Family: PEDALIACEAE

Genus: *Pedalium* D. Royen ex. Linn.

The generic name has been derived from Greek word, Pedalium meaning a rubber; in allusion to the dilated angles of the fruits.

This monotypic genus is known to occur in Africa, Madagascar and tropical Asia including India.

Species: *Pedalium murex* Linn.

Description of Plant

It is a much branched herb with about 15-40 cm height. The stems and branches are often slightly rough with scaly glands. Leaves are opposite, pale glaucous green and fleshy, broadly ovate-oblong, truncate or obtuse, coarsely crenate-serrate of sublobate, 2.5-5.0 x 2.0-3.5 cm and glabrous above with acute base. Flowers are bright yellow and are borne axillary and solitary. Calyx is small, divided more than half way down; lobes 5, linear-triangular and acute and corolla is 2.5cm long; tube 2.0 cm long and slender. The fruits are about 1.2-2.0 cm long and bluntly 4- angled with sharp conical horizontal spines. Seeds are 1 or 2 and are oblong and pendulous.

Flowering and Fruiting

During February-July.

Distribution

It is found distributed along the sea shores, in Saurashtra, Gujarat, Maharashtra, Uttar Pradesh, Chennai and on the river banks in Jammu.

Plant Parts Used

Fruits, leaves, stems and roots of the plant are used.

Uses

Its fruits are considered aphrodisiac, antiseptic, demulcent and diuretic and their decoction is given for incontinence of urine, spermatorrhoea, nocturnal emission and to promote lochial discharges; also as a local application. Its fruits, leaves and stems are steeped in water and an infusion thus, prepared is given as demulcent and diuretic and also to get relieve in strangury and to dissolve the calculi. A powder of its leaves with sugar and milk is given in gonorrhoea and gonorrheal rheumatism. Besides, its leaves along with its young shoots are dipped and kept in boiling milk for few minutes and then are used as an aphrodisiac in seminal debility. A decoction of its roots is also used in antibiliousness.

Folk Uses

In Uttar Pradesh, 3 fruits of this plant are ground and are mixed in 25gm juice of *Acacia nilotica* ssp. *indica leaves* with sugar. This is given 2 times a day for a week in venereal disease, especially in gonorrhoea (Singh and Khan, 1990). In Ranchi, people apply a paste of its fruits locally in the skin-disease- 3-4 times a day till cure of the disease. Besides, an infusion of fresh plant in cold water is employed in dysuria, spermatorrhoea, calculai and burning mictrution (Bhatt and Mitaliya, 1999).

Family: ACANTHACEAE

Genus: *Andrographis* Wall. Ex Nees

The generic name has been taken from Greek words Andros=male and Grapho=to write, but deviation is not known.

The are annual herbs or shrubs. Leaves are entire. Flowers are in lax panicles or in dense subscaptate axillary and terminal racemens and often unilateral. Bracts are small and bracteoles are minute or absent. Fruits are oblong-linear or elliptic capsules. Seeds are 6-12, hard, oblong or subquadrate, glabrous and regosely pitted.

Its 20 species occur in the tropicala Asia, 18 species are found in India, chiefly in the mountains of Southern India.

Species: *Andrographis paniculata* (Burm.f.) Wall. Ex Nees

Description of Plant

It is an erect and branched annual herb with 4-angled branches and about 30-90 cm high. Leaved are 5.0-7.5 x 1.2-2.5 cm, lanceolate, tapering to the bas and acute pale beneath. Flowers are small and solitary and are arranged in lax spreading axillary and terminal racemes or panicles. Capsules are 2.0 cm long and tapering at each end. Seeds are numerous, subquadrate, yellowish brown, rugose and glabrous.

Flowering and Fruiting

During October-January.

Distribution

It is found throughout the plains of India; sometimes cultivated.

Plant Parts Used

Whole dried plant as a panchang is used.

Uses

The entire plan is considered febrifutge, tonic, alterative and anthelmintic and is used in debility, dysentery and dyspepsia; its infusion is given in fever. Besides, the plant is also used in spleen complaints, colic, strangulation of intestine, constipation, diarrhea, cholera, phthisisi, consumption and bite of rabid jackal.

The plant is also used in Unani system of medicine as blood purifier. Its green leaves with that of *Aristolochia indica* and fresh inner root-bark of country sarsaparilla are made into an electuary which is used by Hakims of India as a tonic and alternative medicine in syphilitic chancre and foul syphilitic ulcer.

Folk Uses

It is dried root with *haldi* are taken into equal proportions and made into a paste with water which is applied for itching and skin rash. Its leaves are used in flatulence by the local Nat and Mushar people in eastern Uttar Pradesh.

Preparation

Vishmajwarhar kshar, Sudarshan churna, Kalmesh-navayas churna and Switradilepa.

Genus: *Barleria* Linn

The generic name has been given in honor of J. Barrelier, a French Botanist of the 17th Century.

The genus comprises of shrubs or undershrubs with spines or unarmed. Leaves are opposite and entire. The flowers are showy, sessile and solitary or in dense or rather lax spikes or racemes. The bracts and bracteoles are small, large or absent. The fruit is an ovoid or oblong capsule and 2 or 4-seeded below the middle. Seeds are compressed, ovate and clothed with silky appressed hairs.

About 230 species occur in tropical parts of the world. 26 species are found in India, chiefly in the mountains of Southern and Western India and a few in tropical and subtropical Himalayas.

Species: *Barleria prionitis* Linn.
(*B. pubiflora* (Benth. ex Hohe.); Brummitt and Wood; *B. prionitis auct.* non Linn.)

Description of Plant

It is a spinous undershrub with terete or obscurely 4- angled and glabrous stem. Leaves are petioled, elliptic, acuminate and tipped with a spinule. Flowers are yellow in cymes which are combined into a terminal leafy spike. The fruits are 2-seeded capsules.

Flowering and Fruiting

During August-October.

Distribution

It is known to occur throughout the hotter parts of India.

Plant Parts Used

The leaves, bark and roots are used.

Uses

The decoction of its aerial parts is given in whooping cough tuberculosis and toothache. An infusion of the leaves is applied to soles of the feet to make them harden to tolerate extreme heat, cold and rough soil and the juice of leaves with coconut oil is used for removal of pimples. Its leaves and flowering tops are diuretic and their juice is given in urinary and paralytic affection and stomach disorders (Asolkar *et al.*, 1992). The leaves are also used as alterative in syphilic infections (Watt, 1972). Besides, the bark is known as diaphoretic and expectorant and its fresh juice is given in anasarca.

The paste of its roots is applied to disperse boils and glandular swellings and the root-decoction is used as mouthwash.

Folk Uses

The '*Mushra*' tribals of eastern Uttar Pradesh use its leaves in catarrh with fever, glandular swelling and toothache. Besides, its crushed leaves are also applied on wounds in different parts of the country.

Genus: *Blepharis* Juss.

The generic name has been taken from Greek word, Blepharis meaning eyelash; referring to the fringed bracts and bracteols.

The genus comprises of undershrubs which are usually spinous or with spinous-toothed leaves. Leaves are opposite or 4-nately lobeod. Flowers are purple, blue or white in teminal spikes on suppressed lateral branches. Calyx is 4-partite nearly tobase; segments unequal in opposite pairs and corolla is 2-lipped. Stamens are 4, didynamous and ovary is 2 celled with 2 ovules in each cell.

Its 100 species are found distributed in Paleotropics, Mediterranean region, South Africa and Madagascar. Only 5 of them have been recorded from India.

Species: *Blepharis persica* (Burm. F.) Kuntze (*B. edulis* Pers.)

Description of Plant

This is a grey-pubescent of nearly glabrate undershrub with short, rigid and branched stem. Leaves are in four pairs at the sterile nodes; upper pair 5.0 x1.0 cm., oblong or narrow-elliptic, sessile and spinous-margined and lower pair smaller. Flowers are blue coloured and are borne in 9.0 cm long strobilate inflorescences. Posticous calyx-segment is 1.0-1.5 cm long, broadly ovate and hairy; 2 innermost are 0.5-0.8 cm long. The capsules are 5.0 cm long and 2 seeded.

Flowering and Fruiting

During August-September.

Distribution

It is known to occur wild in Punjab State of the country.

Plant Parts Used

Leaves, roots and seeds are used medicinally.

Uses

Its leaves in ayurveda are considered as acrid, astringent, aphrodisiac, appetizer, alterative, alexiteric and cooling and are used in 'tridosha' diseases, fevers, urinary discharges, leucoderma and mental derangement; while roots in Unani system of medicine, are used as diuretic and to regulate the menstruation. Besides, its seeds are considered to be aphrodisiac, diuretic, expectorant and resolvent (Chopra *et al.*, 1956) and these are used in urinary diseases and inflammations and in the diseases of blood, chest, lungs and liver.

Genus: *Hygropohila* R. Br.

The generic name has been derived from Greek words, Hygros=moist and Phileo=to love; alluding to the habitat of plants.

The genus comprises of erect or procumbent herbs. Leaves are sessile and entire. Flowers are nearly sessile and are borne in spikes or axillary whorls. Calys is 5-lobed or 4-parted; lobes lanceolate, Corolla tube is dialated near mouth; limb 2-lipped, of which upper lip is erect, concave and notched or bilobed. Stamens are 2 or 4 and ovules 2-4 or numerous. The capsules are linear-oblong with 4-8 or more seeds.

About 80 species are known to occur in the tropics of the world. 11 of them have been reported from India.

Species: *Hygropohilla auriculata* (Sch.) Heine (*H. spinosa* T. ander; *Asteracantha longifolia* (Linn.) Nees; *Barleria auriculata* Sch.)

Description of Plant

It is a stout herb usually with unbranched subquadrangular erect stems upto 0.6-.15 m height. Leaves are sparsely hispid on both sides, tapering at base and sessile inverticals of 6 at a node; 2 outer leaves of whorl are 1.8 x 3.5cm, oblong-lanceolate or oblanceolate and 4 inner 3.8cm long; each leaf of 6 leaves is with straight sharp yellow spine in its axils. Flowers are purple-blue in a whorl of 8 in 4 pairs at each node. The capsules are 0.8cm long, linear-oblong and pointed with 4-8 seeds.

Flowering and Fruiting

During March-May.

Distribution

It occurs is moist places throughout India; chiefly in Darbhanga (Bihar).

Plant Parts Used

Leaves, seed and the root are used.

Uses

A leaves of the plant in ayurveda, are known ot possess aphrodisiac, bitter and hyptonic properties and are used in diarrhea, dysentery, thirst, urinary calculi inflammations, biliousness, constipation and anuria. While in Unani system of medicine, these are applied in gleet, lumbago and pain of joints. Its seeds are acrid, aphrodisiac, bitter, cooling and tasty and with milk and sugar are given in spermatorrhoea and gonorrhoea. Besides, leaves, roots seeds are used as diuretic and are employed for jaundice, dropsy, anasarca and diseases of urino-genital tract. A confection of its seeds containing a large number of aphrodisiac, demulcent, nutrient and aromatic stimulant substances, is used in impotence, seminal and other debilities besides, the powder with honey and ghee to be used in asthma.

Folk Uses

The inhabitants of Uttar Pradesh, take a decoction of is roots as diuretic and use its seeds in gonorrhoea and spermatorrhoea.

Preparation

Paushtik churna.

Genus: *Justicia* Linn.

The generic name has been assigned in the honour of J. Justice, a Scotch Horticulturist of the 18th century.

The genus comprises of herbs or shrubs. Leaves are opposite and entire. The flowers are sessile or nearly so, in spikes or panicles and rarely solitary; bracts are small and bracteoles narrow, obsolete or absent. The fruit is ellipsoid or ovoid capsule which bears seeds from the base or narrowed into a solid stalk and 4-seeded at the apex. Seeds are ovoid, compressed, tuberculate and scaly or glochidiate.

About 300 species occur in tropical and subtropical regions of the world. 20 species have been reported from India.

Species: *Justicia adhatoda* Linn. (*Adhatoda vasica* Nees; *A. zeylanica* Med.)

Description of Plant

It is a dense evergreen and erect undershrub or shrub attaining a height upto 1.2-2.5 m. Stem is densely appressed-pubescent on young parts. Leaves are ovate-lanceolate to oblong with cuneate often decurrent base, acuminate, glabrous except hairy nerves beneath, subentirecrenulate and 5.0-15.0x3.0-8.0 cm; petiole is 0.5-3.0 cm long. The flowers are white and are borne in 3.0-15.0 cm long peduncled spikes. The capsules are 2.0 cm long, short hairy and 4 seeded. Seeds are 1-2 in each cell, glabrous and tubercled.

Flowering and Fruiting

Flowering: During December-April and Fruiting: February-May.

Distribution

It is found throughout the sub-Himalayan tracts upto 1219 m.

Plant Parts Used

Leaves, roots and flowers are usable.

Uses

Its leaves are known as antiseptic and are used in cough, chronic bronchitis, asthma and phthisis; also in rheumatism and as insecticidal. A crude extract of its leaves is more useful for respiratory ailments as it exhibits uterine stimulation and bronchoconstrictor as the undesirable effects. The leaves are also used in homeopathy for colds, coryza, coughs, pneumonia, spitting of blood, fever, jaundice, catarrh, whooping coughs and bronchitis.

The roots of the plant are antiseptic and expectorant and are used in cough, asthma and phithisis and are also given in intermittent fever. Besides, according to its flowers are used as antiseptic.

Folk Uses

Its leaves and wood ashes mixed with honey are used for cough and asthma and dried leaves are smoked as cigarettes for rapid recovery from asthma. In Eastern India, 10 gm dried rots and seven black peppers are made into a powder which is given once to expedite delivery by the Kond tribals. Besides, reported that 5.0-10.0 ml juice of its leaves or decoction of the roots is also taken twice daily in '*Swas roga*' by the people in Rajasthan.

Preparation

Vasaka churna, Vasaka kwath, Vasakarista, Vasaka ghirta, Vasa avleha, Vasa kshar, Vas chandanaditaila, Phalatrikadi kwath, Vasadi kwath, Kanakasav, Vunimbadi kwath, Chavyanprash,Geriforte, Shagat aijaz and Bakhor dama.

Species: *Justicia gendarussa* Burm f.

Description of Plant

This is a glabrous undershrub with subterete branches and reaching a height upto 0.5-1.2 m. Leaves are lanceolate or linera-lanceolate and 7.5- 12.5 cm long with dark midribs. Flowers are white, spotted with purple and are borne in spikes, often forming a panicle. Corolla is nearly glabrous and 1.3 cm long. The capsules are clavate, glabrous, about 1.3 cm long and 4-seeded. Seeds are ovoid.

Flowering and Fruiting

During January-March.

Distribution

This is commonly known to occur throughout India in waste places, hedges and village shrubberies.

Plant Parts Used

Whole plant, its leaves, tender stalks and roots are used.

Uses

In ayurveda, the plant is known for its bitter, dry, hot and pungent properties and is used in bronchitis, dyspepsia, eye-diseases, inflammation and vaginal discharges. According to its leaves are considered alterative, antiperiodic, diaphoretic and insecticidal and their juice is used internally to stop internal haemorrhage and in cough and colic of children and also in earache. A decoction of its leaves and tender stalks is used in chronic rheumatism and fresh leaves mixed with oil are applied externally over glandular swellings and leaf-infusion in cephalgia, facila paralysis and hemiplegia. Besides, roots boiled in milk are used in chronic indigestion, dysentery, fever and rheumatism.

Folk Uses

Its leaf juice is dropped in to the ear in earache and into corresponding nostril on the side of head affected with hemicranic in different parts of the country by the native people.

Family: VERBENACEAE

Genus: *Clerodendron* Linn.

The generic name is derived from Greek words Kleros = chance and Dendron= a tree, referring to medicinal properties.

The genus comprises of shrubs or trees which sometimes maybe sarmentose. Leaves are opposite, rarely 3-4 nate, simple and entire or lobed. Flowers are in axillary cymes or terminal panicles and often leafy below. The bracts are conspicuous or small. The fruits are globose and usually succulent drupes which are 4-grooved and separating into 4 pyrenes, of which 1-3 sometimes suppressed.

About 400 species occur in tropical and subtropical regions of the world. 18 species are found in India, mostly in Eastern Himalayas.

Species: *Clerodendrum inerme* (Linn.) Gaertn.

Description of Plant

it is a much branched, straggling shrub upto 1.0-2.5 m high. Leaves are opposite, obovate or elliptic and entire with cuneate base. Flowers are white in axillary and 3-7 flowered cymes. The drupes are spongy, about 1.2-1.5 cm long and separating into 4 woody pyrenes.

Flowering and Fruiting

During March-July.

Distribution

It occurs almost throughout India is waste lands, hedges and along river-blanks.

Plant Parts Used

Leaves, roots and whole plant are used medicinally.

Uses

Juice of its leaves is alterative and febrifuge and a teaspoonful of the juice along with a little castor oil is given in scrofulous and venereal infections. These are also applied in the form of poultice to resolves the buboes. Its roots are boiled in oil to form a liniment which is applied in rheumatism. Besides, a tincture or decoction of whole plant is given in intermittent and remittent fevers. It is also used as a substitute for quinine and Chiretta.

Folk Uses

In Mumbari, juice of its leaves is taken as a febrifuge by the local people (Kirtikar and Basu, 1935)

Species: *Clerodendrum multiflorum* (Burm.f.) O. Kuntze (*C. phlomidis* Linn.f.)

Description of Plant

It is a large bushy shrub. Leaves are ovate-sinuate or serrate, 5.0 x 3.7cm and mature puberulous or pubescent beneath. The flowers are white and aromatic on slender pedicels and are borne in terminal leafy rounded panicle, composed of axillary 3-9 flowered cymes. The fruits are black obovoid with 4-kernels and about 5.0-10.0 cm across.

Flowering and Fruiting

Flowering: During April-May and Fruiting: During May-June.

Distribution

The plant is found throughout India, chiefly in Uttar Pradesh, Bihar, West Bengal and Deccan Peninsula.

Plant Parts Used

The leaves, bark and roots are used.

Uses

The leaves of the plant are alterative and bitter tonic. Their juice is given in syphilis. The root-decoction is slightly astringent, demulcent and stomachic and is used as bitter tonic for children during convalescence from measles; it is also given in gonorrhoea. Its bark and roots are bitter tonic and are used in nervous disorders and debility. Besides, various plant parts are used in fever, post-natal complaints, cholera and anthrax.

Folk Uses

The juice of its fresh roots is used in asthma, cold and cough and that of stem mixed with oil is used in ear pain by the people in Pauri Garhwal.

Preparation

Arq dasmula. It is also used as one of the constituents to prepare kwath for the ayurvedic medicine, Chavyanprash linctus.

Species: *Clerodendrum serratum* (Linn.) Moon

Description of Plant

It is an erect shrub with thick woody rootstocks. Leaves are sessile, usually in whorls of 3, 10.0-20.0 x 4.0-6.0 cm, oblong-elliptic or obovate with a subcordate or rounded base and glabrous. The flowers are bluish or purplish-white and are borne in axillary cymes forming a terminal leafy panicle. The fruit is a succulent drupe which is globose and black when ripe.

Flowering and Fruiting

Flowering: During May-August and Fruiting: In September-November.

Distribution

It is known to grow more or less throughout India.

Plant Parts Used

Its roots are bitter, acrid, thermogenic, anti-inflammatory, digestive, carminative, stomachic, anthelmintic, depurative, expectorant, sudorific, antispasmodic, stimulant and febrifuge and their decoction is given for asthma, bronchitis and other catarrhal affections of the lungs. The leaves are used in fevers and boiled with oil and butter are made into an ointment which is applied externally incephalgia and ophthalmia.

Besides, the plant is also used in droposy, anasarca, rheumatism, hemiplegia, sores, prolapse ani, fistula ani, cholera and rinderpest.

Folk Uses

Its young shoots and roots are ground in equal quantities and little gur is added to make it tasty; about 25 gm of this preparation is given thrice a day for seven days to cure typhoid fever by the tribals in Jharkhand.

Preparation

Bharangi gur, Bharangayadi kwath, Kankasav, Sudarshan churna, Sringayadi churna and Yograj guggulu.

Species: *Clerodendrum viscosum* Veint. (*C. infortunatum* auct, non Linn.)

Description of Plant

It is a large gregarious tawny-villous shrub about 1-2 m high. Leaves are ovate-cordate, acuminate, entire or serrulate and about 10-25x10.20 cm. The flowers are white, tinged with pink and are borne in terminal panicles. Calyx is persistent with the corolla-tube exceeding the glandular calyx-lobes. The fruits are somewhat globose drupes which are borne within enlarged pinkish calyx and black when ripe. Seeds are oblong.

Flowering and Fruiting

Flowering: During January-March and Fruiting: In March-April.

Distribution

It is commonly found in waste places and along the road-sides throughout India.

Plant Parts Used

The leaves, roots and its flowers are used.

Uses

The leaves are known as bitter, tonic, antiperiodic, anthelmintic and cholagogue and their juice is given as a rectal enema for killing ascaris; also in malarial fever. The leaves are employed externally for tumours, some skin-diseases and for scorpion stings.

The roots are externally applied for tumours and skin diseases. Its root-paste in butter milk is given in colic. These are also used to relieve congestion and torpidity of the bowels and in piles, roots are given in doses of about 1.0 gm with water. Its flowers are used in scorpion stings.

Folk Uses

About 5.0 gm root-paste of this plant in mother's milk is given to childredn as an anthelmintic by the tribals of Gandhamardan hills in Orissa. Reported that a paste of its leaves is applied to cure skin diseases by the native people of Tripura.

Genus: *Gmelina* Linn.

The generic name has been assigned in the honour of S. Gottlieb Gmelin, a celebrated German naturalist and traveler of the 18th century.

The genus comprises of trees or shrubs. Leaves are opposite and entire, toothed or lobed. The flowers are large and yellow or brownish and are arranged in panicled tomentose cymes. Fruits are succulent drupes with bony, undivided and 2 or 4 celled endocarps.

About 35 species occur in East Asia, Indo Malayan region and Asia including 2 species from tropical Africa and Madagascar. 5 species have been reported from India.

Species: *Gmelina arborea* Linn.

Description of Plant

It is an unarmed large deciduous tree about 20 m high and pubescent with somewhat corky bark. Leaves are large, tawny tomentose underneath when young, cordate, ovate, acuminate and 10.0-20.0 x 7.5-15.0 cm. The flowers are yellow tinged with brown in terminal and axillary racemiform cymes. Inflorescence, calyx and corolla are toothed with dense soft and tawny tomentum. Fruits are ovoid or pyriform, smooth and 1.7-2.5 cm long drupes and orange-yellow when ripe. The seeds are oblong.

Flowering and Fruiting

Flowering: During February-April and Fruiting: In April-July.

Distribution

It is found along road-sides and in the forests throughout India.

Plant Parts Used

The plant fruits are known as refrigerant and their decoction is given in fever and bilious affections. Its leaves are demulcent and their juice is useful in cough, gonorrhea, foetid discharge and maggot in ulcers. The paste of its leaves is applied on head to relieve headache. It roots are bitter, tonic, demulcent, laxative, refrigerant and stomachic and their decoction is given in anasarca, fever and indigestion. The root-bark also improves this and relieves abdominal pain. Besides, various plant parts are used in swellings of throat, fever dropsy, spleen trouble, pain, colic, rheumatism, convulsions, delirium, smallpox, syphilis, sores, urticaria, dyspepsia, cholera, phthisis, bronchitis, asthma, diarrhea, intoxication, haemorrhage, septicaemia, graved and rinderpest.

Folk Uses

In North-Eastern India, the decoction of its bark is given for stomach ailments and extract of the root as blood purifier. Its bark is also used for poisoning by the Panwaria and Kol tribals of Eastern Uttar Pradesh.

Preparation

Dashmularisht, Kutajarisht, Sriparnyadi kwath, Sriparni taila and Brihat panchmuladi kwath.

Genus: *Lantana* Linn.

Old Italian name is the generic name for the wayfaring tree, Viburnum lantana; applied to this genus on account of resemblances in foliage, flowers and fruits.

A genus comprise of rambling or scandent shrubs or undershrubs which are pubescent or scarbrous. The youngbranches are 4-angeld and sometimes prickly. Leaves are opposite, petioled, simple, crenate and often rugose. The flowers are in peduncled capitate ovoid or cylindric spikes. Calyx is small, membranous and truncate or obscurely 4-5 tooothed. Corolla is slender and cylindric, The stamens are didynamous. Ovary is 2 celled with one ovule in each cell; style is short and stigma is oblique and subcapiatae. The fruit is a fleshy drupe.

About 150 species occur in tropical America, West Indies and tropical South Africa; 8 specis are reported from India.

Species: *Lantana camara* Linn. Var *aculeata* (Linn.) Moldenke (*L. aculeate* Linn.; *L. acmara auct.* non Linn)

Description of Plant

A straggling aromatic shrub and pubescent or scabrous with about 1.2-2.4 m height. The branches are 4- angeled and armed with recurved prickles. Leaves are opposite, simple, ovate, crenate and acute or shortly acuminate. The flowers are white, light purple or yellow and are borne in pedunculate and short-capitate spikes. Calyx is minute. Corolla-tube is curved, spreading horizontally and divided into 4 unequal laobes. Stamens are 4 and didynamous. Ovary is 2-celled with one ovule in each cell. The fruits are glabrous and globose drupes bearing an outer thin epicarp, pulpy mesocarp and hard inner endocarp, about 0.5 cm in diameter and shining black when ripe. Seeds are 2-celled.

Flowering and Fruiting

Almost throughout the year.

Distribution

The plant is native of tropical America, now completely naturalized throughout India and found wild in hedges, railway tracks and in waste places.

Plant Parts Used

The aerial parts of the plant are used.

Uses

The plant is considered antiseptic, antispasmodic, carminative and diaphoretic. The decoction of entire plant is given in tetanus, rheumatism and malaria and for ataxy of abdominal viscera. But the decoction should be used carefully in low doses as it possesses toxic effects. Besides, the bruised leaves are used to promote healing of wounds and in fistula.

Folk Uses

About half cup of plant decoction with a little quantity of 'Kola namak' is taken twice a day till relief in tetanus by the aboriginals of Central India.

Genus: *Premna* Linn.

From Greek word the generic name has been derived Premnon= the stump of a tree; indicating the stunted growth of most of the plant of this genus.

A genus comprises of trees and shrubs, sometimes climbing. Leaves are opposite, entire of toothed. Flowers are borne in corymbose or paniculate, sometimes pubescent cymes. Calyx is small and cup-shaped; limb 2-5 toothed or 2 lipped and corolla-tube is short. Stamens are 4, didynamous and are inserted below the throat of corolla. Ovary is 2 or 4 celled with 4 ovules. The drupes are small, globose or oblong ovoid with hard endocarps.

About 200 species are found in tropical and subtropical Africa and Asia; of which 25 species have been recorded from India.

Species: *Premna corymbosa* Rottl. (*P. corymbosa* (Burm.f.) Miq.; *P. intergrifolia* Linn.; *P. Obtusifolia* R.Br.; *Cornutia corymbosa* Burm.f.)

Description of Plant

A large shrub or small tree with spinous trunk and branches and yellowish bark, reaching a height of 9 m. Leaves are pubescent when young, oblong or ovate, entire or dentate and 5.0-9.0 x 3.0-6.5 cm. Flowers are greenish-white and are borne in dense corymbs. Calyx is 2-lipped or 5-toothed and corolla is twice of the calyx length; outer lobe much larger. Stamens are slightly excerted and ovary glabrous. The drupes are black, globose and 0.4 cm across. Seeds are oblong.

Flowering and Fruiting

Flowering: During May-June and Fruiting: In July-August.

Distribution

It occur wild in Maharashtra, Chennai, Kerala and Andamans.

Plant Parts Used

Roots and leaves are used medicinally.

Uses

The plant roots in ayurveda, are known for their alexipharmic, bitter, pungent, heating, laxatived and stomachic properties and these are used in anaemia, diabetes, inflammations, swelling, bronchitis, dyspepsia, piles, constipation and fever. A paste of its roots with water and butter fat is given in urticaria. A leaves of the plant are considered carminative, galactagogue and stomachic and their decoction or infusion is given internally in colic, flatulence and eruptive fevers. Besides, leaves rubbed along with pepper are administered in colds and fevers and externally are applied on piles and tumours.

Preparation

Agnimantha kasaya, Agnimantha mullaka and Dasmularista.

Genus: *Tectona* Linn. f.

From Malayan name, the generic name has been derived Tekku for teak-wood plant.

It comprises of trees. Leaves are large, opposite or whorled, petioled and entire. The flowers are numerous in dichotomous cymes forming large terminal panicles. Calyx is campanulate and 5-6 lobed. Corolla is white, tube short and limbs are with 5-6 nearly equal spreading imbricate lobes. Stamens are 5 or 6. Ovary is 4- celled with one ovule in each cell and stigma is shortly bifid. The drupes are enclosed in enlarged 4-celled bladder-like calyx.

Its 3 species are known to occur in the Indomalayan region of the world, of which 1 is native of the Philippine islands and 1 is found in India.

Species: *Tectona grandis* Linn. f.

Description of Plant

A large deciduous tree often fluted near the base and attaining a height of 25-50 m. Bark is thin, light brown or grey and peeling off in long thin strips. The branchlets are quadrangular, channeled and stellately tomentose. Leaves are opposite, simple, 30-60 x 25-30cm, ovate, scabrous or subglabrate, smaller and bract-like in the inflorescence. The flowers are white, on short pedicels and are borne in a large dichotomously or trichotomously branched cymose panicles with short lanceolate bracts. The drupes are about 1.25 cm across, globose, more or less indistinctly 4-lobed and covered with dense felt of branched hairs. Sees are 1-4 in number and obovoid.

Flowering and Fruiting

Flowering: During August-September and Fruiting: In November-December but the fruits can be seen retained for long time on the tree.

Distribution

Both cultivated and wild it occurs almost throughout India with greater concentrations in Centrala and Peninsular India, Assam and Bengal.

Plant Parts Used

Wood, bark, seeds, flowers and the roots are usable.

Uses

The wood of the plant is used a astringent, hepatic, diuretic, stimulant, local refrigerant and sedative. The bark is astringent and is sued in bronchitis. A plaster of its powdered bark is used for hot headache and for swellings and internally, the bark powder is given in dyspepsia, burning of stomach and as vermifuge. The ashes of the wood is applied to swollen eyelids and the oily product obtained by distillation of wood chips is on eczema .

Its seeds and flowers are considered diuretic. The flowers are used in biliousness, bronchitis and urinary problems. The roots are used in anuria and retention of the urine. The oil obtained from the seeds is also useful dressing for the cure of skin itches and as a promoter for the growth of hair. When a powder of its seeds is rubbed on the pubis produces a marked diuresis in partial suppression of urine.

Folk Uses

Its roasted seeds are given to women in the month of '*Chaitra*' to avoid pregnancy for one year by Sahariyas in Madhya Pradesh. Besides, a paste of its powdered wood is applied on forehead in headache and swellings.

Genus: *Vitex* Linn.

A genus comprises of shrubs or trees. Leaves are opposite and digitately 3-5 foliolate. The flowers are sessile or in peduncled cymes forming terminal and axillary or wholly axillary panicles or corymbs. Calyx is campanulate and 5-toothed or rarely 3-toothed. Corolla is small and 2-lipped; corolla-tube is short and limb 5-lobed. Stamens are didynamous and usually exserted. Ovary is 2-4 celled with one ovule in each cell and style is filiform. The drupes are globose or ovoid supported by enlarged calyx. Seeds are obovate or oblong.

About 250 species occur in tropical and temperate regions of the world. Of them, 13 species have been reported from India.

Species: *Vitex negundo* Linn.

Description of Plant

A deciduous shrub with thin grey bark. The branchlets and undersides of leaves or inflorescence are hoary with short grey pubescence. Leaves are 3-5 foliolate; leaflets lanceolate, 2.5-12.5 x 0.7-3.3 cm, the lowest pair is smallest sessile or subsessile, the middle pair, if present, more or less distinctly petiolulate and the odd leaflet is largest with a 0.7-1.5 cm long petiolule,entire or crenate, glabrate, dark-green above and pale greenish tomentose beneath. The flowers are small and bluish purple and are borne in lateral cymes forming an elongated terminal thyrsus and often compound at the base. Calyx-teeth are triangular. Corolla is bluish or purplish white and 5-lobed. Stamens are 4, didynamous and excerted. Ovary is 2-4 celled and 4-ovuled. The drupes are succulent and black when ripe. Seeds are obovate or oblong.

Its 250 species occur in tropical and temperate regions of the world. Of them, 13 species have been reported from India.

Species: *Vitex negundo* Linn.

Description of Plant

A deciduous shrub with thin grey bark. The branchlets and undersides of leaves or inflorescence are hoary with short grey pubescence. Leaves ae 3-5 foliolate; leaflets lanceolate, 2.5-12.5 x 0.7-3.3 cm, the lowest pair is smallest sessile or subsessile, the middle pair, if present, more or less distinctly petiolulae and the odd leaflet is largest with a 0.7-1.5cm long petiolule, entire or crenate, glabrate, dark-green above and pale-greenish-tomentose beneath. The flowers are small and bluish purple and are borne in lateral cymes forming an elongated terminal thyrsus and often compound at the base. Calyx-teeth are triangular. Corolla is bluish or purplish white and 5-lobed.Stamens are 4, didynamous and excerted. Ovary is 2-4 celled and 4-ovuled. The drupes are succulent and black when ripe. Seeds are obovate or oblong.

Flowering and Fruiting

Flowering: During June-August and Fruiting: In cold season.

Distribution

Throughout India it is found in scrub-jungles and road-sides in the warmer parts.

Plant Parts Used

Leaves, roots, bark, fruits, sees, flowers and whole plant are usuable.

Uses

The leaves are vermifuge, aromatic, tonic alterative, anodyne, antiparasitic, discutient, anti-inflammatory and antirheumatic. Rheumatic patients are advised to have bath in water boiled with its leaves. A smoke of its dried leaves is taken in headache and catarrh and the decoction of the leaves with long pepper is given in catarrh fever with heaviness of head and dullness of hearing; also as both in the puerperal state of women. The juice of its leaves is used for removing foetid discharges and worms from ulcers. The juice of the leaves with cow's urine is administered in the doses of 25 gm every morning in the enlargement of spleen also. Its leaves are considered very efficacious in dispelling inflammatory swellings of the joints from rheumatism and of the testes from suppressed gonorrhoea or gonorrheal epididymitis and orchitis and also over sprained limbs, contusions and leech bites. For the purposes, the fresh leaves are put into an earthen pot heated over a fire and applied as hot as can be tolerated without pain or the bruised leaves are applied as poultice to the affected part.

The roots of the plant are anodyne, diuretic, febrifuge, expectorant and tonic and these are used in boils, colic, dyspepsia, leprosy and as anthelmintic and also as demulcent in piles and dysentery. Its root-bark tincture is used in rheumatism and irritable bladder. Its fruits are cephalic, nervine tonic and emmenagogue and dried ones are used a vermifuge. It seeds are cooling and are used in leprosy caused by *Mycobacterium tuberculosis* and for other cutaneous diseases. The flowers are astringent, cardiotonic and cooling and are prescribed in cholera, fever, diarrhea, liver complaints and haemorrhage of the digestive system.

The whole plant as astringent, cephalic and stomachic, which is used in asthma, bronchitis, consumption, eye diseases, leucoderma and painful teething. A compound oil is prepared from the juice of the plant and eleven other substances in different properties is thought to be specific for syphilis and other venereal diseases.

Folk Uses

The dried leaves of the plant are smoked in headache and seeds after boiling are eaten by the people in different parts of the country. In Konkan, the juice of its leaves with that of *Eclipta prostrate* and *Ocimum sanctum* is extracted and Ajwan seeds are bruised and steeped in it and this is given in doses of 5.0 gm in rheumatism (Nadkarni, 1954).

Preparation

Nirgundi kalpa, Nirgundi taila, Vishgarbha taila and Safuf fanjkisht.

Species: *Vitex peduncularis* wall.

Description of Plant

A tree with cinerous-pubescent branches and reaching a height of 6-12 m. Leaves are trifoliate; leaflets 11.5 x 2.5 cm, acuminate, lanceolate, entire and densely covered by minute glands beneath. Flowers are borne in long peduncled and many flowered panicles. Calyx is 0.2 cm long, grey-pubescent and corolla 0.5 cm long. The drupes are cuboid-globose and 0.5 cm across.

Flowering and Fruiting

During May-July.

Distribution

It occurs wild in Assam, Andhra Pradesh, Bengal, Bihar and Madhya Pradesh.

Plant Parts Used

Leaves and bark are used.

Uses

An in fusion of its leaves or root bark or stem bark is used in malarial and blackwater fevers. Besides, bark of this plant is also applied externally in the form of a paste in chest pain.

Folk Uses

In Chota Nagpur, a paste of its bark is applied on the chest in pain. The people of Surguja district in Madhya Pradesh, take young leaves and bark of the plant and pound them and the past thus obtained is given in bodyache and also applied topically in morning for 2-3 days in bodyache.

Species: *Vitex trifolia* Linn.

Description of Plant

A small tree or shrub with smooth grey bark. Leaves are simple and trifoliate; leaflets sessile, obovate or obovate-oblong, entire and glabrate above and white-mealy beneath. Flowers are small, palepurple and are borne in pedunculate cymes. Calyx is 0.3cm long in flower and slightly larger is fruits. Corolla is tomentose. Filaments are hairy at the base. The drupes are globose or obovoid and 0.6cm across. Seeds are obovate or oblong.

Flowering and Fruiting

During March-June.

Distribution

Throughout the torpical and subtropical parts in the plains and lower hills in India it is found scattered.

Plant Parts Used

Leaves, flowers, roots are the parts, used medicinally.

Uses

In ayurveda, its leaves are known as acrid, anthelmintic and pungent and are used to improve memory, promote growth of the hair and in pains, inflammation, leucoderma, bronchitis, and fever and also applied externally to contusions, rheumatic pains, sprains and swellings. The plants flowers with honey are prescribed in fevers accompanying vomiting and thirst. The fruits are used in amenorhhoea and as cephalic, emmenagogue and nervine tonic. Besides, roots of the plant are used as anodyne, antiemetic, expectorant and tonic and also successfully in thirst.

As a substitute of Vitex negundo, it is used.

Family: LAMIACEAE

Genus: *Hyssopus* Linn.

The Latinized form of Hebrew name is the generic name Ezoph or Azob meaning holy herb. This aromatic herb was considered sacred and was used in biblical times for cleaning the sacred places.

It comprises of shrubs or undershrubs. Leaves are sessile, obtuse and entire. Flowers are horne in the whorls of 6-15 flowers in axillary and terminal spikes. Calyx is 5 toothed and corolla 2-lipped; upper lip erect, flat and notched. Stamens are 4, exserted and style lobes subequal.

Its 15 species are known to be distributed in South Europe and from Mediterranean region to Central Asia. Only one of them has been reported from India.

Species: *Hyssopus officinalis* Linn.

Description of Plant

It is a brightly coloured shrub or subshrub reaching a height of 30-60 cm. Leaves are sessile, oblong linera or lanceolate and entire. Flowers are borne in the whorls of 6-15 flowered axillary and terminal spikes. Calyx is 0.6-0.8 cm long and corolla is bluish purple. The nutlets are narrow, smooth and triquetrous.

Flowering and Fruiting

During August-October.

Distribution

It occurs wild in Western Himalayas, from Kashmir to Kumaon between 2438-3352 m.

Plant Parts Used

Whole plant and its leaves are used.

Uses

The entire plant is considered anthelmintic, carminative, pectoral and stimulant and is used in the form of an infusion in colds, cough, consumptions and lung complaints. Besides, its leaves are known to possess carminative, emmenagogal, stimulant and stomachic properties and are used in colic and hysteria. A syrup of sap of its leaves is also made with sugar and honey which is given as vermifuge for round worms.

Genus: *Leucas* R.Br.

From Greek word Leukos = white the generic name has been derived, referring to the colour of flowers.

The genus comprises of herbs or undershrubs. The flowers are usually white in axillary or terminal distant whorls. Calyx is tubular, 10 toothed, short, erect and equal or unequal. The nutlets are 4, ovoid, triquetrous, obtuse and dry with small basal scar.

Its 100 species occur in tropical America, West Indies, tropical and South Africa, Arabia, Indo-Malaysia and South China. 40 species are found in India, mostly in Southern and Western parts.

Species: *Leucas cephalotus* Spreng

Description of Plant

It is an erect, simple and branched annual herb with about 30-90 cm height. The stems are patently hairy. Leaves are 3.0-8.0 x 1.0-4.0 cm, ovate or ovate-lanceolate, subacute, crenate-serrate, hairy and land punctuate beneath. The flowers are sessile in large dense globose, terminal whorls and are about 2.5-4.0 cm in diameter. Fruits are 3-gonous, oblong, smooth, 0.3 x 0.12 cm and brown nutlets.

Flowering and Fruiting

Flowering: During July-October and Fruiting later: In December- January.

Distribution

It occurs in the Himalayas from Kashmir to Sikkim upto 1828 m and Southwards throughout India.

Plant Parts Used

Panchang (whole dried plant) and the flowers are used.

Uses

The herb is known as stimulant, diaphaoaretic and insecticidal and its fresh juice is used externally in scabies. The extract of entire plant is dropped in nostrils in headache and also used as *'anjan'* in eye in jaundice. The wounds are washed by its decoction. Besides, a syrup and from its flowers is used as a remedy for cough and colds.

Folk Uses

Its dried leaves mixed with *Nicotiana tobaccum* in the ration of 1:3 are smoked to cure piles by the Bhoxas of Uttar Pradesh. Reported that thick roots of the plant are used a tooth brush by the Irula tribals in South India. It is supposed that its continuous use for 40 days makes one fully resistant to snake poison. Besides, one teaspoonful extract of the seeds is given once a day for three days after menstruation to check conception.

Preparation

Vishmajavaradi kshar and Sudarshan churna.

Species: *Leucas plukenetii* (Roth) Spreng. (*L. aspera* Linn.; *Phlomis plukenetii* Roth.)

Description of Plant

A hispid or scrabid annual with erect and usually much branched stem reaching a height of 15-45 cm. Leaves are subsessile or shortly petiolate, 2.5-7.5 x 0.5-1.0 cm, linear-oblong or oblong-lanceolate, entire or crenate and more or less hairy with tapering base. Flowers are white and sessile or nearly so and are borne in terminal and axillary whorls. Calyx is 0.8-1.5 cm long and contracted above the nutlets and corolla tube 0.5 cm long, enlarged and pubescent above, upper lip 0.3 cm long, densely white woolly. The nutlets are oblong, substruncate at the apex, smooth, brown and about 0.2 cm across.

Flowering and Fruiting

During September-November.

Distribution

It occurs wild more or less throughout India in the plains.

Plant Parts Used

Whole plant, its flowers and leaves are used.

Uses

The whole plant is used as antipyretic and insecticidal and flowers of the plant in colds. Besides, its leaves are said to be useful in chronic rheumatism and their juice is applied in psoriasis and other skin eruptions.

Genus: *Marrubium* Linn.

Based on Hebrew word, Marrob, signifying a bitter juice, referring to bitterness of the plant is the generic name.

The genus comprises of perennial tomentose or woolly herbs. Leaves are villous. Flowers are in axillary whorls. Calyx is tubular, teeth 5-10. Corolla is short and tube naked or annulate within, upper lip erect and lower spreading. Stamens are 4 and are included and style lobes are short and obtuse. The nutlets are obtuse.

Its 40 species are known to occur in temperate Eurasia and Mediterranean region. Only 3 of them have been recorded from India.

Species: *Marrubium vulgare* Linn.

Description of Plant

It is a perennial herb with stout, white woolly and simple or sparingly branched stem reaching a height of 40-60 cm or more. Leaves are 10.0-30.0 cm long, crenate, soft villous, orbicular-ovate orbicular and grayish above and whitish below. Flowers are white and are borne in dense axillary whorls. The nutlets are obtuse.

Flowering and Fruiting

During July-August.

Distribution

It occurs wild in temperate western Himalayas, chiefly in Kashmir between 1524-2438 m.

Plant Parts Used

Whole plant is used medicinally.

Uses

In Unani system of medicine, the plant is known to possess antipyretic, bitter, carminative, expectorant and tonic properties and is useful in pains of the joints, bronchitis, diseases of liver, spleen and uterus and its leaves in eye sores, night blindness and to facilitate the expulsion of the foetus An infusion of the plant in the doses of 30 ml is given orally as alterative, expectorant and stimulant in cough, bronchitis, dyspepsia, jaundice and in chronic rheumatism.

Folk Uses

An infusion of the plant sometimesis sued as a domestic remedy for bronchitis with profuse expectoration in certain parts of the country.

Genus: *Salvia* Linn.

From Latin word the generic name has been derived. Salvus meaning well preserved; in allusion to the qualities attributed to saga plant.

An erect herbs, shrubs or undershrubs. Leaves are variable in size and shape. Flowers are small or large and showy and are borne in racemose or spicate whorls. Calyx is tubular or campanulate and 2 lipped and corolla tube naked or annulate within; upper lip erect, lower 3-lobed and middle usually broader. Stamens are 2, filaments short and ovary is 4- partite. The nutlets are triquetrous and smooth.

Its 700 species are found distributed in tropical and temperate regions of the world; 24 of them have been reported from India, chiefly from temperate and alpine Himalayas.

Species: *Salvia aegyptiaca* Linn

Description of Plant

It is a low much-branched straggling undershrub. Leaves are subsessile or shortly petiolate, 0.6-3.8-0.2-0.8 cm, linear-lanceolate, acute, rigid, crenate and much wrinkled and hairy. Flowers are in the whorls of 2-4 flowers and are borne in long racemes. Calyx in flower is 0.3-0.4 cm long, glandular hairy and in fruits is 0.6-0.8 cm long. Corolla is 0.5 cm long; upper lip oblong, subquadrate and notched, lower 3-lobed and middle one is emarginated and larger. The nutlets are oblong-ellipsoid, bluish-black, smooth and about 0.2 cm long.

Flowering and Fruiting

During the greater parts of the year.

Distribution

In the plains of Punjab it is known to occur wild

Plant Parts Used

Whole plant and its seeds are used medicinally.

Uses

The entire plant is used as a cure for eye diseases. A syrup is made from the different parts of the plant which is used in thirst and to improve the taste of water. Its seeds are used in diarrhea, gonorrhoea and haemorrhoids. Besides, the seeds are collected, roasted and are ground and mixed with water and sugar are used to suit the taste.

Species: *Salvia plebia* R.Br.

Description of Plant

It is an erect, often multicauline, annual-biennial herb reaching a height of about 15-50cm. Leaves are 5.0-10.0 x 2.0-4.5 cm, lanceolate-oblanceolate to oblong with a rounded-sobcordate base, subotuse,crenate, rugose and nearly glabrous. Flowers are borne in paniculate, interrupted, pubescent and spicate racemes. Calyx is 2-lipped, 0.3 cm long and glandular hairy; upper lip shorty 3-dentate and corolla is 0.4-5.0 cm long; upper lip retuse. The onutlets are minute, obovid, smooth, glabrous and about 0.1 cm long and brown coloured.

Flowering and Fruiting

During March-June.

Distribution

It occurs wild almost throughout India.

Plant Parts Used

Seeds are usable.

Uses

The seeds of the plant are used in diarrhoea, gonorrhoea, menorrhagia and piles and also given to promote sexual powers. Besides, there are also given in leucorrhoea and for seminal weakness.

Family: NYCTAGINACEAE

Genus: *Boerhavia* Linn.

The generic name has been assigned in the honour of Herman Boerhave, a famous Dutch Physician of 18th century.

A diffuse or erect herbs with divaricate branches. Leaves are opposite. The flowers are small in panicles, umbels or heads; pedicels jointed.The bracteoles are minute, often deceiduous, rarely whorled and involucrate. Fruits are small, 5-ribbed or angled and viscidly glandular.

It 40 species occur in tropical and subtropical regions of the world; 8 species in drier parts of India.

Species: *Boerhavia diffusa* Linn.
(*B. repens* Linn.)

Description of Plant

It is a diffusely branched and creeping perennial herb with fusiform, stout and woody rootstocks. Stems are 60-90 cm long, slender, prostrate or ascending, swollenat the nodes, minutely hairy and sometimes viscid. Leaves are opposite, 1.2-2.5 cm long, ovate, oblong or suborbicular, green and glabrous above with rounded or sub cordate bark and swubundulate margins; petiole is as long as the blade. Flowers are pink, small or minute and not involucrate and are borne in small umbels which are arranged in axillary and terminal panicles. Fruits are 5-ribbed, 1.3 cm long, clavate, rounded and glandular.

Flowering and Fruiting

During winter season.

Distribution

A common weed found throughout India, especially in West Bengal during rainy season and in cold places.

Plant Parts Used

Whole plant is used.

Uses

The decoction of the entire plant is used in oedema and dropsy. Its roots are expectorant, diuretic and laxative and are used in oedema, jaundice, ascites,anasarca, strangury, gonorrhoea and internal inflammation. The roots and leaf-ash are taken to cure night blindness. A poultice of the herb mixed with palm kernel oil is applied to boils. The leaves of the plant are used in eye wounds, muscular pain, dropsy and in gonorrhoea; also to purify the blood and to hasten the delivery and their paste is taken orally to check bleeding after delivery. Dry powder of its leaves mixed with mustard oil is used externally on itches and eczema. The leaves boiled with rice, garlic and water are rubbed on body in rheumatism.

The powder of entire dried plant is used in abdominal cancer and tumour. Its seeds and flowers are considered contraceptive in ayurveda. Charak, used it in the form of an ointment in leprosy and skin diseases. It is also said to be effective in urethritis.

Folk Uses

Its leaves in West Bengal are used by the natives as appetizer. Besides, reported that about 2.0 gm of its root paste with cow's milk is taken by the women for abortion in Central India.

Preparation

Punarnava mandur, Punarnava-kshar kwath, Punarnava ark, Punarnava ashtak, Punarnava asav, Punarnava churna, Calcury and Jawarish zaruni ambari ka naksha kalan.

Family: AMARANTHACEAE

Genus: *Achyranthes* Linn.

The generic name has been derived form the Greek words Achyron =chaff and Anthos = a flower, referring to the scarious inflorescence.

It comprises of herbs or undershrubs. Leaves are opposite, entire and petioled. Flowers are bisexual in slender and simple or panicled spikes which are soon deflaxed. The bracts and bracteoles are spinescent. Perianth-segments are 4-5, rigid, somewhat connate below, lanceolate and shining. Stamens are 2-5, rigid, somewhat connate below, lanceolate and shining. Staments are 2-5 and anthers are 2-celled. Ovary is oblong, subcompressed; style filformand ovule is solitary. Fruits is an oblong or ovoid utricle. The seeds are inverse and oblong with liner or lanceolate cotyledons.

About 100 species occur in the tropical and subtropical parts, abundantly in Africa and Asia. Of them, 4 species are found in India, mostly in Southern and Western parts.

Species: *Achyranthes aspera* Linn.

Description of Plant

It is stiff effect herb about 30-90 cm high with erect and simple or slightly branched stem. Leaves are 3.8-12.5 x 5.0-7.5 cm, opposite, elliptic, obovatge or suborbicular, usually rounded at apex and tomentose orvelvety on both surfaces. Flowers are bisexual, greenish-white and are borne is long spikes which elongate in fruits and reach upto 50 cm in length. Fruit is an oblong utricle which is enclosed in the hardened perianth and disarticulating easily and carrying spinous bracteoles with it. Seeds are subcylindrical with a truncate brown apex.'

Flowering and Fruiting

Flowering: In winter and Fruiting: In summer season.

Distribution

A common weed found throughout India upto 1000 m usually in dry area and along the road sides.

Plant Parts Used

Whole plant is used.

Uses

The entire plant is considered pungent, purgative, diuretic and astringent and is used in dropsy and haemorrhoids. The decoction of the plant is useful in pneumonia, cough and kidney stones and its ash is given in haemorrhoids. The dried plant is useful in colic and gonorrhoea and also acts as laxative. The ashes of the plant with water and *jaggery* are effective in ascites and anasarca; sesame oil medicated with ashes of the plant is used as eardrops. *Achyrol*, its leaf-extract is used in leporosy and the juice of the leaves is in eczema and leprosy. A paste of its leaves with water is applied to bites of poisonous insects, wasps and bees etc. Its roots are given in stomach pain and their paste is applied to remove the opacity of cornea and to wound as haemostatic besides the decoction in diarrhea. The root-paste is also given to stop bleeding after abortion. A pinch of its root powder in combination with pepper powder and honey forms a good remedy for cough.

Its seed-paste is applied against insect-bites. The seeds are rubbed with rice-water and are prescribed to patients of bleeding piles and their powder is soaked in butter milk which is given in biliousness. Its flowers are ground and mixed with curd and sugar and are given in menorrhagia and the flower tops are used in treatment of rabies. Besides, a decoction of the whole plant is diuretic and efficacious in renal dropsy and in combination with that of Kakjanga (*Leea aequata*) is useful in insomnia. The fresh juice of its leaves thickened into an extract by exposure to the sun and then mixed with a little opium, may be beneficially applied to primary syphilitic sores . The powder of its roots with banagbhasm is used as aphrodisiac to increase the sexual desired.

Folk Uses

The plant root is tied to the ear as a remedy for malarial fever in different parts of the country. In the Vaishnodevi hills area of Jammu Kashmir, a decoction of its roots is given to women after menstruactionas an antifertility agent.

Preparation

Apamargkshar taila,Apamarg asav, Apaamarg avaleh, Cystone, Kushta sam-ul-far, Kustha shangraf and Kushta hartal warqi. Besides, Somal bhasm, Hartal bhasm and Vang bhasm also obtained with the help of apamarg.

Genus: *Amaranthus* Linn.

The generic name is derived from Greek words A = not an Mairaino= to wither or Amaranthos = unfading, in allusion to the ever lasting qualities of its flowers.

This genus comprises of annual erect or decumbent herbs. Leaves are alternate. Flowers are small and monoecious or polygamous in dense, terminal and axillary spikes or panicles; bracts are herbaceous and often persistent and bracteoles 2. Male flowers show perianth fo five, 1-3-membranous, ovate and lanceolate segments. Stamens in the male flowers are usually 5 but may varyfrom 1 to 5 and anthers 2-celled. Female flowers show perianth segments oblong or sphathulate anderect in fruit. Ovary is ovoid and compressed, style short or absent and ovule is solitary, subsessile and erect. Fruit is an ovoid and compressed utricle or circumscissile, membranous or coriaceous capsule. The seeds are erect, orbicular and compressed.

Its 60 species occur in tropical and temperate region of the world; of which, 20 species are reported from India.

Species: *Amaranthus spinosus* Linn.

Description of Plant

It is an erect and spinous herb or under shrub. Stems are 30-60 cm long, glabrous, hard and often tinged with red. Leaves are 3.0-10.0 cm long, ovate or oblong, spine-tipped and glabrous on upper surface. Flowers are unisexual and grayish-green in dense spikes. The male flowers are produced more than female ones. Bracts are setaceous and exceeding the perianth. Stamens are 5. The capsules are dehiscent, ovoid, thickened above and rugose. Seeds are black and shining.

Flowering and Fruiting

During July-December.

Distribution

It is found throughout India, ascending upto 1524 m in the Himalayas; grows abundantly in Malabar and West Bengal. Often troublesome as a weed in cultivated fields, waste places and along road-sides.

Plant Parts Used

Whole plant is used.

Uses

The entire plant is known as diuretic and emollient and is used as good remedy in colic. Its leaves as a poultice and applied to abscess, buboes, wounds and burns to promote suppuration and discharge of pus; also used as enema in stomach troubles, in curing piles and leprosy. It roots are sued in gonorrhoea, eczema, menorrhoea and inflammatory swellings and their powdr is to cure onychia.

Besides, its *avaleh* (linctus) made with honey is used in *raktapitta*.

Folk Uses

Bhuyan (1994), reported that a medium-sized root of the plant is inserted into the vagina with a little bit of *'hing'* after applying it on the top of the root as abortifacient by the tribals in Arunachal Pradesh. It is done inside the vagina for easy abortion. Besdies, the herb is boiled with pulse and is given to cows as a lactagogue for more production of the milk. The people in Ganjam and Phulbani districts in south Orissa, take an aqueous decoction of this roots two or three times a day to check chronic diarrhea.

Genus: *Chenopodium* Linn.

The generic name is derived from Greek words Chen = a goose and Pous=a foot, referring to the shape of the leaves.

This genus comprises of annual or perennial erect or prostrate herbs. Leaves are alternate, entire and lobed or toothed. The flowers are minute and bisexual in axillary clusters or cymes. The bracts and bracteoles are absent. Perianth is 5-lobed; segments concave and incurved. The fruit is a membranous utricle which is enclosed in a erianth. Seeds are usually horizontal with crustaceous or coriaceous testa.

About 100-150 species occur in temperate regions of the world; 7 (8-10, according to Babu, 1977) species are found in India, chiefly in Hiamalayan region.

Species: *Chenopodium ambrosioides* Linn.

Description of Plant

An erect, often much-branched and strongly scented annaual herb with 60-120 cm height. Stems are angular, ribbed and shortly pubescent. Leaves are oblong-lanceolate, glandular, short petioled and 3.0-10.0 x 1.0-3.0 cm; higher ones are small. Flowers are bisexual and minute forming interrupted spikes which are combined into a leafy panicles. The fruits are more or less globular and slightly compressed with a thin pericarp surrounding the seeds. Seeds are small, obtuse, smooth hand lustrous with a bitter pungent taste.

Flowering and Fruiting

During winter season.

Distribution

It is a native of West Indies and South America, now has been naturalized and found distributed in Southern India, Maharashtra, Kashmir, West Bengal and in Northern Himalayas upto an altitude of 900m in Garhwal regions.

Plant Parts Used

The whole plant is used.

Uses

The entire plant is used as a remedy against parasites. Its fruit oil is effective against hook-worm and roundworm infections. The leaf extract is anthelmintic and its shoot and root extract is considered nematicidal. Essential oil obtained form its leaves is considered as a strong antifungal agent against human pathogenic fungi. Its seeds with sugar are also given for wormal diseases.

Family: POLYGONACEAE

Genus: *Polygonum* Linn.

The generic name has been derived from Greek words Poly = many and Gonu= the knee, referring to the stem being often enlarged and bent at the joints.

Usually herbs and rarely of shrubs. Leaves are alternate and entire; stipules membranous and tubular. The flowers are bisexual, small, axillary or terminal and clustered; the cluster is sessile or in spiciform capitate or panicled racemes. The bracts and bracteoles are membranous and ochreate. Perianth is 4-5-cleft and green or variously coloured. Stamens are 5-8 and perigynous. Ovary is compressed or trigonous and 1-celled; ovule is 1 and style 2 or 3 which are free or connate below. The fruits are compressed or trigonous nutlets. The seeds are with lateral or excentric embryo and small cotyledons.

Its 300 species are known to occur cosmopolitcally throughout the world, chiefly in temperate regions; of which 7 species are reported from India.

Species: *Polygonum barbatum* Linn.

Description of Plant

It is an erect and glabrous annual herb, Leaves are 10.,0-15.0 x1.0-2.0 cm, linear lanceolate, acute or acuminate with a rounded or subcordate base and glabrous. The racemes are erect, slender and about 5.0-10.0 cm long.

Flowering and Fruiting

During August-December.

Distribution

This is known to occur throughout the hotter parts of India.

Plant Parts Used

Seeds, shoots and roots are used.

Uses

The seeds of the plant are considered tonic, purgative and emetic and these are used to relieve griping pains of colic. A deceoction of this shorts is sued as a stimulating wash for ulcers. Besides, the roots are used as cooling and astringent.

Species: *Polygonum recumbens* royle

Description of Plant

A puberulous or scabrous herb with ascending, prostrate, woody and much branched stems reaching a height of 30-60 cm. Leaves are closely set, lanceolate and acuminate or truncate. Flowers are small and white or pink in and are colour and are borne axillary. The nuts are black, lustrous and about 0.2 cm long.

Flowering and Fruiting

During July-August.

Distribution

In hilly regions from Eastern Hiamalayan region to Kashmir, it occurs.

Plant Parts Used

Whole plant is used.

Uses

Whole plant is crushed to form a paste and it is applied on boils 3-4 hours before operation so as to reduce bleeding. Besides, its leaf juice is used in blood dysentery.

Folk Uses

The inhabitants of hilly regions in Uttar Pradesh (now Uttarakhand), apply a past of the whole plant on wounds.

Genus: *Rheum* Linn.

The generic name has been derived from Greek word, Rheo meaning to flow; referring to the purgative properties of the plant.

It comprises of stout herbs with large woody roots. Leaves are large, entire, toothed or lobed with scarious stipules. Flowers are usually bisexual and are borne in panicled racemes. Sepals are 5. Stamens are 6-9. Ovary is 2-4 angled and stigma capitate or horse shaped. The nuts are 2-4 winged.

Its 50 species are known to occur in temperate and subtropical Asia. Only 10 of them have been recorded from India, chiefly from Himalayas.

Species: *Rheum australe* D. Don.

Description of Plant

It is a small herb with stout roots and leafy stems upto 1.5-1.8 m high. Radical leaves are cordate, long petioled, orbicular or broadly ovate and about 60.0 cm in diameter with 30.0-45.0 cm long petioles. Flowers are small, pale-red or dark purple in axillary till panicles. The nutlets are ovoid-oblong and about 1.3 cm long with cordate base and notched apex.

Flowering and Fruiting

During July-August.

Distribution

It is found in the Himalayas, from Kashmir to Sikkim between 3300-5200 m; also cultivated in Assam.

Plant Parts Used

Roots of the plant are used.

Medicinal Uses

A roots of the plant are considered astringent, purgative and tonic and combined with ginger in the pill forms of 0.3-7.0 gm, are used in bowel complaints. The stimulating effect of its roots combined with aperient properties finds useful application in atonic dyspepsia. Besides, a decoction of root bark in the doses of 15-30 gm is also prescribed in bowel complaints. Roots cause a deep tinge to the urine when given internally but it need no alarm and misconception however, it is considered detrimental to patients, suffering from gout, rheumatism, epilepsy and uric acid troubles.

Folk Uses

Gangwal tribe in Garhwal Himalayas, apply a paste of this roots in tumours, wound and headache. Besides, the paste is also applied on curts, sprains and swellings in certain parts of the country.

Preparation

Its roots are used in baby's tonic antidiarrhoetic and antidywsenteric preparations like Grape water and Ghuttis etc. and Lavanbhaskar churna.

Species: *Rheum webbianum* Royle

Description of Plant

A small herb with leafy stems and about 0.3-1.8m height. Leaves are 10.0-60.0 cm in diameter, orbicular-cordate or reniform and long peptioled. Flowers are pale-yellowish or pink and smaller in axillary and terminal panicles. The fruits are oblong or orbicular, about 0.8 cm in diameter and notched at the both ends with wings.

Flowering and Fruiting

During July-August.

Distribution

It is known to occur wild in Kashmir, Himachal Pradesh, Uttar Pradesh and Sikkim between 3048-4267 m.

Plant Parts Uses

Roots of the plant are used medicinally.

Uses

Its roots are used in form of decoction or in powder form along with ginger, in the doses of 0.3-0.7 gm in bowel complaints.

Besides, the roots are used as a substitute for Rheum australe.

Folk Uses

Its roots are used on cuts and wounds by the tribal people in hilly regions of Uttarakhand.

Family: ARISTOLOCHIACAE

Genus: *Aristolochia* Linn.

The generic name has been derived form Greek words, Aristos=best and Lochia=birth, with reference to the curved shape of the flower resembling the human foetus and hence commonly called 'Birth wort'.

A prostrate or twinging shrubs or perennial herbs. Leaves are alternate, entire of lobed and often with a stipule-like leaf in axil. Perianth is coloured; tube inflated below and then contracted and hairy within. Stamens are 6 or rarely 5 or more. Ovary is inferior and more or less 6-celled with many ovules. The capsules are lantern-liked with many seeds.

Its 350 species are known to occur in tropics and temperate regions of the world. Out of which only 15 species have been recorded form India.

Species: *Aristolochiabracteolata* Lamk. (*A. bracteata* Retz.)

Description of Plant

A slender, decumbent and glabrous perennial herb with 30-45 cm height. Leaves are entire or undulate, usually obtuse, cordate at the base, reniform or broadly ovate and long peptioled. Flowers are solitary. Perianth is 2.5-5.0 cm long with cylindric tube and trumpet shaped mouth and dark purple lip. The capsules are 1.2-2.0 cm long, oblong-ellipsoid and glabrous. Seeds are deltoid with a cordate base and 0.6 cm long.

Flowering and Fruiting

During rainy season.

Distribution

It occurs in the plains of northern India from Haryana and Uttar Pradesh south wards to Peninsula India upto Andhra Pradesh and Maharashtra.

Plant Parts Uses

Whole plant, its leaves and roots are used.

Uses

The plant in ayurveda is known for its anthelmintic, bitter and purgative properties and it is used to cure *'vata'* and *'kapha'* diseases, painful joints and is also applied to kill the maggots on sores, The juice of its leaves is applied to foul and neglected ulcers and bruised ones with castor oil to cure eczema on children's legs. Besides, its powdered root sin the doses of 80-150 gm are used to increased the contraction of uterus during labour and with castor oil are used in colic and torming, amenorrhoea, dysmenorrhoea, tedious labour, intermittent fever and worms.

Folk Uses

Tharus people in Gorakhpur (Uttar Pradesh), take orally its root paste along with black peoper and leaf juice at 30 minute intervals as antidote in snake-bite. Its leaf juice is also used to rinse oral cavity to cure apthous ulcers by the inhabitants of Trimula hills in Andhra Pradesh. Besides, in Bhavnagar (Gujarat), a paste of its leaves is applied on bois, wounds and ecezema by the local tribal people.

Species: *Aristolochia indica* Linn.

Description of Plant

A perennial climber with greenish-white woody stems. Leaves are 10.0-12.5 x 7.5-8.0 cm, entire with more or less undulate margins, cordate, acuminate, obovate-oblong to sub-pandurate and glabrous. Flowers are borne in few-flowered axillary racemes. POerianth is upto 4.0 cm long with a pale-green cylindric tube; mouth and lip clothed with purple tinged hairs. The capsules are 3.8-5.0 cm long and oblong or globose-oblong with flat, ovate and winged seeds.

Flowering and Fruiting

Flowering: During September-October and Fruiting: In February-March.

Distribution

It is known to occur wild throughout the low hills and plains of India.

Plant Parts Used

Roots and leaves of the plant are used.

Uses

In ayurveda, the roots of this plant are described as alexiteric, bitter, emmenagogue and pungent and these are used to cure *'tridosha'* diseases, pains of the joints and bowel problems of children. Its roots are rubbed with honey and are given in white leprosy and dropsy and after macerated with black pepper corns, in cholera and diarrhea. A decoction of its roots in the doses of 30-60 ml is given as stimulant and febrifuge. Besides, the leaf juice forms a valuable remedy incholera, diarrhea, intermittent fever and in bowel complaints.

Folk Uses

In Mumbai, its leaf juice is applied externally to the abdomen in bowel complaints and also given internally for the same.

Family: LAURACEAE

Genus: *Litsea* Lamk.

The generic name *Litesea* is a Japanese name for this genus.

It comprises of trees and shrubs. Leaves are alternate, rarely opposite or subopposite and penninerved. Flowers are small, dioecious and usually borne sessile or shortly pedunculate in 4-6 flowered umbels. Perianth tube is ovoid, campanulate or very short; lobes 6 or 4, equal or unequal. Male flowers; Stamens 9 or 12 in trimerous and 6 in 2 merous. Female flowers: ovary enclosed in perianth-tube or free, style short or long and stigma irregularly lobed. The fruits are drupes or berries, resting on unchanged perianth or partly clasped at base by much enlarged discoid or copular perianth.

Its 400 species are found in the warmer parts of Asia from North to Korea and Japan, Australia and America. Only 46 of them have been recorded from India.

Species: *Litsea glutinosa* (Lour.) C.B. Robinson (*L. chinensis* Lamk; *L. sebifera* Pers.; *Sebifera glutinosa* Lour.)

Description of Plant

A small evergreen tree upto 25 m in height and 1.5 m in girth with a clean bole of 6 m length and pale brown, roughish and corky bark and yellowish-grey to grayish brown wood. Leaves are 10.0-25.0 x 5.0-10.0 cm, aromatic, elliptic ovate or oblong lanceolate and pubescent. Flowers are yellowish, 8-12 together in umbellate heads.

Perianth lobes are generally wanting and stamens 20 or more. The fruits are globose, black-purple and about 0.6-0.7 cm across.

Flowering and Fruiting

Flowering: During summer and Fruiting: In rainy season.

Distribution

It is found throughout India ascending upto 1350 m in the Himalayas.

Plant Parts Used

Roots, leaves, bark and fruits are the parts, used medicinally.

Uses

Its roots in ayurveda, are considered aphrodisiac, cooling, galactagogue and sweetish and are used in biliousness, burning sensations, consumption, bronchitis and in leprosy and in Unani system of medicine, are considered aphrodisiac, astringent, bitter, expectorant and tonic and are useful in inflammation, pain of the joints, thirst and diseases of spleen and in paralysis. Their decoction is employed as emmenagogue. The leaves of the plant are mucilaginous and they are bruised and are applied in the form of a poultice on wounds. The bark of the plant is known as aphrodisiac, anodyne and demulcent and is used I the treatment of diarrhea and dysentery and externally, fresh ground bark is used as emollient application to bruises, sprains, rheumatic and gouty joints; also as antidote to bites of venomous animals. Besides, the soil obtained from its fruits is used in rheumatism

Folk Uses

In Eastern Ghats (Andhra Pradesh), local people take a paste of its stem bark and made pills from it; 2-3 pills of which are administered orally in leucorrhoea, twice a day for 9 days. Besides, 2-3 spoonfuls of stem-decoction is given for chest pain by Porja people twice a day till cure of the disease in Andhra Pradesh.

Species: *Litsea monopetala* (Roxb.) Pers. (*L. polycantha* Juss.)

Description of Plant

A small or medium sized evergreen tree reaching a height upto 21 m and a girth of 1.8 m with pale brown bark and olive-grey to yellowish wood. Leaves are alternate, oblong, glossy and 7.5-23.0 x 3.5-12.5 cm. Flowers are sot, pubescent and brown and are borne in corymbs or umbels. Perianth is 0.2 cm long and stamens 9-13. The fruits are globose-ellipsoid, 0.1 cm long and black when ripe.

Flowering and Fruiting

Flowering: During July-August and Fruiting later: In October-November.

Distribution

It occurs throughout north, east and central India upto 1200m.

Plant Parts Used

The bark and roots are used.

Uses

Its bark as astringent, stimulant and stomachic which is used in diarrhea. After brushing, this is applied on contusions and in powdered form, is dusted over affected parts of the body in pains arising from bellows or bruises or due to hard work. Besides, its roots are also applied externally on pains, bruises and contusions.

Family: PIPERACEAE

Genus: *Piper* Linn

The generic name is the Latin name of the Pepper plant.

It comprises of perennial herbs or shrubs, often climbing, usually glandular and aromatic with swollen nodes. Leaves are entire, alternate, simple and unequal sided at the base. Stipules are membranous, enclosing the buds. The flowers are minute, dioecious and rarely bisexual and spicate. The spikes are usually leaf-opposed. Perianth is absent. Stamens are 2-4, filaments short and anthers are 2-celled. Ovary is 1-celled, style is short, conic and beaked or none and ovule solitary and erect. The fruit is a small and ovoid or globose 1-seeded berry. Seeds are usually globose.

Its about 2000 species are known to occur in tropical parts of the world; of which 50 species occur in India, mostly in tropical and subtropical Eastern Himalayas and Southern and Western India.

Species: *Piper longum* Linn.

Description of Plant

The plant is a slender, aromatic and climbing herb. Stems are creeping, jointed and become attached to other plants. Leaves are 5.0-9.0x3.0-5.0 cm, ovate, cordate, shortly acuminate, glabrous and membranous; the lower long petioled and upper sessile. The spikes are pedunculate and upright; male larger and slender and female 1.3-2.5 cm long and 0.4-0.5 cm in diameter. The fruits are very small and ovoid and sunk in solid fleshy spike which is 2.5-3.8 cm, erect, blackish green and shining.

Flowering and Fruiting

Flowering: During rainy season and Fruiting: In autumn.

Distribution

The plant is found distributed in hotter parts of India from Central Himalayas to Asaam, Khasi and Mikir hills, lower hills of West Bengal and evergreen forest of Western ghats from Konkan to Kerala. Also recorded from Car Nicobar islands.

Plant Parts Used

Fruits and roots are used.

Uses

The unripe fruits are used as alterative and tonic and mature fruits in the diseases of respiratory tracts, bronchitis, asthma and cough as carminative, as sedative in insomnia and epilepsy, as general tonic, as cholagogue, in bile and gall bladder obstruction, as emmenagogue, abortifiacient and anthelmintic; also applied on sprains and muscular pains. The roots are said to be used as antidote to snakebites.

It is an important ingredient in the preparation of medicated oil, used externally in sciatica and paraplegia. A mixture of long pepper, its root, black pepper and ginger in equal parts is a useful preparation for colic, flatulence, cough, coryza and hoarseness. Long pepper in combination with black pepper is used in the preparation of irritating snuff which is used in coma and for drowsiness. In the form of a powder, it is suspended in warm water and is given to women after parturition to check piles and fever. As vermifuge, it is prescribed for colic in children. Pipla mul in combination with borax and *baiberang* in equal doses is used as contraceptive.

Folk Uses

The natives of India, use a powder of long pepper mixed with honey in cold, cough, asthma and hiccough. This is also used in hoarseness and for gout and rheumatism (Ahuja 1965).

Preparation

Gurpippali, Pippalikhand, Pippalasav, Pippalimuladi Kwath, Videngadi-louha, Chandraprabha vati, Gokshuradi kwath, Kukwadi churna, Khadiararista, Aragbadhyadi kwath, Amlakayadi churna, Sudarshan chruna, Sarvajwarhar-louha, Amritarista, Anandbhairav ras, Hingavastak churna, Lavanbhaskar churna, Rasonvati,Agnitundi vati, Sanjivani vati, Punarnava mandur, Nayas louha,Rakta pittantak louha, Tlisadi churana, Sitopaladi churna, Trikatu churna, Kaphketu ras, Sringadi churna, Kanakasav, Karanjadi yog, Yograj guggulu, Narachras, Itriphal fauladi, Angaruya-i-kabir and Majunkhadar.

Family: SANTALACEAE

Genus: *Santalum* Linn.

The generic name has been derived from Sanskrit name Chandama, for sandal wood tree.

The genus comprises of trees or shrubs. Leaves are opposite, rarely alternate and coriaceous. The flowers are bisexual, axillary or in terminal trichotomous paniculate cymes. The drupes are subglobose with ovoid or globose seeds.

Its about 25 species are found in East Malaysia and Australia to Eastern Polynesia. Out of which only one species has been reported from India.

Species: *Santalum album* Linn.

Description of Plant

A small, evergreen glabrous and semi-parasitic tree with slender branches attaining a height upto 18m with dark grey or nearly black or reddish and rough bark. Sapwood is unscented and white but heartwood is scented and yellowish-brown or dark brown. Leaves are opposite, ovate or ovate lanceolate, glabrous, 1.5-8.0 x 1.5-3.0 cm or larger and thin. The flowers are brownish purple, violet or straw-coloured, unscented and are borne in terminal and axillary paniculate cymes. The fruits are globose drupes, 1.2 cm across, purple black with hard ribbed endocarp. Seeds are ovoid or globose.

Flowering and Fruiting

Flowering: During June-September and Fruiting: In November-February.

Distribution

It occurs from Vindhya mountains southward, particularly in Karnataka and Tamil Nadu and ascending upto 1200m. Also introduced in Rajasthan, Uttar Pradesh, Madhya Pradesh and Orissa, where it has become naturalized.

Plant Parts Used

The wood and sandal wood oil are used.

Uses

Its wood is ground up with water into a paste which is applied to the temples in headache, fevers and local inflammations and to skin diseases to allay head and pruritus; also used a sdiaphoretic. A decoction of the wood mixed with that of dried ginger is used in haemorrhoids. The sandalwood oil and oil derived from its heartwood are considered as cooling, diuretic, diaphoretic and expectorant. This is used in symptomatic treatment of dysuria, gonorrheal urethritis and cystitis. It is mixed with double quantity of mustard oil and is used for pimples on the nose.

The plant is also used in Unanai medicine as germicide, blood purifier, antiseptic and fungicide. Ilaj-ul-gurba recommends a paste made of equal parts of sandal oil and borax with sufficient quantity of water as useful application in pityriasis-versicolor and similar affections.

Folk Uses

In different parts of the country, a paste of its wood with water is applied externally in inflammated swellings, prickly neat and to skin diseases to allay itching, inflammation, heat and headache. Besides, the sandal oil is applied over skins in burns.

Preparation

Chandanadi vati, Chandanadi churna, Chandanasav, Sudarshan churna, Pippalasav, Khamirasandal sada, Khamirasandal tursh tilawala, Juavarish sandain, Majun sandal, Sharbet sandal, halwa-i-supari pak, Khamira abrisham-shira-i-unabwala and Majun helila.

Family: THYMELACEAE

Genus: *Aquilaria* Lamk.

Description of Plant

An evergreen large tree upto 15-25 m in height. Leaves are alternate, linear-lanceolate to lanceolate and obovate, oblong and coriaceous. Flowers are white or green and bisexual and are borne in terminal umbellate cymes. Perianth is 0.5 cm long, slightly hairy outside and densely villous inside. Ovary is tawny-tomentose. The capsules are obovate-cuneate, slightly compressed and about 2.5 cm long.

Flowering and Fruiting

Flowering: During May- June and Fruiting: In July-August.

Distribution

It occurs in eastern parts of India in the forests of Arunachal Pradesh and hilly tracts of Assam, Manipur, Nagland and Tripura.

Plant Parts Used

Wood is used medicinally.

Uses

The wood of the plant is knowns in ayurveda, for its alterative, bitter, carminative,oleaginous, pungent and tonic properties and is used successfully in '*kapha*' and '*vata*' diseases, skin diseases, leucoderma and eye problems while in Unani system of medicine, it is considered aphrodisiac, carminative, diuretic, laxative, stomachic and tonic and is used in chronic diarrhea, diseases of liver and intestines, bronchitis and vomiting and to stabilize the foetus in uterus.

The famous agar wood or eagle wood of commerce is derived form the fungus-infected tree through-wounds caused by *Aspergillus, Fusarium, Penicillium* and Fungi *Imperfectii* and as a liniment, it si applied in various skin diseases; also in rheumatismand paralysis.

Preparation

Agruadi taila.

Family: EUPHORBIACEAE

Genus: *Acalypha* Linn.

The generic name is derived from Greek name Akalepohe for common nettle plant.

The genus comprises of herbs, shrubs or trees. Leaves are alternate, ovate, long stalked and toothed or crentate. Flowers are usually monoecious and minute in short axillary spikes or females 1-2 in pedunculate solitary bract or casually dioecious. Male flowers minute and female at the base of large accrescent leafy bracts, low on the male spike or in separate spikes. Male flowers: sepals 4, valvate; petals absent; disk absent and stamens usually 8 or more. Female flowers: sepals 3-4, minute; petals absent; disk absent and ovary is 3-celled with solitary ovule in each cell. The capsules are concealed in bracts and consist of 2 valved cocci. Seeds are 3 and subglobose.

There are 450 species found distributed in tropical and subtropical parts of the world. About 10 of them have been reported from India.

Species: *Acalypha indica* Linn.

Description of Plant

It is an annual erect herb upto 30-90 cm in height with numerous, long, ascending and angular branches. Leaves are long petioled, 2.5-7.5 x 2.0-4.5 cm, ovate or rhombic-ovate and crenate serrate. Flowers are borne in numerous, lax, erect, elongated and axillary spikes; male flowers are minute and clustered at the top; while female flowers are scattered in accrescent, broad and leafy bracts. The capsules are concealed by bracts and are small and hispid usually with single seed. The seeds are acute, ovoid and smooth.

Flowering and Fruiting

Almost throughout the year; chiefly during rainy season.

Distribution

It occurs throughout the hotter parts of India.

Plant Parts Used

Whole plant, its leaves and roots are used medicinally.

Uses

The whole plant as emetic and expectorant which is used as a remedy for asthma, bronchitis, pneumonia and rheumatism. The leaves of the plant are considered emetic and laxative and their powder or a decoction, after mixing with garlic is used as anthelmintic, An expressed juice of its leaves is sued as expectorant in smaller doses and also in chronic bronchitis, asthma and consumption and in larger doses (one teaspoonful) as emetic for children. The decoction of the leaves is used in ear ache as instillation and as fomentation round the aching ear. Their fresh juice is used in scabies and other skin diseases and with lime and onion, as astimulating application in rheumatism. In cases of obstinate constipation of children, the balls are prepared from its leaf paste which are introduced into the rectum to produce free motions. Besides, the roots are bruised in hot water and are employed as a cathartic. It is used as substitute for Polygala senega (snake root).

Folk Uses

Irula people in Tamil Nadu, use the decoction of its leaves for ear pain, snake bite and scabies.

Species: *Acalypha recemosa* wall. ex Benth (*A. paniculata* Miq.)

Description of Plant

A small, much-branched shrub. Leaves are numerous, alternate, ovate or oblong-ovate, acuminate, serrate and 3.2 x 7.5 cm with long petiole. Flowers are small and monoecious; males in small clusters in axillary spikes and females are shortly pedicelled on the branches of large, erect, lax, slender and terminal panicles. The capsules are small and glandular hairy.

Flowering and Fruiting

During July-October.

Distribution

It is known to occur wild in South India.

Plant Parts Used

Whole plant is used medicinally.

Uses

This plant is considered emetic and expectorant and is used a good substitute of *Acalypha indica*. Its roots are used as cathartic and leaf decoction or their powder as a laxative and in various skin diseases. Besides, juice of entire plant is used in erysipelas and haemorrhage.

Genus: *Balaiospermum* Blume

The generic name has been derived from Greek words, Balios=spotted and Sperma=seed; referring to the spotted seeds of the plants.

It comprises of shrubs. Leaves are alternate, minute-toothed or lobed with 2 glands at the base. Flowers are small, monoecious or dioecious and are arranged in panicles or racemes. Male flowers: sepals 4-5, membranous, orbicular and concave; petals absent; disk is 4-6 glands and stamens 10-30. Female flowers: sepals 5-6, lanceolate, entire or toothed; petals absent; disk entire and ovary is 3-celled with single ovule in each cell. The fruit is a capsule which consists of 2 valved crustaceous cocci. The seeds are ovoid.

Only 6 species are known to occur in South-east Asia,Malaya Peninsula, Java, Sumatara and India.

Species: *Baliospermum montanum* (Willd.) Muell.-Arg. (B. axillare Blume; *Jatroopha Montana* Willd.)

Description of Plant

A leafy stout shrub about 1-2 m in height. Leaves are sinuate-toothed; lower large, 1.5.0-30.0 cm long, oblong-ovate or rounded or palmately 3-5 loboed; upper small. 5.0-8.0 cm long and lanceolate; petiole with a pair of sitpular glands. Flowers are borne in axillary racemes; all male or with a few female below. Male flowers: calyx globose, 4-5 partite, glabrous or pubescent; disk of 6 glands and stamens 20. Female flowers: sepals ovate-lanceolate, pubescent and ovary densely strigose. The capsule are obovoid, usually hairy and consisted of o2-valved crustaceous cocci. The seeds are oblong, mottled and 0.8 x 0.5 cm.

Flowering and Fruiting

Almost throughout the year; chiefly during March-May.

Distribution

It is found in tropical and subtropical Himalayas from Kashmir to Arunachal Pradesh upto an altitude of 1000 m and southwards to Peninsular India ascending upto 1800 m in Kerala hills.

Plant Parts Used

Its roots, seeds and leaves are used.

Uses

The roots of the plant in ayurveda, have been described as anthelmintic, alexiteric, diuretic and purgative and these are used in skin-diseases, wounds, piles, spleen enlargement, inflammation, anaemia, leucoderma and jaundice. Seeds of this plant are purgative and externally they are used a rubifacient and stimulant. An oil obtained from its seeds, is a powerful hydragogue and is applied externally in rheumatism.

Its seeds are used as a substitute or adulterant for that of Jamal ghota (Croton tiglium).

Preparation

Dantyarista, Dantyadi churna, Dantiharitaki, Dantyadi lep. Abhyariasta, Kaisor guggulu, Laghu vis garva taila and Punarnava mandur.

Genus: *Drypetes* Vaho.

The generic name has been taken from Greek word Drypto = to lacerate or to tea,a referring to the spiny nature of some of the species.

By handsome trees the genus is represented. Leaves are alternate, entire or serrulate, penninerved and reticulate. The flowers are monoecious or dioecious and in axillary pedicelled. The males are clustered and the females are subsolitary. Petals and disk are absent. The fruit is an ovoid or globose drupe. Seeds are solitary and ovoid, with fleshy albumen and broad and flat cotyledons.

Its 200 species are distributed in the tropical parts, South Africa and subtropical East Asia. 2 species have been reported to be found distributed in India, especially in Assam.

Species: *Drypetes roxburghii* (Wall.) Hurusawa (*Putranjiva roxburghii* Wall).

Description of Plant

It is a handsome evergreen tree usually with pendent branches and dark-grey bark and about 12 m high. Leaves are alternate, 5.0-10.0 cm long, simple, glabrous, obliquely ovate or ovate-lanceolate, serrulate and shining. The flowers are dioecious and small. The males are in head like dense axillary clusters oand females impairs or solitary. The drupes are ellipsoid-ovoid, white-tomentose and about 1.2 cm long with pointed apex.

Flowering and Fruiting

Flowering: During April-May and Fruiting: During cold season.

Distribution

It is found wild or cultivated throughout tropical parts of India.

Plant Parts Used

Generally the leaves, seeds and fruits are used.

Uses

A decoction of the leaves, fruits and stones of fruits is given in cold and fever and that of leaves alone is used to wash the infected eyes. Its juice is used in filarial. The fruits are used as anodyne and the seeds as antidysenteric, stimulant and tonic.

Folk Uses

The native women strung up its nuts in rosaries whose male child die immediately after birth in different parts of India.

Genus: *Emblica* Gaertn.

The generic name has been derived from its Sanskrit name Amlika.

It comprises of trees. Leaves are alternate and distichous; the branchlets resemble pinnate leaves and stipules are narrow or absent. The flowers are small, monoecious and are axillary or on the old wood. Capsules are 3-crustaceous or coriaceous 2-valved cocci; berries or drupes with 3-4-celled stone. The seeds are trigonous.

Its only 4 species are known to occur in Madagascar, East Asia and Indo-Malayan region. 2 of them are reported to grow in India.

Species: *Emblica officinalis* Gaertn. (*Phyllanthus emblica* Linn.; *Cicca emblica* Kurz.)

Description of Plant

The plant is a moderate-sized deciduous tree with greenish-grey or red bark which is peeling off in scales and long stripes. Leaves are pinnate, distichously closest, linear-oblong and obtuse. Flowers are greenish-yellow and are borne in axillary clusters along the branchlets; males are on slender pedicels and females are sub-sessile and few. The fruits are of three 2 valved cocci, globose, fleshy and obscurely 6-lobed when ripe. Seeds are trigonous.

Flowering and Fruiting

Flowering: During February-May and Fruiting: In October-March.

Distribution

It is found wild or cultivated throughout tropical parts of India.

Plant Parts Used

Fruits, seeds, flowers, leaves, bark and roots are sued.

Uses

The fruits are antidiarrhoeal, astringent, antidysenteric, antiscorbutic, carminative, cooling, stomachic and tonic and unripe fruits are aperient. These are useful in urinary problems and are prescribed in anaemia, jaundice and dyspepsia in combination with iron. A fermented liquor prepared from its fruits is sued in jaundice, dyspepsia and coughs. The seeds are used as a collyrium in eye complaints and their infusion is given in asthma, bronchitis and fever. The infusion of its seeds is also useful in diabetes. Seeds fried in ghee and ground in '*conjee*' are applied as a *lep* on the forehead to stop nose-bleeding.

Its flowers are used as cooling, refrigerant and aperient. The juice of this leaves is applied externally too ulcers and their infusion mixed with fenugreek seeds is given inchronic dysentery. The bark and roots are used as astringent. The juice of its bark combined with honey and turmeric is used as a remedy for gonorrhoea. Besides, its root bark rubbed with honey is used in aphthous stomatitis.

Folk Uses

A sharbat prepared from its fresh fruits with or without raisins and honey is drunk as a cooling drink and sherbet with lemon juice is taken for bacillary dysentery.

Its fruits and leaves are ground to make a paste which is given orally about 100 gm thrice daily for 5 days in Sun stroke and as a preventice drug locally. The galls are collected from it stem along with its bark and ground, then, these are boiled with the common pulses in 1:2 ratio and taken as a food every day for two months to cure night blindness.

People of Phulbani district in Orissa, make a powder by pounding its dry seeds and take this powder twice a day for a week to cure leucorrhoea.

Preparation

Chavyanprash, Brahma rasayan, Dharti louha, Triphala, Amalkyadi churna,Sudarshan churna, Sarvajawar louha, Agnitundi vati, Sanjivani vati,Phaltrikadi kwath, Punarnava mandur, Navayas louha, Rakta pittantak louha, Pippalayasav, Sringayadi churna, Karanjadi yog and Itriphan ustkhudus.

Genus: *Euphorbia* Linn.

The generic name has been assigned in honour of Euphorbus, a Physician of Juba, King of Mauritania.

It comprises of herbs, shrubs or trees with milky juice, Stems are slender and leafy, thick and fleshy, sometimes leafless. Leaves are alternate with or without stipular prickles, sometimes opposite or whorled especially on flowering branches, Flowers are in heads, resembling single flower, consisting of a cup shaped involucre which encloses many male flowers and one central female flower. Male flowers are with simple pedicelled stamens and female ones with 3-celled pedicelled ovary. Fruit is a capsule, dehiscing ventrally or both ventrally and dorosally. The seeds are albuminous with broad and flat cotyledons.

About 2000 species occur in the world, chiefly in sub-tropical and warm temperate regions. 60 species of them are met in India, mostly in Southern and Western parts and the Hiamalayas.

Species: *Euphorbia hirta* Linn.
(*E. pilulifera* Linn.; *Chamaesvce hirta* (Linn.) Millsp.)

Description of Plant

An annual herb, about 15-20 cm high, erect or ascending, often branched from the base and sometimes with a woody base, hispid with long yellowish crisped hairs. Stem is usually terete. Leaves are opposite, 1.0-4.0 x 0.5-1.5 cm, obliquely oblong-lanceolate, acute, toothed or serrulate. Inflorescence is axillary cyme, peduncled and pubescent. The fruits are depressed-globose and keeled. Seeds are reddish brown, 0.08 cm long and ovoid-trigonous.

Flowering and Fruiting

Almost throughout the year.

Distribution

A common in waste places, road sides and gardens throughout India.

Plant Parts Used

The whole plant is used.

Uses

The whole plant is used for the disease of children in worms, bowel complaints and cough. Its decoction is given in asthma and bronchial infections and the juice in dysentery and colic. The latex of the plant is used as an application to warts and in diseases of urino-genitary tracts. It is also used in postnatal complaints, failure of lactation, breast pain and skin eruptions.

Folk Uses

Santhals, give its roots to allay vomiting and the entire plant to nursing mother when the supply of the milk is deficient or fails. Besides, its about 20 leaves are crushed and their extract is given orally with honey once a day in the morning for leucorrhoea for a month by the people in Orissa.

Genus: *Mallotus* Lour.

The generic name has been from Greek word Mallotos = wooly, referring to the woolly nature of its fruits.

It comprises of trees or shrubs. Leaves are opposite or alternate, entire, toothed or 3-lobed, often gland-dotted beneath and sometimes with glandular areas at the base on the lower surface. The flowers are small, dioceious or monoecious and apetalous and are arranged in axillary simple or branched spikes or racemes. The males are fasicled and females solitary in the bracts. The capsules are with ovoid, oblong or globose seeds.

Its 2 species occur in tropical Africa and Madagascar and 140 species in Eastern and South-eastern Asia, Indo-Malaya to New Caled and Fiji and in Northern and Southern Australia. About 20 species of them have been reported from India.

Species: *Mallotus philippensis* Muell.-Arg.

Description of Plant

A small much-branched evergreen tree with about 5-9 m height. Its branchlets, young leaves and inflorescence are rusty or tawny pubescent. Leaves are alternate, variable in size and shape, ovate or rhomboid, acute or acuminate, glabrous above, pubescent and with many close-set orbicular reddish glands beneath. The flowers are small and dioecious. The male flowers are clustered in racemes while the females are solitary. The capsules are densely covered with crimson powder. Seeds are globose, smooth, 4.0 cm in diameter and black.

Flowering and Fruiting

Flowering: During September-November and Fruiting: In February-May.

Distribution

It is found widely distributed throughout the tropical parts of the country.

Plant Parts Used

Fruits and seed-oil are used.

Uses

The crushed glandular hairs of its fruits knows as *'kamela'* powder are used as anthelmintic, cathartic and styptic. This powder mixed with sweet oil, is employed as ointment to ring worm, pityriasis and freckles. It is used internally by the Arabianas in leprosy and snake-bites. It is also applied over syphilitic ulcers and is used in herpes. It is also effective in weeping eczema and in low doses, is given internally toreduce the fertility in females.

The kamela is chiefly adulterated by the pounded bark of *Casearia tomentosa* and a powder prepared from the red fruits of *Ficus bengalensis*. Its seed oil forms a good substitute of Tung oil, used as hair fixers and in ointments.

Folk Uses

Kamela, the fruit powder of this plant alongwith curd is given as wormicide and with 'whey' in stomach ache by the tribal people in Dehradun area.

Preparation

Krimighatini-vatika, Qurs didan, Itgriphal-i-didan, Roghan quba and Zimad jarb marham kharish jaded.

Genus: *Phyllanthus* Linn.

The generic name has been derived from Greek words Phyllon = a leaf and Anthos = a flower, referring to the appearance of flowers on leaf-like branches.

It comprises of herbs, shrubs or trees. Leaves are alternate and distichous, the branchlets with their leaves resembling pinnate leaves; stipules are narrow or none. The flowers are small, monoecious and axillary or on the old wood. Sepals are 4-6, free or shortly connate, imbricatre and more or less biseriate. Petals are absent. Male flowers: disk glandular, rarely none, stamens 3 or 4-5, pistillode none. Female flowers: disk is glandular, ovary 3-celled, style free or connate, 2 ovules in each cell. The fruit forms a capsule with 3 crustaceous or coriaceous 2-valved cocci or a berry or a drupe with 3-4 celled stone. Seeds are trigonous.

About 600 species occur in tropics and subtropics of the world excluding Europe and North Asia, About 40 species of them have been recorded form India, chiefly from Southern and Western parts and the Eastern Himalayas.

Species: *Phyloanthus urinaria* Linn.

Description of Plant

An annual, erect and decumbent herb with about 10-30 cm height. Leaves are variable in size, distichously imbricate, sessile or nearly so and 0.6-1.0 x 0.2-0.6 cm with minute stipules. Flower are very small, yellowish, solitary and are borne axillary. Sepals are oblong and rounded. The capsules are globose, echinate, scarcely lobed and 0.3 cm in diameter. Seeds are about 0.1 cm long, trigonous and transversely furrowed.

Flowering and Fruiting

During July-December.

Distribution

It is found throughout the hotter parts in India.

Plant Parts Used

Whole plant, its fruits and leaves are used.

Medicinal Uses

In ayurveda, the whole plant and fruits of this plant are known for their acrid, alexipharmic, bitter, cooling and sweetish properties and these are used in thirst, bronchitis, leprosy, anuria, biliousness, asthma and hiccough. Besides the whole plant is also used successfully in dropsy, genitor urinary problems, gonorrhoea and jaundice.

Its leaf juice mixed with coconut milk is given as a good appetizer to children.

Folk Uses

In Chota Nagpur, the roots of the plant are given to sleepless children to induce the sleep.

Genus: *Ricinus* Linn.

The generic name is the Latin name for the Castor oil plant.

This monotypic genus is a native of Africa that got naturalized in tropical countries including India.

Species: *Ricinus communis* Linn.

Description of Plant

A tall glabrous and glaucous annual shrub reaching a height upto 4.0 m. Leaves are palmate, 7-many lobeod; lobes are oblong to linera, acute or acuminate and gland-serratted. The flowers are large and monoecious in terminal subpanicled racemes. Male flowers: calyx membranous, spilitting valvately into 3-5 segments, the stamens are many in a dense globose head of branched filaments and anthers. Female flowers: calyx is spathaceous and caduceus, style spreading and highly coloured and ovary 3-celled. The capsules are of 3-2 valved cocci, globosely oblong and smooth or echinate. Seeds are oblong, smooth and mottled.

Flowering and Fruiting

During the greater parts of the year, chiefly in winter.

Distribution

It is a native of Africa, now runs wild in waste lands throughout India; also cultivated in gardens and the fields.

Plant Parts Used

The seeds, leaves and root-bark are used.

Uses

The seeds are known as purgative and counter irritant and are said to be used in scorpion-stings and as fish poison. An oil, derived from its seeds known as castor oil, in a doses of 5-20 ml is used as purgative A poultice of its seed-paste is applied on sores and gout or rheumatic swellings. The leaves are anodyne and galactogogue and externally, are applied to boils and sores in the form of a poultice. Besides, the root-bark is considered emetic and purgative and is used in lumbago and in skin diseases. The root bark is also effective in reducing the fat of the body.

Folk Uses

Its warm leaves are tied over the effected parts of the body in sprains in different parts of the country. A past of its leaves with luke warm ghee is applied on the swollen joints; also tied on the boil or on the carbuncle in order to burst it by the tgribals in jammu region. Besides, the pulp of one unshelled seed is given to women just before intercourse as contraceptive by aboriginals in various parts of the country.

Preparation

Erandpak, Erandmuladi kwath, Erandsaptak kwath, Rasnadi kwath, Rasna panchak kwath, Vishgarbh taila and Jimade sheeraeshutur.

Family: ULMACEAE

Genus: *Holoptelea* Planch

The generic name has been taken from the Greek words Holos = entire and Ptelea = the elm, but deviation is not known.

It genus comprises of large deciduous tree. Leaves are alternate, penninerved and entire, the stipules are lateral and scarious. The flowers are polygamous or bisexual; male without rudimentary ovary and are arranged in fascicles at the scars of the previous year's shoots which are scaly but leafless. The fruits are dry. Samaroid or flat and indehiscent. Seeds are flat and exalbuminous with longitudinally and complicate cotyledons.

About 2 species are confined only to tropical Africa and Indomalayan region; of which one is known to be found in India.

Species: *Holoptelea integrifoilia* Planch.

Description of Plant

A large deciduous tree with light brown or grey bark. Leaves are alternate, 7.5-12.0 x 5.0-7.5 cm, elliptic or broad-oblong, acuminate, entire, coriaceous and glabrous or pubescent beneath with rounded or acute base; nerves 5-8 paired. The stipules are ovate and membranous. The flowers are 0.4 cm across. Poerianth is pubescent and their segments are 4-5 partile. Fruit is an oval or orbicular samara, about 2.5 cm long; wing reticulately veined and notched at the apex.

Flowering and Fruiting

Flowering: During March-April and Fruiting in May-August.

Distribution

The tree is found along Sub-Himalayan tracts, Ajmer, Bundelkhand, Bihar, Assam and Western Peninsula; also planted along the road-sides.

Plant Parts Used

Its bark and the seeds are used.

Uses

The juice of its boiled bark is applied to rheumatic swellings, and the paste of stem-bark and sees in ringworm disease. Its stem bark alone is also used in scabies. Besides, the seeds soaked in water are applied over swellings.

Folk Uses

The tribals of Varanasi, use its leaf-paste externally for the treatment of skin diseases like ringworm and itches.

Family: CANNABACEAE

Genus: *Cannabis* Linn

The generic name has been taken from its ancient Greek name, Kannabis.

This montypic genus is known to occur wild in central Asia and throughout India.

Species: *Cannabis sativa* Linn.

Description of Plant

A tall annual herb about 1.2-4.9 m high with angular stem. Leaves are palmate, alternate or lower one opposite and 5.0-20.0 cm long; the lobes of the upper leaves are 1-5 and in lower 5-11, linera laneeolate, sharply toothed, long-pointed and narrowed at base. The flowers are small, greenish and unisexual; males are borne in long dropping panicles and females in short axillary spikes. The achenes are 0.3 cm across, ovate and flat and are enclosed in persistent perianth.

Flowering and Fruiting

During March-May.

Distribution

It is a native of Central and Western Asia, now naturalized in the sub-Himalayan tract and abundant in wastelands from Punjab Eastwards to Bengal and Bihar extending Southwards.

Plant Parts Used

Its seeds, leaves, dried flowering or fruiting top of the female plant and resionous exudates, are used for medicinal purposes.

Uses

The plant is considered as a tonic, analgesic and antiseptic and is useful in gonorrhoea, menorrhagia, diarrhea, cholera, hydrophobia, tetanus and rheumatism. Its leaves are used in ear troubles and their juice is applied on cuts after mixing with sugar. The juice of leaves is also used destroys worms and vermin; crushed ones are applied on skin in skin diseases. The chronic use of cannabis is useful inpatients of diabetic hypertension. The leaves are also used internally about 2.5 gm per day to relieve pain and in swelling orchitis.

Narcotic drugs Bhang, Ganja (Marijuana) and Hashish (Charas) are prepared from this plant. Bhang, deep green in colour, consist of its dried leaves and Ganja is prepared from those dried flowering or fruiting top of female plants from which no resin has been removed. *Charas* is the resinous exudates of its leaves which has a powerful smell and dark green or brown in colour. Bhang and Ganja are used as appetizer and nervous stimulant and in bowel complaints. Bhang is made into a paste with water then mixed with milk and sugar and is taken as intoxicating liquor producing cuphoria.Charas is sued for soothing activity in cases of mania and hysteria and also in asthma and tetanus. Like opium, it does not produce loss of appetite or constipation.

Folk Uses

Its female flowering tops are used in treatment of hydrocele and prolapose of outerus locally by the people in North-West Himalayan region. The treatment consist in blanching of coarsely pounded leafy flowering tops of female plants and wrapping the affected testicles with the warm pounded mass at bed time. Though, the swellings of testicles starts reducing within a week or ten days, but the treatment should be continued for 2-3 months. A bed time in case of prolapse of uterus, the leaf material prepared as above is put in soft muslin cloth and kept inside the vagina at bed time. This treatment is very effective in the initial stages of uterus prolapse. Besides, the leaves of this plant are bruised and are smoked as narcotic throughout the different parts of the country.

Preparation

Madnanand modak, Jati-phaladi churna, Lai churna and Majun-phalakser. Besides, In homeopathy, flowering tops of freshly harvested plants are used.

Family: MORACEAE

Genus: *Ficus* Linn

The classical name of the cultivated plants is the generic name.

This genus comprises of tree or shrubs sometimes scandant or epiphytic with milky sap. Leaves are alternate, opposite, entire, toothed or lobed; the stipules are enveloping the buds and re caduceus. The flowers are minute and unisexual on the inner walls of a fleshly receptacles, the mouth of which is closed by imbricate bracts often mixed with bracteoles of four forms male, female, gall and rarely neuter. The fruit is an enlarged hollow cup-shaped and closed receptacle. Seeds are pendulous with membranous testa and curved embryo.

In the warmer parts of the world its 800 species are known to occur in the warmer parts of the world, chiefly in Indomalayan region. From India out of which about 70 species have been reported.

Species: *Ficus bengalensis* Linn.

Description of Plant

A large evergreen tree upto 30 m height and extending laterally by sending sown aerial roots. Aerial roots are many and sometimes develop into accessory trunks and assist the lateral spread of the tree indefinitely. The bark is light grey-white, smooth and about 1.25 cm in thickness. Leaves are coriaceous, 10.0-20.0 x 5.0-2.0 cm, ovateelliptic with subcordate or rounded base. Male flowers are many near the mouth of the receptacle: sepals 4; gall flowers: perianth as in male and fertile flowers: perianth shorter than in male, style elonagate. The male and female flowers are in same receptacle. In pairs the fruits are sessile in pairs, puberulous, sub-globose, 1.3-2.0 cm across and red when ripe.

Flowering and Fruiting

Flowering: During summer Fruiting: During rainy season.

Distribution

Throughout the plains it occurs and forest tracts of India. In avenues for shades also planted.

Plant Parts Used

Generally, It seeds, latex, bark and the tender ends of its aerial roots are used.

Uses

As cooling and tonic its seeds are used. Its leaves as poultice are applied to abscesses and tender leaves pasted with honey are considered beneficial in '*raktapitta*'. The milky juice obtained from its branches is applied externally for pains, in rheumatism and lumabago, sores and ulcers and soles of the feet when cracked or inflamed. It is also used in toothache and mixing with the sesame oil is applied on burns and is genital diseases. Its root fibres are used in gonnrrhoea and tender ends of hanging roots as antiemeric. A paste of the roots is applied to scalp to grow hair long and for menorrhagia. An infusion of the bark is considered to be a good tonic, effective in diabetes, dysentery, gonorrhoea and in seminal weakness. The bark with black pepper is used in snakebites. An aqueous extract of its bark and that of leaves is used as a depressant on uterine and cardiac muscle and also on cholinergic-blocking of smooth and skeletal muscle. A concentrated juice in combination with its fruit is taken as aphrodisiac and also is of great value in spermatorrhoea and gonorrhoea.

Its bark along with the bark of *F. virens, F. religiosa, F. racemosa* and *Azadirachta indica*, pass by the name Panchaval kala, is used for gargle in salivation, as a wash for ulcers and as a injection in leucorrhoea.

Folk Uses

A small quantity of its milky juice dropped in 'batasha' is taken daily to as aphrodisiac and alone also in dysentery throughout the country. The milky juice of the branches is also applied on cracked heels. The tips of its adventitious roots are crushed and boiled in cow's milk and the decoction thus obtained, is strained and served hot in piles. A powder of its fruits in shade is prepared, which is taken with honey in the morning and evening for a week for spermatorrhoea.

Preparation

Nyagrodhadi churna, Nyagrodhadi ghrita, Nyagrodhadi kwath and Majun bargad.

Species: *Ficus racemosa* Linn. (*F. gloomerata* Roxb.)

Description of Plant

A medium–sized or large tree with spreading branches and about 9.0-12.0 m height. The bark is smooth and reddish brown. Leaves are alternate, 10.0-17.5 cm long, obovate-oblong or oblong- lanceolate, glabrous above when full grown and 3-nerved. Male sepals are inflated. Gall and female perianth is toothed. Fruits are borne in large clusters on short leafless branches derived from the trunk and the main branches and are subglobose, 2.5-5.0 across and when ripe.

Flowering and Fruiting

Flowering: During spring season and Fruiting: In April–July.

Distribution

Throughout India, it occurs along the sides of ravines and bank of streams; also occurs on rocky slopes, sometimes gregariously. It is also grown along road sides and in villages.

Plant Parts Used

The leaves, fruits, latex, bark and roots are used.

Uses

The leaves in powdered form are mixed are mixed with honey and are given in bilious affections and their decoction in bronchitis. The fruits are astringent, stomachic, carminative and are used in menorrhoea and haemoptysis. Its roots and fruits are considered to have hypoglycaemic activity against diabetes. The juice of its roots is used in dysentery and that of fruits is given as a vehicle to medicine in diabetes and urinary complaints (Chatterjee and Pakrashi, 1991). The bark is astringent and is given to cattle when suffering from rinder–pest. Its milky juice is administrated in piles and diarrhea and with sesame oil in cancer (Watt, 1972). Besides, the various parts of the plant are used in small pox, muscular pain, adenitis, scabies, spermatorrhoea, orchitis, epididymitis, hydrocele and failure of lactaion (Asolkar *et al.*, 1992).

Its bark along with that of *Ficus virens, F. religiosa, F. bengalensis* and *Azadirachta indica,* pass by the name Panchaval kala, is used for gargle in salivation, as a wash for ulcers and as a injection in leucorrhoea.

Folk Uses

The small blister-like 'galls' common on the leaves, are soaked in milk and mixed with honey are given to prevent pitting in small pox. Its dried fruits are pounded and their powder in the doses of 10-20 gm is given for body strength in Surguja district of Madhya Pradesh. Besides, the people in North east Karnataka, use leaf extract of this plant with that of Cuminium, mixed with water in empty stomach in stomach complaints.

Preparation

Udumbar sar.

Species: *Ficus religiosa* Linn.

Description of Plant

A large or medium-sized deciduous tree with spreading branches and about 10-12 m high. Bark is grey with brownish specks, smooth and 1.3 cm thick. Trunks are irregularly shaped. Leaves are alternate, 10.0-17.5 x 7.5-12.5 cm, broadly ovate or rotund, caudate and somewhat pendulous, upper surface glaucous, 5-7 veined and long petioled. Male flowers are very few and sessile; sepals 3, stamen I and filament short. Female and gall flowers; sepals 5, style short and lateral. Fruits are sessile in axillary pairs and are depressed-globose, 1.3 cm across and dark purple when ripe.

Flowering and Fruiting

Flowering: During summer and Fruiting: In rainy season.

Distribution

It is found wild or cultivated throughout India; also grown as avenue tree, or along road-sides.

Plant Parts Used

The fruits, seeds, leaves and its bark are often used.

Uses

The bark is known as astringent and is used in gongrrhoea; pulverized one is applied externally on unhealthy ulcer and wounds to promote granulation and found very efficacious when rubbed with honey to apthous sores of children. The dried bark in powdered form is used in anal fistula and in the form of a paste, as an absorbent in inflammatory swellings. Its tender and fresh leaves are beneficial when used along with butter fat to cover the inflammatory ulcers. An oil medicated with its leaves is used as eardrops in earache. The leaves and its twigs are also used as laxatives. The fruits are used as mild laxative and digestive. The seeds are laxative and in powder form, are given for three days during menses and as to sterilize women if given for long time. Besides, various parts of the plant are used in otitis media, suppurativa, mouth sores, atrophy, emaciation or cachexy, rheumatism, small pox, carbuncle, mucus in urine, spermatorrhoea, gravel, cholera and rinderpest.

Its bark along with the barks of *F. virens*, *F.* racemosa, *F. bengalensis* and *Azadirachta indica*, pass by the name Panchaval kala, is used in combination for garagle in salivation, as a wash for ulcers and as a injection in leucorrhoea. Water, in which its freshly burnt bark has been steeped is said to cure the case of obstinate hiccup. Besides, the powder of the dried bark is also used in anal fistula. For the purpose, hakim introduced a metallic tube something like a blow- pipe into the fistula and putting a small quantity of the powder into it, blew the same into the fistula.

Folk Uses

Native people of India, take about 5.0 gm of its leaf–ash with water to remove urinary calculi. Bhil tribals in Madhya Pradesh, insert a leaf periole in both the ears to suck the poison from body. About twenty pairs of leaves are used alternatively to take out the poison from the body by the people.

Preparation

Hartal bhasm and Sharbat peepar.

Species: *Ficus virens* Ait.
(*F. lacor* Buch.-Ham.; *F. infectoria* sensu Roxb.)

Description of Plant

It is a large and spreading tree about 12-5 m high with greenish-grey and smooth bark. Sometimes, sending down a few aerial roots. Leaves arealternate, 7.5-12.0 cm long, oblong-ovate or ovate and shortly acuminate with subundulate or entire margins; petioles are 3.5–5.0 cm long and channelled. The male flowers are few and sessile: sepals 4, stamen 1 and anther broad ovate. Female and gall flowers: sepals and stigma elongate. Fruits are usually sessile in axillary pairs and are sub-globose and white when ripe.

Flowering and Fruiting

Flowering: In rainy season and Fruiting: During February-May.

Distribution

In Northern India, Madhya Pradesh and Peninsular India. It is also frequently planted as an avenue and ornamental tree.

Plant Parts Used

The bark, leaves and fruits are used.

Uses

Its bark along with the barks of *F. bengalensis, F. racemosa. F. religiosa* and *Azadirachta indica,* pass by the name, *Panchava*l kala, which is used for gargle in salivation, as a wash for ulcers and as a injection in leucorrhoea. Its pulverized bark mixed with honey os applied externally in the diseases of female generative organs. The leaves of the plant are considered to be useful diet for those who suffer from *'raktapitta'* disease. Its fruits are used in bronchitis and scabies and their juice as cardiotonic and to increase the appetite.

Folk Uses

The bark dust is mixed with coconut oil and smeared on skin affections like dermatitis, ringworm and ulcers in different parts of the country.

Genus: *Streblus* Lour.

The generic name has been derived from Greek word Sterblos = crooked, indicating the straggling twiggy branches.

The genus comprises of shrubs or trees. Leaves are alternate, penninerved and scrabid; the stipules are small and lanceolate. The flowers are axillary and usually dioecious; the males in peduncled spikes and females are solitary or 2-4 together and stalked. The fruits are membranous, straight, subglobose and laxly covered by the persistent perianth. Seeds are globose.

Its 22 species occur in Madagascar, South 'East Asia and Indo-Malayan region. Only 4 species have been reported from India.

Species: *Strevlus asper* Lour.

Description of Plant

It is an evergreen green tree with light grey bark. Leaves are alternate; elliptic-rhomboid or obovate and acute or acuminate with more or less toothed margins towards the apex and rough on both the surfaces. The flowers are usually dioecious; males in short peduncled heads or spikes and solitary or few together in the axils of the leaves. The berries are globose, about 1.2 cm long and yellow when ripe.

Flowering and Fruiting

Flowering: During January-February and Fruiting: In May-July.

Distribution

It is found in the drier parts of India from Rohilkhand, Eastwards and Southwards to Travancore and in Andamanns islands.

Plant Parts Used

Bark and roots are usable.

Uses

The decoction of its bark is given in fever, dysentery and diarrhoea. Besides, the roots are used as an application to unhealthy ulcers and sinuses. Its roots are ground in water and as an antidote are given orally to snake-bitten person.

For tea its leaves are used as a substitute.

Folk Uses

About 100 gm of stem-bark is ground and boiled in water in 1:4 ratio till the contents reduced to 25 gm and left over night. The remainder is given once a day for two months to cure filarial in Ranchi. However, drug should not be taken on empty stomach.

Family: MYRICACEAE

Genus: *Myrica* Linn.

On its Greek name the generic name is based, Myrike for tamarisk plant. In Greek, Myrio means to flow and name is in allusion to flowing of wax from the berries.

This genus comprises of aromatic and glandular shrubs. Leaves are alternate. Flowers are unisexual and borne in cylindric bracteate catkin like spikes. Male spike is sometime fascicled or panicled and female always solitary, occasionally a few flowers at the top of male spikes. The drupes are small, succulent and ovoid or globose.

Thirty five species occur cosmopolitically throughout the world except in North Africa, Central and South east Europe, Southwest Asia and Australia. Only 1 species has been recorded from India.

Species: *Mryica esculenta* Buch-Ham.ex D. Don. (*M. nagi Hook. F.*)

Description of Plant

It is small of medium sized, evergreen tree with rough and brownish-grey bark reaching a height of 3-5 m. Leaves are oblong-obovate, lanceolate, serrate, crowded at the end of branches and 7.5-15.0 x 2.5-5.0 cm. Flowers are unisexual in axillary spikes are 0.7 cm long and arranged recemosely on a common axillary stalk and female spikes are erect, axillary and 1.3-2.5 cm long. The drupes are ellipsoid or ovoid, reddish or cheese coloured when ripe and composed of spindle shaped fleshy fibres radiating from rugose nut.

Flowering and Fruiting

Flowering: During October–November and Fruiting later.

Distribution

In subtropical Himalayas this is found from Ravi eastwards to Assam as will as in Lushar, Khasi, Jaintia and Naga hills berween 900-2100 m.

Plant Parts Used

The bark, Seeds and fruits of the plant are used medicinally .

Uses

In ayurveda, the bark is described as acrid, bitter and pungent and is used to cure 'kapha' and 'vata' diseases' asthma' fever, piles bronchitis, anaemia, dysentery and ulcers and in Unani system of medicine, the bark is known to posses astringent, bitter, carminative and tonic properties and is used for liver complaints, headache, inflammations, piles, gleet and in chronic bronchitis; also as uterine inflammations, piles, gleet and in chronic bronchitis; also as uterine stimulant. It is chewed to get relief in toothache and a lotion prepared from it, is used to wash the sores. An oil is also prepared form its bark, which is dropped into ear in earache. Besides, pessaries made of its bark are used to promote the menses., stated that a paste of its seeds with stimulant balsams is mixed with ginger and is used externallyas a rubefacient application to fore arms, calves and extremities during collapse stages of cholera and with asafetida and camphor is applied over piles successfjlly. Besides, fruits yield a wax which is externally used for healing ulcers.

Preparation

Katphaladi kwath, Kathphaladi churana, Devdaradi Kwath and Dulm-el-kanduh.

MONOCOTYLEDONS

Family: ORCHIDACEAE

Genus: *Eulophia* R. Br. ex. Lindl.

The generic name has been derived from Greek words, Eu = good and Lophos = crest; in allusion to the beautiful labellumn bearing ridges.

This genus comprises of terrestrial herbs with leafy tubers or rhizomes. Leaves are appeared with or after the flowers and are long, narrow and plicate. Flowers are borne in racemes or rarely in panicles on a large, erect and sheathed lateral scape.Sepals are free, spreading and subequal and petals subsimilar. Lip is adnate to columon base of its foot; side lobes erect and embracing the columon and midlobe spreading or recurved. Anthers are terminal and pollinia 2, globose and are attached by a caudicle to flat gland of rostellum.

Its 200 species are known to occur in Pantropics of the world. Of which 26 species have been recorded from India; chiefly from Himalayas and Western ghats.

Species: *Eulophia dabia* (D.Don) Hochr. (*E. campestris* Lindl.; *Bletia dabia* D. Don.)

Description of Plant

It is a terrestrial herb with fleshy, irregularly oblong and lobed tubers. Leaves are 2, arise from apex of a slender sheathing pseudostem and develop long after the plant has flowered. Scape is 15-30cm long and sheathed at intervals by loose membranous bracts. Flowers are drooping, subsecund and yellowish or green with pink or purple markings. Sepals are slightly attached to base of lip and are linear- lanceolate and acute or acuminate. Petals are spreading, oblanceolate and narrower than sepals. Lip is as long as the sepals without a foot; pollinia broad. The capsules are ellipsoid and about 2.0 cm long.

Flowering and Fruiting

Flowering: During March-May and Fruiting: Later in July- August.

Distribution

It is found throughout the greater parts or India; mostly in plains.

Uses

Its tubers are considered aphrodisiac, astringent, nutritive and tonic and are used in stomatitis, purulent cough and heart problems. These are also used in scrofulous diseases of the neck both externally as well as internally and also given for intestinal worms as anthelmintic.

The tubers of the plant are used as a substitute for Salep (*Orchis latifolia*).

Preparation

Sometimes, Vrihat varunadi kwath is prepared for use in piles and scanty urine but generally it is sold raw in the marker under trade name 'Paushtik churna'.

Genus: *Orchis* Linn.

The generic name is taken from Greek word, *Orchis* meaning testicles; in allusion to the appearance of tuberous roots.

The genus comprise of terrestrial, erect leafy herbs with enter, oblong or palmately lobed tubers. Leaves are sheathing. Flowers are borne in racemes or spikes. Sepals are free and subequal; lateral spreading or conniving in a hood with petals. Lip is shortly adnate to column, spreading or pendulous and entire ro 3–lobed. Column is short; anther adnate to the face or coloumn and pollinia 2.

Its 35 species are found distributed in Madeira, temperate Eurasia to India and Soutwest China. Only 6 of them have been recorded from India.

Species: *Orchis latifolia* Linn.

Description of Plant

It is a terrestrial, erect and leafy herb with palmate tubers. Leaves are many, erect, oblong, linear-oblong or lanceolate with flat or concave tip. Flowers are borne in dense–Flowered, cylindric and 2.5-15.0cm long spikes. Sepals and petals are acute or obtuse; lateral sepal ovate and reflexed. Lip is oblong or rhomboid, crenate, entire or obtusely 3-lobed. Column is very short. Anthers are adnate to coloumn face; pollinia 2.

Flowering and Fruiting

During June–August.

Distribution

In Western temperate Himalayas from Kashmir to Sikkim upto 3657 m., it occurs.

Plant Parts Used

Its tubers are used.

Uses

Salep has long been esteemed in India as a restorative and invigorator and as a tonic and aphrodisiac in the diseases characterized by weakness or loss of sexual powers. It is also mentioned in Sushrut samhita, and is placed with the group of drugs which pacify '*kapha*' and has been used in erysipelas etc.. It is stated that its tubers ate considered as astringent, expectorant and nutrient is most important constituent of tubers, is mucilage or starch which is obtained after boiling with water or milk to yield a thick jelly which is highly nutritious and forms a best article of diet for weak or connalescent persons and in cases of chronic diarrhoea and dysentery.

Genus: *Vanda* W. Jones ex R. Br.

For *Vanda roxburghii* plant, the generic name is the Sanskrit name.

This genus comprises of ephiphytic herbs with leafy stems. Leaves are thickly coriaceous or fleshy, flat and keeled or terete. Flowers are large and showy, axillary and are borne in simple lax or dense racemes or sometimes solitary. Sepals are spreading or connivent and narrow at the base. Lip is large, usually saccate or spurred at base; disk usually ridged or lamellate or carunculoate. Columon is short and stout with or without a shot foot. Pollinia are 2, globose, ovoid or obovoid and caudicle short and broad or long.

Its 60 species are found distributed in China. Indomalayan region and in Marianna island. Of which only 10 species have been reported from India.

Species: *Vanda tessellatum* (Roxb.) G. Don. (*V. roxburghii* R. Br.; *Epidendrum tessellatum* Roxb.)

Description of Plant

An epiphytic herb with leafy stem and thickly coriaceous, 15.0-20.0 x 1.5-2.0 cm, recurved, narrow and keeled. Flowers are greenish-yellow in 6–10–flowered racemes. Sepals are yellow and tessellated with brown lines and white margins; lateral sepal 2.5 x 1.5 cm, obovate with subcuneate bases and wavy margins. Petals are yellow with brown lines and shorter than sepals. Lip is 0.1cm long and bluish dotted. Column is very short; pollinia subglobose or ellipsoid. The capsules are clavateo-oblong with acute ribs and about 7.5-9.0cm long.

Flowering and Fruiting

Flowering: During May-June and Fruiting: Later.

Distribution

It is found wild in Assam, Bengal, Bihar, Kerala, Madhya Pradesh, Chennai and in Maharashtra.

Plant Parts Used

Roots and leaves of the plant are used medicinally.

Uses

In ayurveda, the roots of this plant are well known for their alexiteric, antipyretic and bitter properties and are used in bronchitis, dyspepsia, rheumatic pains and hiccough. A paste of its leaves is applied successfully over the body during fever and their juice is introduced into the *aural meatus* as a remedy for *otittis media*.

Folk Uses

In Bihar, the whole plant is boiled in 'Karanja oil' in 1:2 ratio and is filtered. It is massaged gently on the back of patient suffering from backache problem twice a day for 5 days.

Preparation

Rasnasaptak kwath, Rasna guggulu, Rasnadasmul kwath, Rasnapanchak kwath, Mahamash taila, Maharasnadi kwath and Kukuvadi churna.

Family: DIOSCOREACEAE

Genus: *Dioscorea* Linn.

In the honour of Dioscorides, the generic name has been assigned a native of Anazauba in Cilicia in the age of Nero.

The genus comprises of perennial herbs with slender twining stems often bearing bulbils the axis of its leaves. The root-tubers are solitary and large. Leaves are alternate or sometimes opposite, entire or lobed and digitately 3-9–foliate. The flowers are small, bracteate, unisexual and usually dioecious. Fruit is a three winged, loculicidal capsule. The seeds are compressed and often winged.

Its 600 species are distributed in tropical and subtropical parts of the world. Out of which 50 species have been reported from India.

Species: *Dioscorea bulbifera* Linn. (*D. sativa* auct. Non Linn.)

Description of Plant

It is twining herb with glabrous stem. Leaves are alternate, ovate-triangular to suborbicular with a deeply cordate base, basal lobes rounded, simple, glabrous and 10-20 x 8-15cm. Male flowers are in raceme-like, pendent, axillary, solitary or fascicled, simple or paniculate spikes of 5.0-8.0 cm length. Female flowers are in pendulous fascicled spikes of 10-20 cm length. The capsules are oblong, winged and 2.0-2.5x 1.0-1.5 cm. Seeds are winged at the base.

Flowering and Fruiting

Flowering: During July–September and Fruiting in December–March.

Distribution

Throughout India, it is common ascending upto 1828 m in the Himalayas.

Plant Parts Used

Generally, its tubers are used.

Uses

The tubers are used in piles, dysentery and syphilis and are also applied to ulcers after being dried and powderd. Besides, ethanol (50 per cent) extract of its aerial parts is also used as diuretic.

Preparation

Chavyanprash.

Family: ZINGIBERACEAE

Genus: *Alpinia* Roxb.

In the honour of Prospero Alpini (1553-1617), the generic mane has been assigned an Italian Botanists, who made collection of the plant in tropical Africa.

The genus comprises of herbs with leafy stems and horizontal rootstocks. Leaves are oblong or lanceolate. Flowers are borne in terminal racemes or panicles. Calyx; is short, tubular and 3-toothed and corolla- tube is perfect; anther lobes divided by broad connective. Lip is spreading with recurved margins, sometimes with 2 subulate processes at base. Ovary is 3-celled. The fruits are globose, dry or fleshy and usually indehiscent. Seeds are globose or angled.

Its 250 species occur in the warmer parts of Asia and Polynesia. About 5 species of them have been recorded from India.

Species: *Alpinia galangal* Swartz

Description of Plant

It is an aromatic and perennial herb with tuberous rootstocks. Leaves are oblong-lanceolate, acute, glabrous, green above and paler beneath and 23.0-45.0 x 3.5-11.0 cm. Flowers are greenish–white and are borne in 15-30 cm long dens–flowered racemes. Calyx is tubular and 3-toothed and corolla is 3.2 cm long; lobes obtuse, oblong and subequal. Stamen is 2.0 cm long. The fruits are orange-red coloured.

Flowering and Fruiting

Flowering: During May-June and Fruiting: Later.

Distribution

Throughout India it is found chiefly in Eastern Himalayas and Southwest India.

Plant Parts Used

Tuberous rootstocks are used medicinally.

Uses

The tuberous or rhizomes of plant in ayurveda, are known for their bitter, heating, pungent and stomachic properties and are used to increase appetite, taste and voice and in '*vata*' disease, bronchitis and heart problem while in Unani system of medicine, these are considered aphrodisiac, carminative, diuretic and expectorant and are used successfully in headache, sore throat, diabetes and kidney problems. Vaidyas prescribe this drug to be taken in combination with liquorice root, long pepper and tail pepper in bronchitis and as a stomachic tonic. A paste of whole plant in honey is given in whooping coughs to children and also to diminish the quantity of urine in urinary disease. Besides, this is also used to destroy bed smell of mouth and to improve voice in throat affections.

Folk Uses

In Manipur and North-eastern parts of India, its bulbous roots and flowers are eaten as aromatic for various purposes.

Genus: *Headychium* J. Koenig

The generic name has been derived from Greek words, Hedys= sweet and Chion = show, referring to the sweet scented, snow-white flowers of *H. coronarium*, which is commonly known as Ginger–lily.

The genus comprises of tall leafy, perennial herbs with tuberous root stocks. Leaves are distichous and oblong and subcoriaceous. Calyx is tubular and 3-toothed and equal. Perfect stamen is only one. Lip is large and 2-fid. Ovary is 3-celled with numerous ovules; stigma projects just beyond the anther. The capsules are globose and 3-valved. Seeds are numerous and small.

Its 50 species are known to be distributed in Madagascar, Indomalayan region and Southwest China. Of which 25 have been reported from India; chiefly from Himalayas.

Species: *Hedychium spicatum* Buch.-ham. Ex J.F. Smith

Description of Plant

It is a perennial herb. Leaves are oblong or oblong- lanceolate and 30.0 cm long spikes. Corolla–tube is 5.0-6.0 cm long; segments linear. Lip is cuneate, deeply 2- fid and nearly sessile with rounded lobes. Filament is pale rid; anthers linear. The capsules are glabrous and globose.

Flowering and Fruiting

During August-October.

Distribution

It meets wild in subtropical Himalayas and Kumaon between 1524-2133 m.

Plant Parts Used

Its tuberous rootstocks are used.

Uses

The roots of this plant in ayurveda, have been described to possess acrid, astringent, bitter and heating properties and in Unani system, as carinative, expectorant, emmenagogue and laxative. A decoction or powder of its roots in the doses of 30-60 gm is given as bitter, carminative, stimulant and stomachic; also in dyspepsia.

Folk Uses

Hilly tribal people in Uttar Pradesh, use its roots in asthma, bronchitis, blood purification, gastric problems and as antiemetic.

Preparation

Satyadi churan, Satyadi varga, Satyadi kwath, Sudarshan churna and Chandraprabha vati.

Family: HYPOXIDACEAE

Genus: *Curculigo* Gaertn.

The generic name is based on Latin word curculio = a weevil; referring to beaked ovary of the plants.

The genus comprises of herbs with a tuberous rootstock or tunicate corn. Leaves are lanceolate and plicate or linear and flat. Flowers are spicate racemose or subcapitate; lowers usually bisexual and upper male. Perianth is 6-partite. Stamens are 6 and adnate to perianth lobe base; filaments short and anthers linear and erect. Ovary is inferior and 3-celled with 2 or more ovules in each cell. Seeds are subglobose and often beaked.

Its 10 species (6 according to Santapau and Henry, 1973) are known to occur in the tropics of world. Of which 3 species have been recorded from India.

Species: *Curculigo orchioides* Gaertn.

Description of Plant

It is a perennial herb with stout, short or elongated rootstocks. Leaves are basal, sessile or narrowed into a short petiole, 15.0-45.0x 1.2-2.5 cm and glabrous or thinly hairy. Scapes are short, clauate and hairy. Flowers are bright yellow and sessile. Perianth tube is produce above the ovary and filiform and hairy; segments 6, lanceolate-oblong, acute and ciliate at the top. Stamens are small; anthers linear. Ovary is lanceolante; style stout and stigma 3, 0.6-0.8 cm long. The capsules are 1.3 cm long and 1-4–seeded with a slender beak.

Flowering and Fruiting

During July–December

Distribution

In subtropical Himalayas, Uttar Pradesh, Bihar, Assam and Bengal.

Plant Parts Used

Its roots and leaves are used.

Uses

The roots of plant in ayurveda, are known for their aphrodisiac, appetizer, bitter, fattening, heating and sweet properties and are used in piles, *'vata'* diseases, biliousness, fatigue and blood–diseases while in Unani system of medicine, these are considered aphrodisiac, antipyretic, carminative and tonic and are used in bronchitis, opthalmia, diarrhea, gonorrhea and pain of the joints. Its roots usually combined with bitters and aromatics in the form of electuary, are prescribed in dysuria, gonorrhoea, menorhagia, leucorrhoea and menstrual disorders in the doses of one teaspoonful twice a day; also given with warm milk and sugar for the same purpose. Besides, its leaves are used in whitlows.

Folk Uses

Local people in Jharkhand take its 1-2 rootstocks and ground with water and give once daily for one week to cure dysentery with blood. In Koraput (Orissa), Bondo tribals apply its root paste with salt on boils.

Preparation

It is one of the constituents of various Majuna and *Paks* and aphrodisiac preparations; important one is Musalyadi churana.

Family: LILIAECEAE

Genus: *Asparagus* Linn.

The generic name has been taken from the Greek word Sparasso meaning a sprout or shoot, referring to the edible shoots of the plant.

This genus comprises of erect or straggling herbs or shrubs with a stout creeping root stock. Leaves ate reduce to scales or spines with a tuft or green needle-like or flattened rudimentary branchlets. Flowers are usually bisexual, small or minute, axillary, solitary fascicled or in racemes and pendulous. Perianth is 6- partite. Stamens are 3. Fruit is a globose berry. Seeds are few or solitary by abortion.

About 300 species occur in the old world, mostly in dry places. 20 species are found in India, chiefly in tropical and subtropical Himalayas and Southern and Western India.

Species: *Asparagus adscendens* Roxb.

Description of Plant

It is a suberect, prickly shrub with white tuberous roots. Stems are tall, stout, stout suberect, terete, smooth, white and much-branched; branchlets are ascending, white, grooved and angled and spines stout, straight and 1.2-2.0cm long. Flowers are white, 0.6-0.7 cm across and are borne in 3.0-5.0 cm long racemes. Pedicels are 1-2nate, jointed above the base and 0.4 cm long. Tepals are nearly equal and 0.2 cm long; anthers reddish brown. Style is 0.03 cm long and ovules many. The berries are ovoid, 0.6-0.7 cm across and 1-seeded.

Flowering and Fruiting

Flowering: During September–October and Fruiting: In November–December.

Distribution

It is found in Western Himalayas and Punjab to Kumaon upto 1615 m.

Plant Parts Used

The roots of the plant are used medicinally.

Uses

Its roots are considered demulcent, galactagogue and tonic and these are used in diarrhoea, dysentery and general debility. These are boiled in milk and added with sugar, also in general weakness.

Its roots are used as a substitute for Salep.

Preparation

Muslipak and Musalyadi yog.

Species: *Asparagus racemosus* Willd.

Description of Plant

A much branched perennial herb with a tuberous rootstock. Stems are triquetrous with 0.5-1.2 cm long patent or recurved spines. Cladodes are 2-3, which are arranged in a tuft and falcate. Racemes are 2.5-5.0 cm long and pedicels are jointed at the middle and slender below the joint. The flowers are white. Perianth is 0.3-0.5 cm long. Style is short. Fruits are red to or reddish violet and 1-3 seeded berries. Seeds are ellipsoid–globose.

Flowering and Fruiting

Flowering: In October-November and Fruiting: During cold season.

Distribution

In tropical and subtropical India the plant is found and upto 1219 m in the Himalayas from Kashmir eastwards.

Plant Parts Used

Roots and leaves are generally used.

Uses

Its roots are refrigerant, demulcent, diuretic, aphrodisiac, antiseptic, alterative, antidiarrhoeal, antidysenteric and galactagogue and are used in fever, rheumatism and as sexual tonic. A decoction of its roots is given for fever and their extract as antifungal. Its boiled leaves are smeared with butter–fat and are applied on boils, small-pox and to prevent confluence.

Folk Uses

About 8 teaspoonful infusion of its root is taken during bedtime as anthelmintic by the people throughout the greater part of the country.

Preparation

Shatavari ghirit, Phalghiri, Shatmulyadi louh, Shatavari panak, Narayan taila, Dhatupaushtik ras, Phalkalyan ghirit, Saraswatrisht, Geriforte, Hala-i-supari pak, Halwa-i-ghaikwar, Majun zanjibil, Majun shir bargadh wali and Majun musli pak.

Genus: *Colchicum* Linn.

On Greek name, the generic name is based Kolchis, an ancient place in Turkey near Caucasus mountains. According to Greek mythology, it was the home Maeda, the sorceress and this plant grew out of a drop of a brew which sorceress Maeda was concocting for witchcarft.

The genus comprise of corm coated herbs. Leaves are radical, linear of lanceolate. Scape is very short, sessile amongst the leaf-sheath and 1-3 flowered. The flowers are large and erect. Perianth is funnel shaped; tube very long and slender and lobes 6, subequal and suberect. Stamens are 6 which are inserted at the bases of segments; anthers dorsifixed. Ovary is sessile and 3-celled with many ovules in each cell. The capsules are septicidal. Seeds are subglobose.

Its 65 species are known to occur in Europe, Mediterranean to Central Asia and North India. Only 2 species have been recorded from India.

Species: *Colchicum luteum* Baker

Description of Plant

It is corm bearing herb. Leaves are few, lorate, linear- oblong or oblaneolate, obtuse and appearing with the flowers; short at the time of flowering and about 15.0-30.0 x 8.0-12.0 cm at the fruiting. Flowers are yellow, 1-2 and 2.5-4.0 cm across when expand. Perianth is golden yellow; segments oblong or oblanceolate, obtuse and many- nerved. Stamens are shorter than perianth; filaments very much shorter and style filiform and longer than perianth. The capsules are about 2.5-3.5 cm long with recurved beaks. Seeds are subglobose.

Flowering and Fruiting

Flowering: During August-September and Fruiting: In October-November.

Distribution

In Western temperate Himalayas from Kangra and Kashmir to Chamba between 914-2133 m., it is found.

Plant Parts Used

The corms of the plant are used.

Uses

There are two varieties of Colchicum, commonly sold in the marker; sweet and bitter. The sweet variety is *'surinjan-i-sbrin'* and bitter variety is *'surinjan-i-talkh'* which is distinguishable due to its bitter taste, smaller size and dark colour.

Its corms considered as aphrodisiac, alterative, aperient, carminative and laxative and these are used in gout, rheumatism and diseases of liver and spleen. Besides, these are also applied externally in the form of a paste on lesson, inflammation, in pains and to the heal wounds.

Its corms are used as a good substitute for Meadow saffron (*Colchicum autumnale*).

Genus: *Drimia* Jacq. Ex Willd.

By herbs with tunicate bulbs the genus is represented. Leaves are radical and linear or lorate. The flowers are racemose or on a long leafless scape and often appearing before the leaves. The fruits are oblong and tri-querrous loculicidal capsules. Seeds are many in each cell, compressed and marginally winged and white fleshy albumen.

About 45 species are found in tropical parts and South Africa. 6 species of them have been reported from India.

Species: *Drimia indica* (Roxb.) Jessop (*Urginea indica* (Roxb.) Kunth; *U. coromandeliana* (Roxb.) Hook. *F.; U. wightiana* Hook. *F.; U.gobindappae* Boraiah and Fatima; *Scilla indica* Roxb.)

Description of Plant

It is a herb with 5.0–10.0cm ovoid and tunicate bulbs.Leaves are appearing after flower, 15-45 cm long, linear, acute and nearly flat. The scape is erect, 30-45 cm long and brittle. The flowers are distant, drooping or spreading and greenish–white; bracts are minute and soon falling. Perianth is campanulate and their segments are oblong–lanceolate and obtuse. Stamens are 0.6 cm long and filaments. The capsules are oblong, triquetrous and loculicidal and about 1.2–1.8 cm long. Seeds are black and flattened.

Flowering and Fruiting

Flowering: During March–May and Fruiting: In July-August.

Distribution

Throughout the plains of India it is found abundantly.

Plant Parts Used

Its bulbs are used.

Uses

Its bulbs are cardiac stimulant and diuretic and in form of a syrup, are useful as an expectorant. However, in large doses, these are emetic and cathartic and may cause cardiac depression. Alcoholic extract of the bulbs possesses anticancereous activity against human epidermoid carcinoma of the nasopharynx and the extract also employed as deobstruent, in dropsy, rheumatism, skin troubles and in burning sensation of soles. Its bulbs are also useful in asthma, cough, bronchitis and paralytic affections.

For Digitalis the bulbs are used as a substitute.

Folk Uses

The extract of its bulbs is applied in burning sensation of the soles of the feet by the native people in various parts of the country.

Preparation

Panak and tincture. Various other preparations are also made of the drug.

Genus: *Gloriosa* Linn.

From the Latin word Gloria, the generic name has been derived = splendour, referring to the beauty of the flowers.

The genus comprises of climbing herbs. Leaves are alternate, opposite or 3-nately whorled, lanceolate and strongly nerved with a long spiral tendril-like apex. The flowers are large, showy, solitary and axillary. The pedicels are reflexed. The fruits are large and coriaceous septicidal capsules. Seeds are subglobose.

Its 5 species occur in tropical Africa and Asia. Only one species is found distributed throughout India, upto 1650 m in the Himalayas.

Species: *Gloriosa superva* Linn.

Description of Plant

It is a climbing glabrous herb with leafy tendrils and fleshy cylindric tubers. Leaves are sessile, opposite or alternate or verticillate, ovate-lanceolate with cordate base, narrowed into a coiled tendril at apex, glabrous and 10.0-15.0 x 2.5.0 cm. Flowers are axillary, solitary and forming a terminal corymb. Capsules are oblong and about 4.5 cm long.

Flowering and Fruiting

During June-October.

Distribution

Throughout India, it occurs ascending upto 1650 m in the Himalayas.

Plant Parts Used

Root tubers, leaves and whole plant are used.

Uses

Its root tubers are tonic, antiperiodic, cholagogue, alterative and purgative. The white flour prepared from its root tubers is given with honey in gonorrhoea and alone in leprosy, colic and intestinal worms. For promoting labour pain, a paste of its root tubers is applied over the supra-pubic region and the vagina and the warm poultice is used in rheumatism and neuralgic pains. Its leaves are antiasthmatic. Their juice is used for killing lice in the hairs. Besides, various parts of the plant are used in spleen complaints, tumours, erysipelas, sores and syphilis and ethanol (50 per cent) extract of the plant as spasmolytic and central nervous system depressant.

The colchicines obtained from the plant can be used as a substitute for Colchicum autumnale.

Folk Uses

Its root stocks are cut into pieces, meshed and boiled in sesame oil for about half an hour, then the oil is filtered and is applied twice a day followed by massage on painful joints. In Orissa, its root paste is applied to the abdomen after applying turmeric paste for three days in *'pirhi'* disease, boil-like thing developed inside the abdomen below rib-cage on left side in children.

Preparation

Kasisadi taila, Languli rasayan and Laghuvis garva taila.

Family: AREACACEAE

Genus: *Borassus* Linn.

From Greek word Borassos, the generic name has been derived which is used for the fruits of the palm.

Tall dioecious palm trees, the genus comprises of the trunk is stout and unarmed. Leaves are terminal, fan-shaped and plicately multifid, sides of lobes induplicate in vernation; petiole spinous and ligule short. The spadix is large and interfoliar and peduncle sheathed with open spathes. The male spadix is with stout cylindric branches, densely clothed with closely imbricating bracts and enclosing the spikelets of flowers. The female spadix is sparingly branched bearing a few scattered flowers. The fruits are large and subglobose drupes and composed or 1-3 obcordate compressed pyrenes. Seeds are compressed and quadrate with 3-lobed tops.

Its 8 species are distributed in the paleo-tropical parts of the old world. Only one of them is reported from India.

Species: *Borassus flabellifer* Linn.
(B. *flabelliformis Roxb.*)

Description of Plant

A tall palm attaining a height upto 30 m. The trunk is black, swollen above the middle and again contracted upwards and is covered with dry leaves or bases of petioles when young and leaving the black scars of petioles in the old age. Leaves are 0.9–1.5 m across, palmately fan-shaped, rigidly coriaceous and many-cleft to lanceolate or linear with bifid lobes; segments are 60-80 and shining. The petiole is 60-120 cm long, stout and semiterete; edges with hard horny spinescent serratures and ligule is short. The male spadix is simply branched and sheathed with many imbricated spathes. Female spadix is simple spike and terminating the branches of the spadix. The drupes are distinctly trigonous when young but when old, the pulp round the pyrenes swells to give globose appearance to the fruits and 15-20 cm across. Pyrenes are 3-1, obcordate and fibrous outside. Seeds are oblong.

Flowering and Fruiting

During December–July.

Distribution

It is found more or less all over India in the dry parts; common along the coastal areas of the Peninsula, Bihar and Bengal. Also grown as ornamental plant.

Plant Parts Used

Roots, flowers, fruits, gum, pulp, bark, spathe and whole plant are usable.

Uses

Its roots considered as cooling and restorative and their decoction is used in gastritis and hiccups along with juice of young terminal buds and leaf stalks. The roots are also used as anthelmintic and as a cure for gonorrhoea. Fermented juice is used as tonic, fattening, expectorant, to allay thirst and in the scalding of urine. In Unani system of medicine, the fruits are given to purify the blood and a slightly fermented juice in diabetes. The sap is prescribed in digestive problems, chronic gonorrhoea and in Hansen's dressed disease.

The pulp of the fruits is used as demulcent and nutritive. The powder of its burnt bark or its decoction is used in dentifrice. Its flowers are used as sinapism in tumours of uterus (Asolkar *et al.*, 1992). An extract of central axis of male inflorescence with mustard oil is used in rheumatism and the gum as efficient in indurations. The ash of the spathe along with other demulcent is applied for enlarged liver and spleen and also considered good as antacid in heartburn.

Besides, the plant is used in heatstroke, headache, earache, pain, epilepsy, convulsions, adneteis, scrofulosa, colli, sores, scabies, burns, syphilis, ulcers of palate, nausea, vomiting, spiderlick, minorrhagia, haemorrhage and septicaemia.

Folk Uses

The fresh-drawn toddy from the plant, is added to rice-flour to make toddy poultice, till it has the consistence as soft and then subjected to a gentle fire for fermentation. Then, this is spreaded on a clothe and applied to the affected parts as a stimulant application to gangrenous ulceration, carbuncles and indolent ulcers in certain parts of the country.

Genus: *Phoenix* Linn.

From Greek name, the generic name is taken Phoinkos for the date plam.

The genus comprises of tall, simple or soboliferous palms. Leaves are pinnatisect; segments are lanceolate or ensiform with induplicate margins. The flower are small, dioecious, yellowish and coriaceous on several branched, interfoliar, erect or drooping spadices. The spathe is basilar, complete and coriaceous. Male flowers: often approximate in pairs, calyx is cup shaped and 3-toothed, petals 3, obliquely ovate and valvate, stamens 6 and anthers are erect and dorsifixed. Female flowers: sepals are 3 and connate in a globose accrescent calyx, petals are 3, rounded and imbricate, staminodes 6 and free or connate in a 6-toothed cup and the carpels are 3 and free, stigmas are sessile and ovules erect. The fruit is single oblong and 1-seeded berry. Seeds are oblong and ventrally grooved.

Its 17 species are known to occur in warmer parts of Africa and Asia. Out of which 7 species have been reported from India.

Species: *Phoenix dactylifera* Linn.

Description of Plant

This is a tall tree upto 30-35 m in height with persistent bases of petioles and foot often surrounded with a dense mass of root suckers. Leaves are grey and longer; pinnae20.0-40.0 cm long, regularly distichous and forming a very acute angle with the petiole. Male panicles are white, compact and 15-30 cm long on a short peduncule; flowers are 0.6-0.8 cm long and sweet scented. Female pecuncles are 0.8-1.2 cm broad. The fruits are oblong, reddish or yellowish brown when ripe and 2.5-7.5 cm long with fleshy pulp. Seeds are cylindric with a longitudinal furrow in front.

Flowering and Fruiting

During March–July.

Distribution

It occurs wild in Punjab, Rajasthan, Gujarat, Uttar Pradesh, Madhya Pradesh, Andhra Pradesh and Delhi etc.

Plant Parts Used

Fruits, juice and gum, obtained from the stem and seeds are used.

Uses

The fruits of this plant in ayurveda, are known for their aphrodisiac, alexiteric, cooling, flattening and sweetish properties and these are used in asthma, bronchitis, fatigue, tuberculosis, leprosy, fever, thirst, vomiting and loss of consciousness and in Unani system of medicine, these are known to possess aphrodisiac, diuretic and sweetish properties and used in bronchitis and to enrich the blood. The fresh juice obtained from its stems, named *laghi* is used as a demulcent, diuretic and refrigerant in genitourinary affections. The gum of the plant is also used in diarrhea and in the diseases of genitor-urinary system. Besides, a paste made of its ground seeds is applied over eyelids in opacity of cornea and to head in headache.

Preparation

Khajurasav.

Species: *Phoenix sylvestris* Roxb.

Description of Plant

A tall and erect palm tree it is reaching a height upto 18 m with thick hemispherical crown. Trunk is solitary and clothed with the persistent bases of the petioles. Leaves are pinnate, grayish-green, 2.0-3.6 m long and glabrous; leaflets are 30.0-60.0 x 1.7-2.5 cm, fascicled, 2-4-farious and rigid and spines are 10 cm long. Male flowers: white, scented spadix 60-90 cm long, peduncle much compressed, sphathes are 30-40 cm long, scurfy and separating into 2 boat shaped valves. Female flowers: spadix and spathe as in male; fruiting peduncles are about 15 cm long; fruiting spadix 90 cm long, nodding and much compressed. The fruit is oblong-ellipsoid, orange-yellow to reddish-brown and edible berry. Seeds are rounded at ends, grooved to centre on one face, pale brown and about 1.6 cm long.

Flowering and Fruiting

Flowering: In April–May and fruiting later during September- October.

Distribution

It is tolerably common throughout the drier parts of India, wild or more often cultivated upto 1500m.

Plant Parts Used

Fruits and roots are used.

Uses

Its fruits considered as tonic and restorative and these are eaten. A fresh juice of its stem is used as cooling beverage. The sap from the trunk called *Nira*, is cooling, slightly narcotic and digestive. Its roots are used in toothache and also given in nervous debility. A paste of its–kernels with roots of *Achyranthes aspera* is eaten with betel leaves as a remedy for ague. Besides, central tender part of the plant is used in gonorrhoea and gleet.

Folk Uses

Its fruits are pounded and mixed with almonds, quince seeds, pistachio nuts, spices and sugar to form a *'paushtik'* or restorative remedy which is used in vogue by the natives. The *'patali gur'* made from the sap of plant is used by the diabetic patient as sweetener. Its root are also used in snake bite by the tribals of Bastar district. For the purpose, its roots are kept in water, during soaking a sap exudes out from its roots in water which is given 4-5 teaspoonful to bitten person.

Preparation

Khajurasav.

Family: ARACEAE

Genus: *Acorus* Linn.

The generic name is the latinzed form of Greek name, Akron, given to *Iris pseudacorus* plant.

The genus comprises of aromatic and marshy herbs with creeping rootstocks. Leaves are distichous and ensiform with equitant base. Peduncle is lea-like; and continued into ensiform spathe. Spadix is sessile, cylindric and dense-flowered. Sepals are 6, orbicular and concave with incurved tips. Stamens are 6; anther renifrom. Ovary is conical and 2-3-celled with numerous ovules; The berries are many seeded.

About 2 Species to occur in North temperate and tropical regions of the world including India.

Species: *Acorus calamus* Linn.

Description of Plant

A semi-aquatic plant with underground stem and rootstocks. Leaves are bright green, acute, thickened in the middle with wavy margins and 0.9- 1.8 cm long. Spathe is 15-75 cm long and spadix is 5.0-10.0 x 1.3-2.0 cm, obtuse and slightly curved. Sepals are as long as the ovary and scarious; anthers yellow. The fruits are turbinate and prismatic with puramidal tips. Seeds are oblong.

Flowering and Fruiting

Flowering: During May-June and Fruiting: Later in July-August.

Distribution

This is found almost throughout India in marshy places, ascending in the Himalayas upto 1828 m in the Sikkim.

Plant Parts Used

Its rhizomatous rootstocks are used medicinally.

Uses

In ayurveda, its rootstocks are known for their anthelmintic, bitter, carminative, diuretic, emetic, laxative and pungent properties and are used to improve appetite and in abdominal pains, inflammations, fevers, epilepsy, bronchitis, delirium, hysteria and dysentery and in Unani system of medicine, these are described as alexiteric, carminative, expectorant and laxative and are used for general weakness, stomatitis, leucoderma and kidney problems An infusion of rootstocks of this plant is given in dyspepsia, flatulence and in loss of appetite; also in hysteria and neuralgia and as antispasmodic in tertiary fevers. This is administered with liquotice in cough, fever, capillary bronchitis and colic. In case or irritation of throat and cough, its root are simply chewed to copious salivation. Besides, the roots with *bhang* and *ajowain* in equal parts are used as a fumigation to get relieve in painful piles.

Folk Uses

In Ranchi (Jharkhand), its about 10 gm roots are ground and given thrice daily for 2-3 days to children in fever, particularly during summer (Singh and Khan, 1990). The people of Surguja district in Madhya Pradesh, take the pound roots of plant and apply on the body from head to toe in fever. Besides, its dried roots are also chewed in coughs.

Preparation

Vachayog, Medhya arasayn, Saraswati churna, Sudarshan churna,Sanjivani vati, Devdaradi kwath, Karanjadi yog, Yograj guggulu, Laghu vis garva taila, Kukuvadi churna and Chandraprabha vati.

Genus: *Pistia* Linn.

The generic name is derived form Greek word, Pistos meaning water; in allusion to aquatic habit of the plant.

This monotypic genus is found in tropical and subtropical parts of the world including India.

Species: *Pistia stratiotes* Linn.

Description of Plant

It is a gregarious, floating and stemless stoloniferous herb with simple white fibrous rootstocks. Leaves are 3.0-10.0 cm long, obovate-cuneate, rounded or retuse at apex and densely pubescent on both surfaces. Spathe are 0.1 cm long, obliquely campanulatge, white, gibbous and closed below and contracted in the middle. Male inflorescence consists of a whorl of a few sessile connate stamens below the apex of spadix and female inflorescence constis of a solitary 1-celled ovary with many ovules. The fruits are ovoid with numerous oblong or obovoid seeds.

Flowering and Fruiting

Flowering: During May-June and Fruiting: In July-August.

Distribution

It is known to occur throughout India in still water.

Plant Parts Used

Whole plant, leaves and roots of the plant are used.

Uses

In ayurveda, the plant is described as bitter, cooling, laxative and pungenet and is used in '*tridosha*' diseases, fevers, tugberculosis and in blood-diseases. A pou tice of its leaves is apaplied in piles and mixed with sugar and rose water, the lea es are given internally in asthma and cough and with rice and coconut milk in dys ntery. An oil obtained by boiling its leaf juice in coconut oil is used externally in chr nic skin-diseases. Besides, roots of the plant possess bitter and diuretic properties and are used in inflammation and burns.

Scindapsus Schott

On Greek name, the generic name is based Skindaspos for an ivy like plant.

The genus comprises of stout and climbing shrubs. Leaves are distichous, entire or pinnatifid and deciduous. Spathe is coriaceous, ovate, acuminate and deciduous. Spadix is sessile and produced above the spathe. Flowers are crowded and bisexual; perianth absent; stamens 4-6 with short flattended filaments and ovary is 1-celled with a single basal ovule.The berries are confluent. Seeds are exalbuminous.

Its 40 specis are found distributed in South China, Southeast Asia and Indomalayan region. Only 2 of them have been recorded from India.

Species: *Scindapsus officinalis* Schott

Description of Plant

It is a large ep;iphytic climber. The stems are thick as the little finger and branches are wrinkled when dry. Leaves are 12.5-25.0 x 6.0-15.0 cm, ovate, elliptic-ovate or orbicular, caudate-acuminate and dark green with rounded or slightly cordate base. Peduncles are solitary and terminal. Spathe is 10.0-15.0 cm long, subcylindrical and spadix is elongated in fruits and greenish yellow. The berries are few and fleshy. Seedsd are ovate-cordate.

Flowering and Fruiting

During almost throughout the year.

Distribution

It is found in tropical Himalayas from Sikkim east wards, Bengal, Chittagong and Andaman islands.

Plant Parts Used

Fruits and the roots of the plant are used medicinally.

Uses

In ayurveda, the fruits of the plant have been described as anthelmintic, appetizer, aphrodisiac, galactagogue and pungent and are used in asthma, dysentery and in the problems of the throat. A decoction of tis sliced fruits in the doses of 150-900 ml is given to cure asthma and diarrhea and the past is applied externally in rheumatism. Besides, its root-decoction is used in asthma, bronchitis, diarrhea and as an expectorant.

Family: CYPERACEAE

Genus: *Cyperus* Linn.

The Greek name given for one or more species of this genus is the generic name.

This genus comprises of annual of perennial herbs with stolons or tuberous roots. Leaves are usually only at the base of stem and sometimes reduced to sheaths. Spikelets are in solitary, globose or umbellate heads or spiles; bracts on or more and leafy below the inflorescence and bracteoles usually similar but smaller below the secondary divisions of the inflorescence. Spikelets are composed of several distichous glumes, the two lowest being empty and differ in size and shape from the rest which are equal and deciduous from below upwards, the uppermost 1-3 are sterile or empty. The fruits are trigonous, rarely compressed or globose nuts.

About 550 species are found in tropical and warmer temperate regions of the world. Out of which 100 species have been reported from India, mostly from the plains and abundantly from Eastern India.

Species: *Cyperus rotundus* Linn.

Description of Plant

It is an erect and perennial glabrous herb with woody subterranean stoloniferous rhizome which is clothed with fibrous remains of leafsheaths. Stems are nodose at base, 3-gonous and 10-60 cm high. Leaves are basal and usually shorter than the stem and are linear, tapering in the upper part to a slender acuminate apex and 0.3-0.5 cm broad. Inflorescence is an umbel of more or less condensed spikes. The flowers are reddish-brown. Bracts are 3 and unequal; the longest is upto 15cm long. The nutws are obovoid, glabrous and 0.1 cm long.

Flowering and Fruiting

Almost throughout the year, chiefly during July-December.

Distribution

This is known to occur throughout India; commonly in waste grounds, gardens, and along roadsides in open spots upto an elevation of 1828 m.

Plant Parts Used

Its tubers are used.

Uses

The tubers are known as diuretic, emmenagogue, anthelmintic, diaphoretic, astringent and stimulant and are used in the disorders of the stomatch and irrigation of the bowles. Its paste is apapleid in healing wounds are sores etc. and also used in intestinal diseases beside, as *'anjan'* (collyrium) in eye-diseases. An ethanol extract of root tubers is used as diuretic and the aqueous extract is useful in conjuctivities. Besides, its bulbous roots are scraped and pounded with green ginger and after mixing with honey, are given in dysentery.

Folk Uses

In Konkan, its fresh tubers are applied to the breast in the form of 'lep' (malagma) as galactagoue. The stem or its leaves are crushed and are applied for jaundice, disorders of stomach and irrigation of bowels and also to heal the wounds by the inhabitants of Pauri Garhwal. Besides, in rheumatoid arthritis, tribals of Santhal Pargana in Bihar, take its tubers and crush them with the rhizomes of Panicum repens and amake the pills of 2.0-3.0 gm each, which are taken orally on empty stomach.

Preparation

Mustakadi Kwath, Mustakarisht, Mustadi chruna, Mustadileh, Shandgapaniya, Jawarish Jalinush and Majun-murrawal-ul-anwah.

Family: POACEAE

Genus: *Bambusa* Schrb.

The generic name has been derived from its Malayan name Bambu.

The genus comprises of large bamboos with woody rhizomatous rootstocks and persistent culms. The culms-sheaths are broad and blade is often triangular. Leaves are linear or oblong-lanceolate, acuminate and shortly petioled. Inflorescence is a large leafless panicle bearing heads of spikelets. The spikelets are many-flowered. Grains are furrowed on one side.

About 70 species occur in tropical and subtropical Asia, Africa and America; 17 of them are found inIndia, chiefly in tropical Eastern Himalayas and Eastern India.

Species: *Bambusa arundinacea* Roxb. (*B. bamboos* Druce)

Description of Plant

It is a tall thorny bamboo with crowded culms arising from branching of its root stock. Culms-sheaths are rounded at apex and hairy with dark-brown hairs on the back and glabrous within. Leaves are linera or linera-lanceolate and glabrous or hairy beneath. The Inflorescence is an enormous panicle, often comprising of the whole culm. The sp;ikelets are lanceolate and acute. Grains are ending in a short beak and are about 0.5-0.8 cm long.

Flowering and Fruiting

Flowering: In summer and Fruiting later during September-October.

Distribution

It is wild throughout the greater part of India, especially in the hill forests of Western and Southrn India, ascending upt o914 m on the Nilgiris, also in Orissa, Assam and Eastern Bengal. Also cultivated in the lower Himalayas and the valleys of the Ganga.

Plant Parts Used

The leaves, young shorts, stems, and bamboo manna are used.

Uses

Its leaves are emmenagogue and are used in haematemesis. A decoction of its leaves is given to induce lochia after childbirth and as anthelmintic, these are particularly use ful of threadworms. The juice of the leaves with aromatics is given inblood vomiting and their infusion internally for gonorrhoea. The young shoots are stomachic and stimulant andare used in respiratory diseases and its pickled or cooked tender shoots aid digestion hence are used as appetizer.

The young shoots of bamboo are used as poultice on the infected lesions and ulcers and are very efficacious in dislodgement of worm-infected lesions and ulcers and are very efficacious in dislodgement of worm-infested wounds. A decoction of the joints of bamboo stem is used as emmenagogue and also as an abortifacient it is applied to inflamed joints to get relief. Besides, its aerial parts including stem and leaves are used by Ayurvedia physicians as blood purifier and in leucoderma and inflammatory conditions. These are burnt and applied locally to ringworm infection.

Banslochna or tabashir or bamboo manna, a crystalline secretion accumulate in the internodes of the female plant, is considered febrifuge, expectorant, tonic, aphrodisiac, demulcent and pectoral and is used in hepatic fever, phthisis, asthma and paralytic complaiants.

Folk Uses

Green bamboo is scratched with a knife and the powder thus obtained is applied externally to heal the fresh wounds by the tribal people in Bastar district in Andhra Pradesh.

Preparation

Sitopaladi churna, Talishadi churna,Sudarshan churna, Chandraprabha vati and Kurs tabashir.

Genus: *Cymbopogon* Sporeng

The generic name has been derived from Greek words Kymbe = a boat and Pogon = a beard, referring to the boat-shaped glumes.

These are tall perennial grasses, usually aromatic and densely tufted. Leaves are coarse, Panicles are much comound, contracted and spatheate. The spikelets are paired, sessile and pedicelled in shrot spikes, which germinate and usually divaricate on a slender and short peduncle and is sheathed by a spathe. One or more of the sessile spikelets at the base of the spike are different from all the others. Ist glume is flat and with inflaxed margins and often winged keels, sometimes grooved or pitted and awnless, IInd is cymbiform and keeled, IIIrd oblong and hyaline and IVth is narrow, hyaline, 2 cleft and awned, pale and minute or none. The pedicelled spikelets are male or neuter. Grains are oblong.

About 60 species occur in the tropical and subtropical regions of the world; 20 species grow in India, mostly in Southern parts and tropical and subtropical Himalayas.

Species: *Cymbopogon jwarancusa* (Junes) Schutt. (*Andropogon jwarancusa* Jones)

Description of Plant

A tall, perennial grass upto, 1.8 m height with aromatic and densely tufted roots. Leaves are linera, flat and 10cm long and 0.5cm broad, Ligue is oblong, membranous and ciliolate. Sheaths are glabrous. Panicles are long, narrow and interrupted and spatheole 1.2-1.5 cm long. Racemes are unequal, 1.5-1.8 cm long and often 5-jointed. Sessile spikelets are ovate or linear-lanceolate and 0.5cm long; lowest pair of sessile raceme is neuter and of pecuncled is heterogamous. Pedicelled spikelet is equal or rather longer than sessile, narrowly lanceolate and purplish; lower involucral glume 7-9 nerved.

Flowering and Fruiting

During July-December.

Distribution

It is found wild in Himalayas from Kashmir to Assam upto 3290 m and in the north western plains reaching down to Mumbai.

Plant Parts Used

Leaves and flowers are used.

Uses

The grass in ayurveda, is known to possess alexiteric, appetizer, astringent, bitter, cooling and stomachic properties. An infusion of its leaves is used as atniperiodic, sudorific and stimulant and also in catarrh and added with purgatives is given in rhueumatism. Besides, the flowers (calyx) are used as haemostatic.

Species: *Cymbopogon nardus* Linn.

Description of Plant

A tall and sweet scented gras reaching a height of 1.5-2.0m. Culms are upto 1.0 cm across at base, solid and pale polished with black and pubescent or glabrescent nodes. Lower leaves are 1.5cm wide and upper cauline over 0.8 cm wide, narrowed at base with filiform apex. Ligule is scarious, glabrous or ciliate and 0.2 cm long. Spikes are 1.0-1.5 cm long, unequally pedicelled and villous. Sessile spikelets are 0.4-0.5 cm long; glume 1, oblong-lanceolae, flat or slightly concave below, hyaline and nerveless or with 2 green nerves between the keels.

Flowering and Fruiting

During July-December.

Distribution

It occurs wild throughtout the hotter parts of India: also cultivated.

Plant Parts Used

Leaves and oil. Obtained from the grass is used medicinally.

Uses

Its leaf-infusion is used carminative and stomachic. Besides, the grass yields an oil which is commercially known as *'citronella oil'* and is used as antiseptic, carminative, diaphoretic, rubifacientm, stomachic and sudorific.

Sketches & Photographs

Adhatoda zeylanica **Medic.**

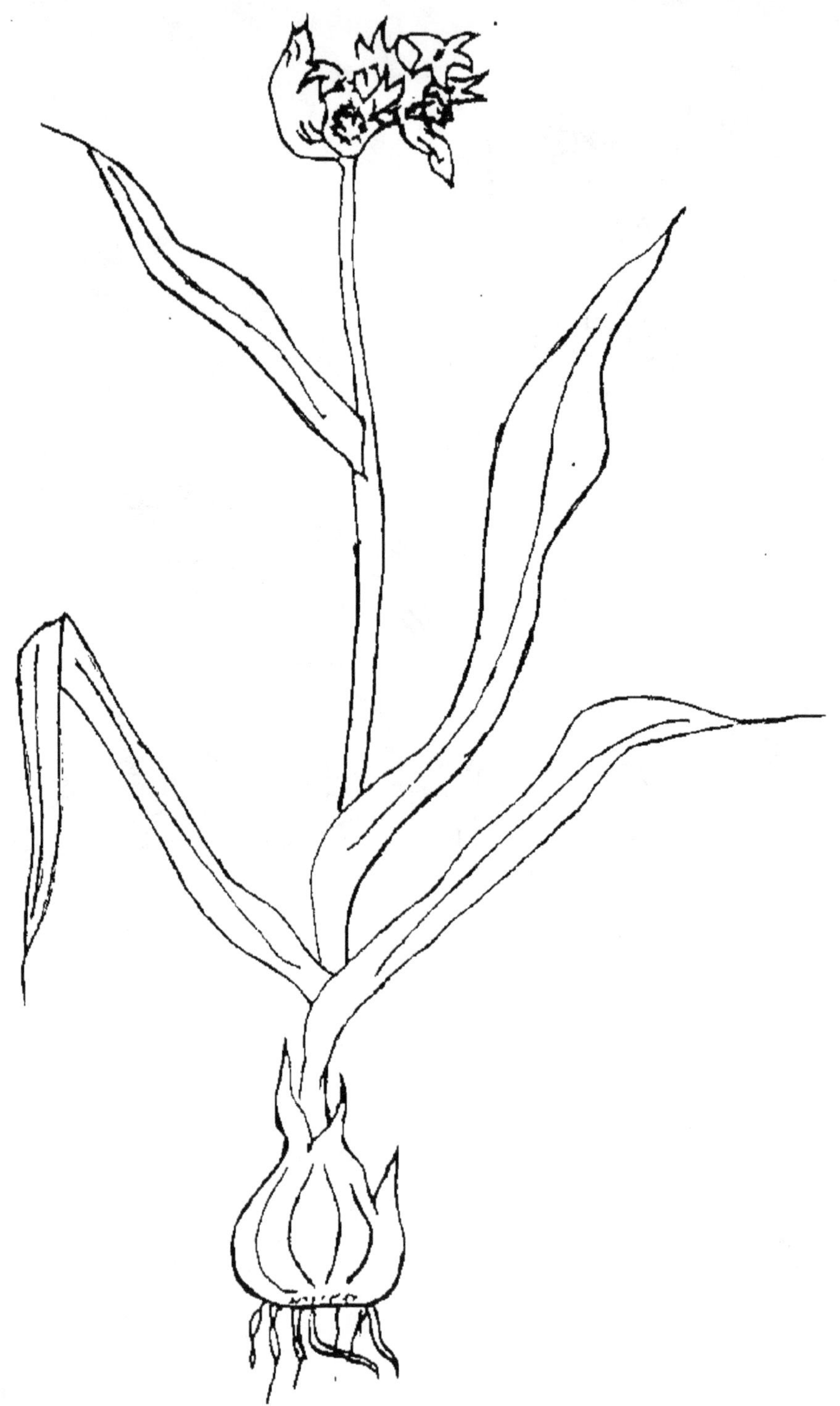

Allium sativum **Linn.**

Boerhavia diffusa **(L)**

***Catharanthus roseus* (L) G. Don**

***Chenopodium album* Linn.**

***Datura metal* Linn.**

Daucus carrota **Linn.**

***Emilia sonchifolia* (L)**

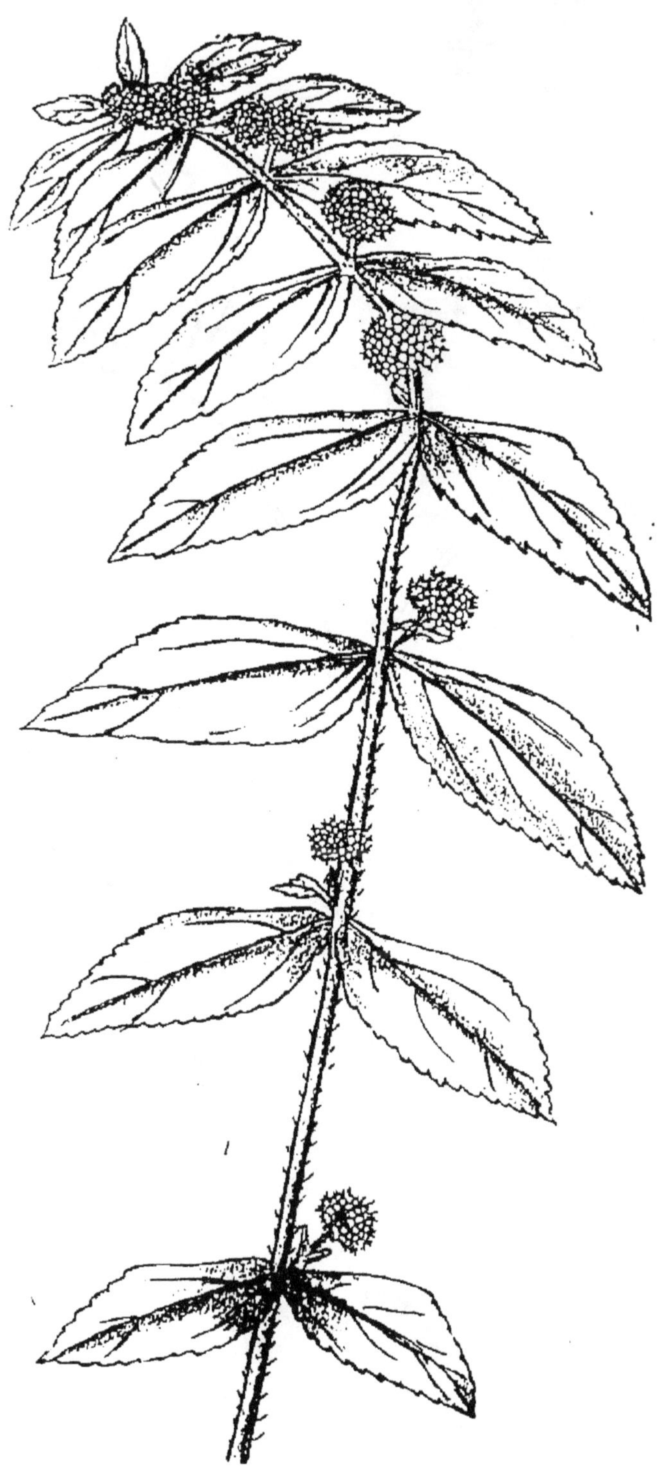

Euphorbia hirta **Linn.**

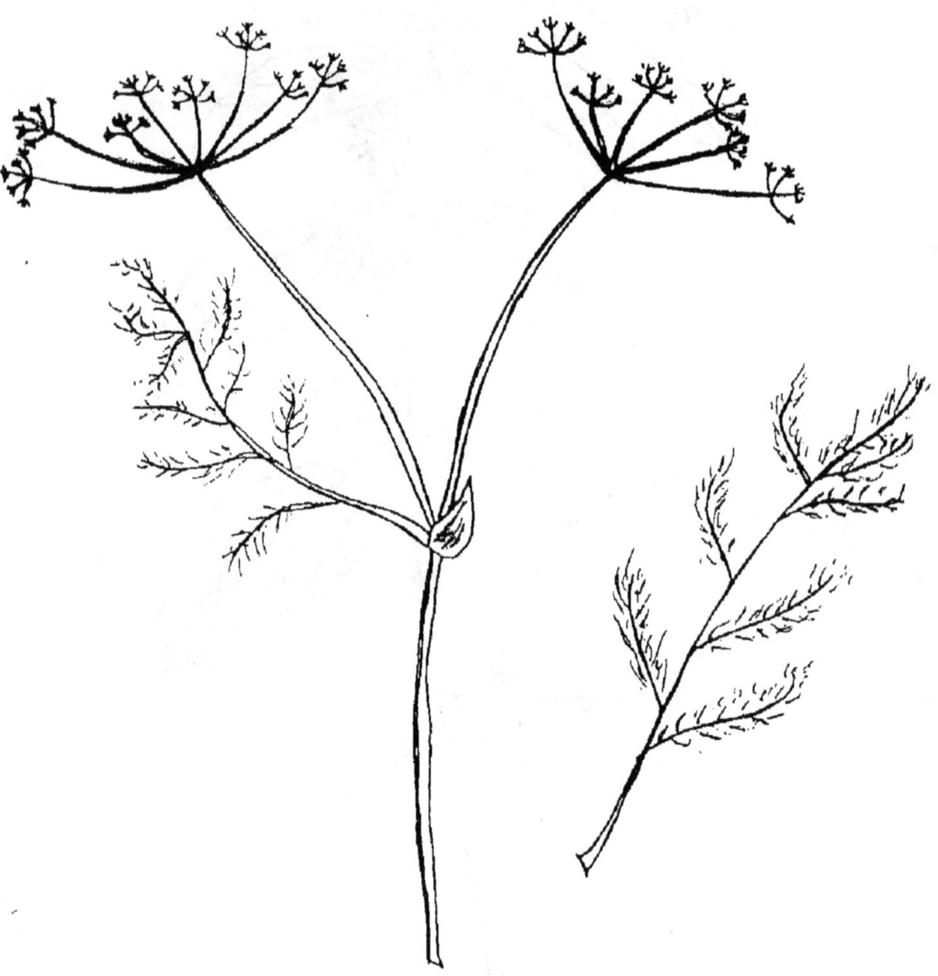

***Foericulum officinalis* Mill**

Gloriosa superaba **Linn.**

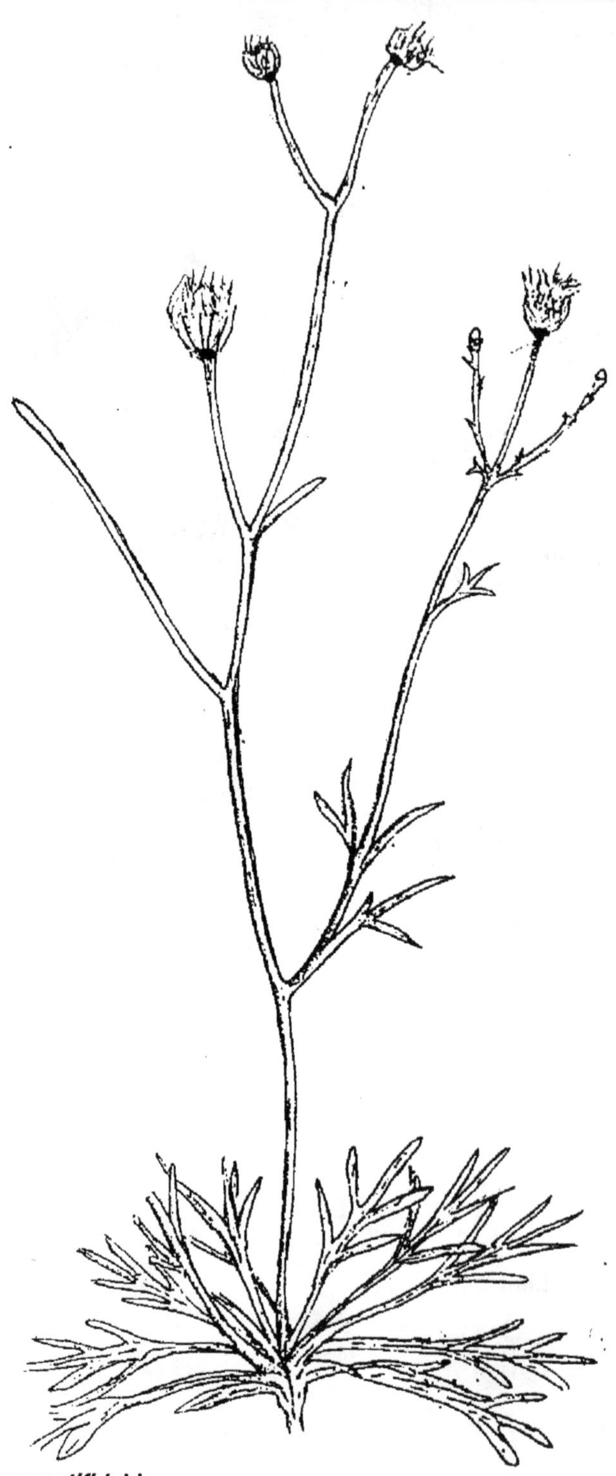

Glossogyne pennatifida Linn.

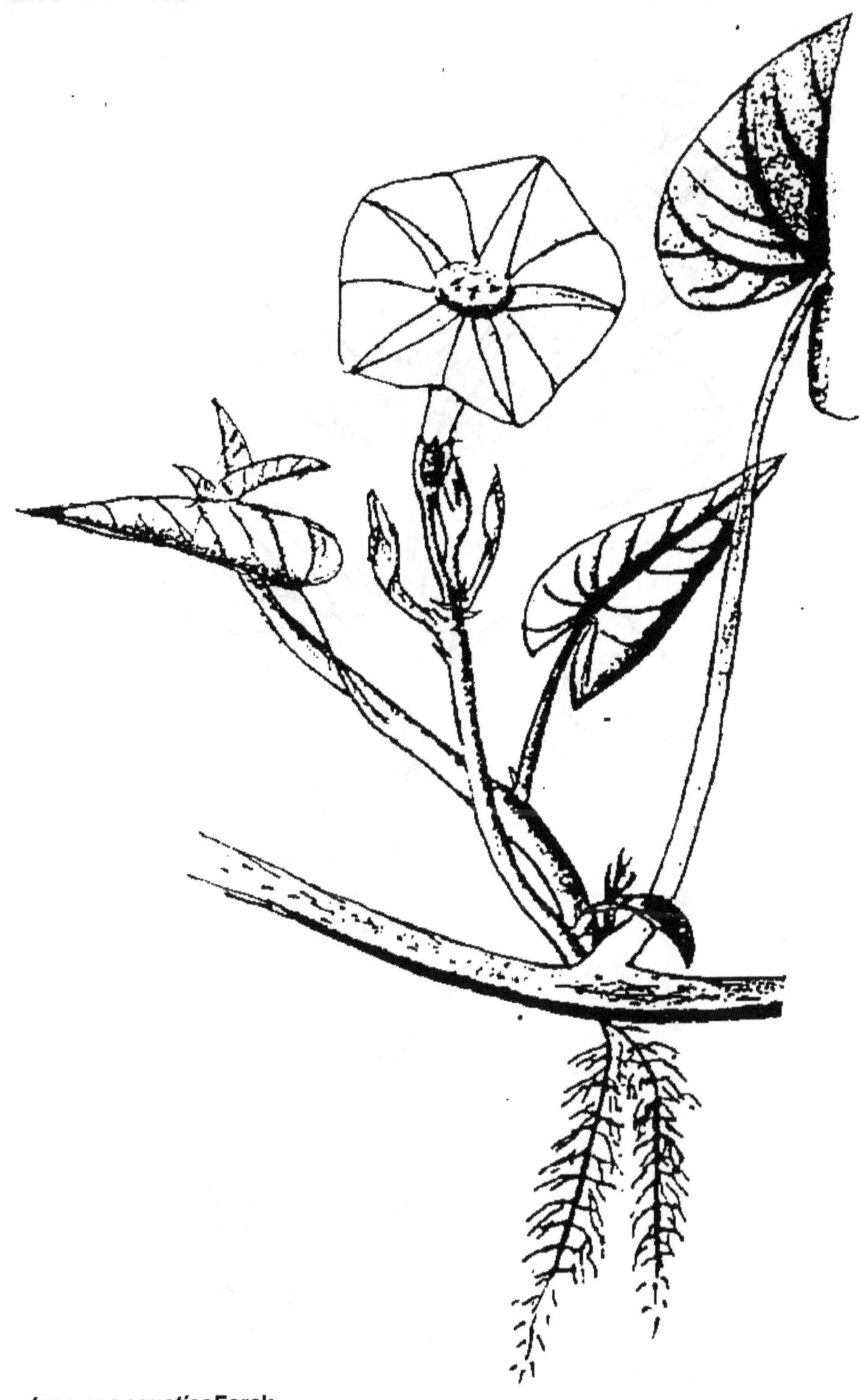

Ipomoea aguatica **Forsk.**

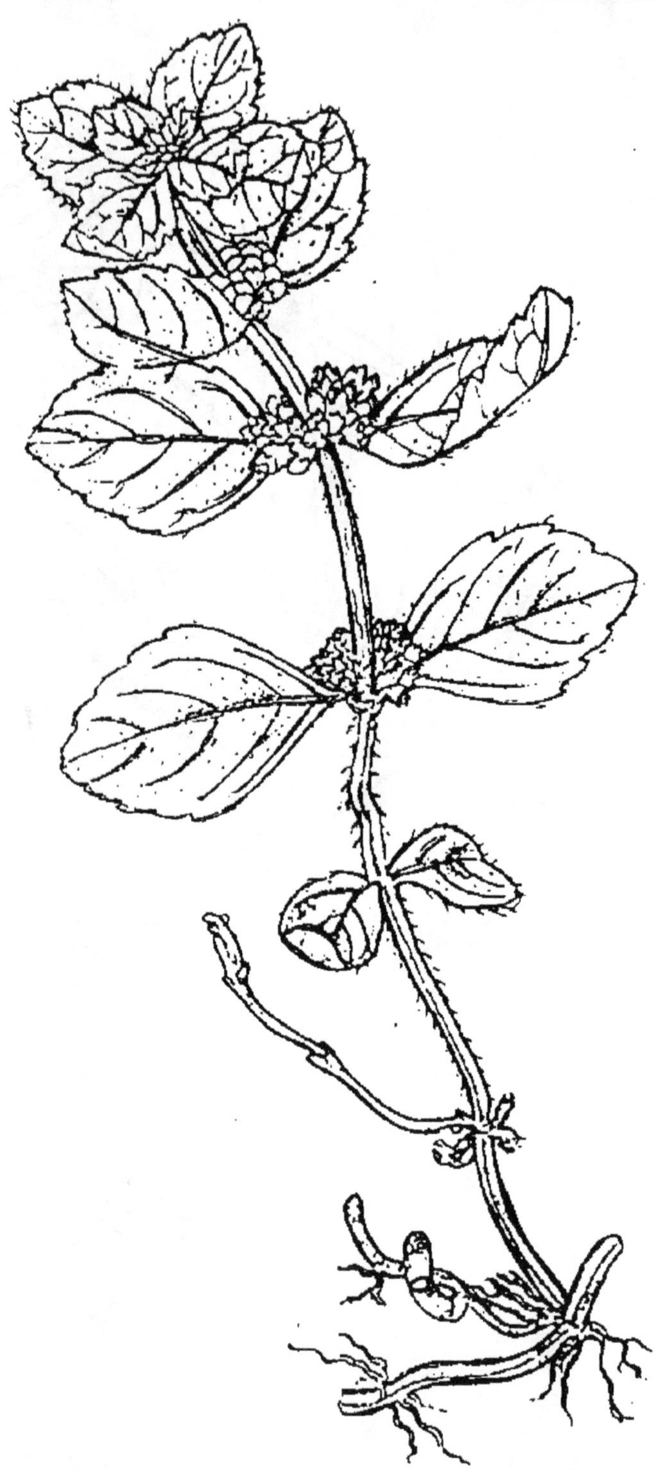

Mentha arvensis **Linn.**

Plumeria indica **Linn.**

Scilla indica **Linn.**

Tribulus terrestris **Linn.**

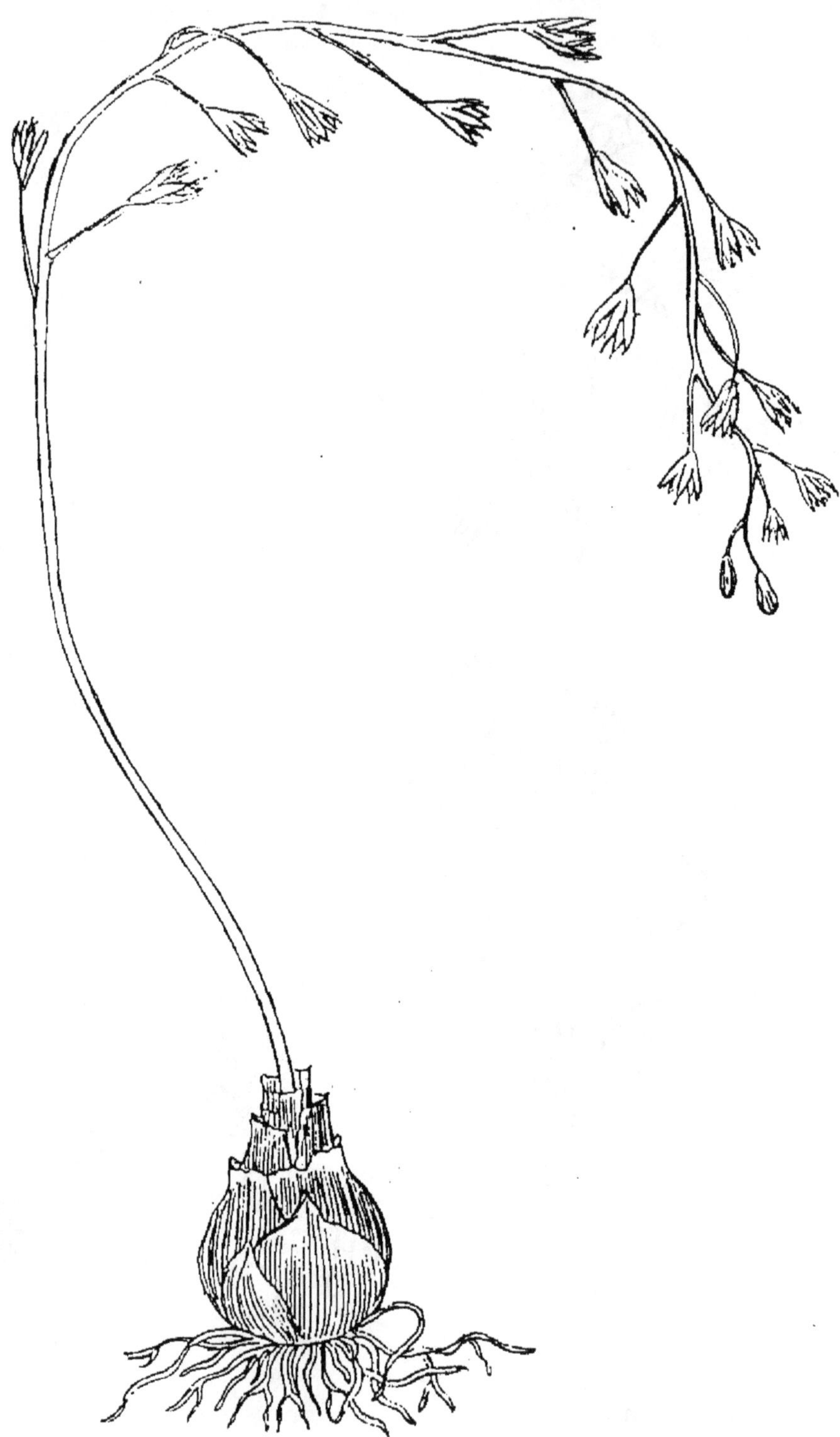

Urgenia indica **Roxb.**

Vanda roxburghii **Roxb.**

Withania somnifera **Danal**

Abrus pecatoricus **Linn.**

Asparagus racemesus

Alstonia scholaris

Atopra belladonna

Barleria prionitis

***Cardiospormum helicabum* Linn.**

Celasterus paniculata Wild.

Cordia dichotoma

Cissampelos pariera

***Centella asiatica* Linn.**

***Datura metal* Linn.**

Dioscroea bulbifera

Disopyros malabarica

Drosera indica

Hemidesmus indicus **Linn.**

Hieracium pilosella **Linn.**

Mallonus philippensis

Moringa oleifera

Randia uliginosa DC.

Rauvolifa serpentina

Saraca asoca

Solanum surrettense

Solanum torvum

Spermocoea hispida **Linn.**

Spheranthus indicus **Linn.**

Taraxacium officinale **Weber**

Trichodesma indicum **Roxb.**

Vanda parviflora

Vitex penducularis **Linn.**

Appendices

Appendix–I
Plant Classification in Ancient Systems

A. CLASSIFICATION BASED IN VEDIC SYSTEM

In vedic literature various classification of the plants are given. The most authentic and widely accepted classification divides the plants into four types: *viz.*, Vanaspati, Vanaspatiya, Aushdhi and Veerudh. *Vanaspati* means the large trees. According to Manu, Ahcarya Chakra and Acharya Susruta, these plants bear fruits without going into flowering condition. Vanaspatiya are those plants which bear both flowers and fruits. For small trees in Atharvaveda, this word is used for small trees. Aushdhi are the plants which destroy after harvesting. It includes the grasses of family Gramineae (Poaceae). The creeping plants are veerudh.

In the vedas, many other systems of classification are also given and Besides later in vedic literature. In view of flower formation the plants are divided into *Pushpini* or *Pushvati* (those possess flowers) or *Apushpa* or *Prasuvari* (flower-less plants) and on the basis of fruit formation, these are divided as *Falini* (plants possess fruits) and *Afala* (Fruit-less plants). Another classification is based on the media in which the plants are borne. On this basis, the plants may be *Sthalaj* (plants grow on plain soil) and Jalaj (plants grow in water). Another classification of the plants is based on the of colour of plants. According to this classification of the plants may be divided as *Babhru* (brown coloured plants), Shukra (white coloured plants), Rohini (red coloured plants), Prishnaya (spotted plants), Asikini (blue coloured plants) and Krishna (black coloured plants). The other classification of the plants in vedic literature is based on the shape and appearance of the plants. This classification has divided the plants into seven group which include, *Prastrinati* (branches spreaded just above from the roots), *Stambini* (One bundle type branch arises from nodulatd root), *Ekshringa* (one apical bud), *Pratanvati* (spreading creeper), *Anshumati* (having spikes), *Kandini* (with nodes and internodes) and *Vishakha* (branchless plants). Further on the basis of the vital properties, the plants are tried to divide into five types as; *Jeevalam* (the plants which gave life), *Pushpam* (the plant which bear flowers), *Madhumatim* (the plants which contain sweet juice), *Unnayatim* (the plants which possess seman properties) and *Aushdhim* (the plants which bear antifever properties).

B. PLANT AND DRUGS: IN CHARAKA'S CLASSIFICATION

As follows, Acharya Chakra has divided the plants into 50 groups:

1. Jeevaneeyam (Promoting Life): Jivaka, Rishabhaka, Meda, Mahaamedaa, Kaakoli, Ksheerakaakoli, Maashaparni, Mudgaparni, Jeevanti and Yastimadhu.
2. Brimhaneeyam (Promoting Growth): Ksheerini, Raajakshavaka, Bala, Kaakoli, Ksheerakaakoli, Vatyaaayani,Bhadroudani,Bhaaradwaaji, Payasyaa and Rishyagandhaa.
3. Lekhaniyam (Reducing Growth): Mustaa, Kushta, Haridraa, Vachaa, Daaruharidraa, Ativishaa, Katukarohini, Chitraka, Karanja and Haimavati.
4. Bhedaneeyam (Proroting Evacuation): Trivrit, Arka, Eranda, Agnimukhi, Danti, Chitramula, Chirbilwa, Sankhini, Katukarohini and Brahmadandi.
5. Sandhaaneeyam (Promoting Union): Yastimadhu, Guduchi, Prisniparni, Ambashtaa, Samangaa, Mocharasa, Dhaataki, Lodhra, Priyangu and Katphala.
6. Deepaneeyam (Promoting Appetite): Pippali, Pippalimoola, Chavya, Chitraka, Sunthi, Amlavetasa, Maricha, Ajamodaa, Bhallathakaasthi and Hingu.
7. Balyam (Promoting Strength): Indravaaruni, Vrishabha,a Sataavari, Maashaparni, Vidaari, Aswagandhaa, Sthiraa, Rohini, Balaa and Atibalaa.
8. Varnyam (Promoting Complexion): Chandana, Tunga, Padmaka, Useera, Yashtimadhu, Majaistaa, Saaribaa, Ksheera-kaakoli, Sitaa and Lataa.
9. Kanthyam (Promoting Voice): Saaribaa, Ikshumoola, Yastimadhu, Pippali, Draakshaa, Vidaari, Kaayaphala, Hansapadi, Brihati and Kantakaari.
10. Hridyam (Promoting Happy feeling): Aamra, Aaamraataka, Lakucha, Karamarda, Vrikshaamla, Amlavetasa,Kuvala, Badara, Daadima and Maatulunga.
11. Triptighnam (Destroying Satisfaction): Sunthi, Chitramoola, Chavya, Vidanga, Murva, Guduchi, Vachaa, Musta, Pippali and Patola.
12. Arsoghnam (Destroying Piles): Kutaja, Bilwa, Chitraka, Sunthi, Ativisha, Haritaki, Dhanvayaasaka, Daaruharidra, Vachaa and Chavya.
13. Kustaghnam (Destroying Skin Diseases): Khadira, Haritaki, Aamalakia, Haridraa,Bhallaataka, Saptaaparna, Aaragwadha, Karaveera, Vidanga and Jaathi Pravaala.

14. Kandughnam (Destgroying Itching): Chandana, Jataamaansi, Aaragwadha, Naktamaala, Nimba, Kutaja, Sarshapa, Yastimadhu, Daaruharidra and Mustaa.
15. Krimighnam (Destroying Parasites): Sigru, Maricha, Gandira, Kebuka, Vidanga, Nirgundi, Kinihi, Gokshura, Bhaarangi and Aakhuparnika.
16. Vishaghnam (Destroying Poison): Haridraa, Manjistaa, Raasnaa, Sukshmaila, Syaamaa, Chandana, Kathaka, Sireesha, Sindhuvaara and Sleshmaataka.
17. Sthanyajananam (Producing Milk): Virana, Saali, Shastika, Ikshumoola, Kusamoola, Darbhamoola, Kaasamoola, Gundra, Utkatamoola and Katrinamoola.
18. Sthanyasodhanam (Purifying Milk): Paathaa,Sunthi, Devadaaru, Mustaa, Murvaa,Guduchi, Kutajabeeja, Bhunimba, Katukarohini and Saaribaa.
19. Sukrajananam (Producing Sperm): Jivaka, Rishabhaka, Kaakoali, Ksheerakaakoli, Mudgaparni, Masshaparni, Medaa, Sataavari, Jataamansi and Karkatakasringi.
20. Sukrasodhanam (Purifying Sperm): Kushta, Elavaaluka, Katphala, Samudraphenma, Kadambaniryaasa, Ikshukanda, Ikshuraka, Vasuka and Useera.
21. Smehopagam (promoting Lubrication): Draakshaa, Yastimadhu, Guduchi,Medaa, Vidaari, Kaakoli, Ksheerakaakolia, Jivaka, Jivanti and Saalaparni.
22. Sedopagam (Promoting Sweet): Sigru, Erandamula, Arka,Vrischira, Punarnava, Yava, Tail, Kuluttha, Maasha and Badriphala.
23. Vamanopagam (Promoting Vomiting): Madhu, Yastimadhu, Kovidaara, Karbudaara, Neepa, Vidula, Bimbi, Sanapushpi, Arka and Pratyakapushpi.
24. Virechanopagam (promoting Purgation): Draakshaa, Kaasmari, Parushaka, Abhayaa, Aamalaki, Vibheetaki, Kuvela, Badara, Karkandhu and Pilu.
25. Aasthaapanopagam (Useful for Non-oily Enemata): Trivrit, Bilwa, Pippali, Kushta,Sarshapa, Vachaa, Kutajabeeja, Satapushpaa, Yastimadhu and Madanaphala.
26. Anuvaasanopagam (useful for Oily Enemata): Raasnaa, Devadaaru, Bilwa,Madanaphala, Satapushpaa, Punarnavaa, Vrischeera, Gokshura, Agnimantha and Syonaka.
27. Sirovirechaniyam (Purging Doshas in the Head): Jyotishmati, Kshavaka, Maricha, Pippali,Vidanga, Sigru, Sarshapa, Apaamaarga, Sweta and Mahaaswetaa.
28. Chardinigrahanam (Controlling Vomitting): Jambu, Aamrapallava, Maadiphala, Badara, Daadima, yava, Yastimadhu, Useera, Mrith and Laaja.
29. Trishnaanigrahaanam (Controlling Thirst): Sunthi, Dhanvayaasa, Mustaa, Parpataka, Chandana, Bhunimba, Guduchi, Useera, Dhaanyaka and Potola.

30. Hiccaanigrahanam (Checking Hiccough): Sati, Pushkaramoola, Badarabeeeja, Kantakaari, Brihati, Vriksharuha, Haritaki, Pippali, Duraalabhaa and Karkatakasringi.
31. Purisha Sangrahaneeyam (Reducing Faecal Matter): Priyangu, Ananthaa, Aamraasthi, Katwanga, Lodhra, Mocharasa, Samangaa, Dhaataki, Bhaarangi and Padmakesara.
32. Purisha Virajaneeyam (Pruifying Faecal Matter): Jambu, Sallakitwak, Kachchura,Yastimadhu, Mocharasa, Devadaaru-niryaasa, Bhrustamrit, Payasyaa, Utpala and Tilakana.
33. Mootra Sangrahaneeyam (Reducing Urine): Jambu, Aamra, Plaksha, Vata,Kapitana, Udumbara, Aswattha, Bhallataka, Asmantaka and Somavalka.
34. Mootra Virajaaneeyam (Purifying Urine): Padma, Utpala, Nalina, Kumuda, sougandhika, Pundareeka, Satapatra, Madhuka, Priyangu and Dhaathaki-pushpa.
35. Moorta Virechneeyam (Increasing Urine): Vrikshaadani, Swadamastra, Vasuka, Suryaavarta, Paashaanabhedi, Darbha, Kusa, Kaassa, Mustaa and Utkatamoola.
36. Kaasahara (Relieving Cough): Draakshaa, Haritaki, Aamalaki, Pippali, Duraalabhaa, Karkatasringi, Kantakaari, Punarnavaa, Taamalaki and Vruschira.
37. Swaasahara (Relieving Dyspepsia): Sati, Pushkaramula, Amlavetasa, Elaa,Hingu, Agaru, Surasaa, Taamalaki, Jeevanti and Chandana.
38. Swayathuhara (Relieving Swelling): Paatala, Agnimandha, Syonaka, Bilwa, Kassmari, Kantakaari, Brihati, Saalaparni, Prisniparni, Gokshura and Dasamoola.
39. Jwaraharam (Relieving Fever): Saaribaa, Sarkaraa, Paathaa, Manjistaa, Draakshaa, Peelu, Parushaka, Haritaki, Aamalaki and Vibheetaki.
40. Sramaharam (Relieving Exhaustion): Draakshaa, Kharjura, Piyaala, Badara, Daadima, Phalgu, Parushaka, Ikshu, Yava and Shastika.
41. Daahaprasamanam (Relieving Burning Sensation): Laaja, Chandana, Kaasmari, Madhuka, Sarkaraa, Neelotpala, Useera, Saaribaa, Guduchi and Hribera.
42. Seethaprasamaaanam (Relieving Cold): Tagara, Agaru, Dhaanyaka, Sringabera, Ajmadaa, Vachaa, Kantakaari, Agnimandha, Syonaaka and Pippali.
43. Udardaprasamanam (Relieving Rashes): Tinduka, Piyaala, Badara, Khadira, Kadara, Saptaparna, Aswakarna, Arjuna, Asana and Irimeda.
44. Angamardaprasamanam (Relieving Body Pains): Vidaarigandhaa, Prisniparni, Brihati, Kanatakaari, Eranda, Kaakoli, Chandana, Useera, Elaa and Madhuka.

45. Sulaprasamanam (Relieving Colic): Pippali, Pippalmoola, Chavya, Chitraka, Sunthi, Maricha, Ajmadaa, Ajagandha Jeeraka and Gandeera.
46. Sonitaasthaapanam (Resatoring Blood): Madhu, Madhuka, Kunkuma,Mocharasa, Loshta,Lodhra,Gairika, Prenkhana, Sarkaraa and Laaja.
47. Vedanaasthaapana (Relieving Sufferings): Arjuna, Katphala, Kadamba, Padmaa, Kumuda, Mocharasa, Sirisha, Vanjula, Elaavaluka and Asoka.
48. Samgnaasthaapana (Restoring Consciousness): Hingu, Katphala,a Irimeda, Vachaa, Sati, Braahmi, Golomi, Jataamaansi, Guggulu and Katurohini.
49. Prajaasthaapana (Fixing Pregnancy): Indravaaruni, Braahmi, Durvaa, Swetadurva, Paatali, Aamalaki, Haritaki, Katurohini, Balaa and Prenkhana.
50. Vayasthaapana (Fixing Youthfulness): Guduchi, Haritaki, Aamalaki, Raasna,Swetaa, Jeevanti, Sataavari, Braahmi, Sthiraa and Punarnavaa.

C. PLANTS AND DRUGS: IN SUSRUTA'S CLASSIFICATION

As follows, Acharya Susruta has divided the plants into 38 ganas (groups):

1. Vidaarigandhaadigana: Vidaarigandhaa (Saalaparnee), Vidaaree, Viswadevaa (Gangerukee), Sahadeva (a variety of Balaa), Swedamshtraa (Gokshura), Prithaakparnee (Prisniparnee), Sataavaree, Saaribaa, Krishna saaribaa, Jeevaka and Rishabhaka, Mahaasahaa (Maahaparnee), Kshudasahaa (Mudgaparnee), Brihatee, Kantakaaree, Punarnavaa, Eranda, Hansapaadi, Vrischikaalee and Rishabhee.

   The Vidaarigandhaadigana counteracts Vaata and Pitta. It is useful in Sosha (emaciation) Gulma (tumour), Angamarda (body soreness), Oordhwaswaasa (a kind of Seassa) and Kaasa (Cough).

2. Aaragwadhaadigana: Aaragwadha, Madana, Gopaghonta (Karkotee), Kantakee (Vikamkata), Kutaja, Paathaa, Paatalaa, Moorvaa, Indrayava, Saptaparna, Nimba, Kurantaka, Daaseekurantaka, Gudoochee, Chitraka, Saarngeshtaa (Kaakaajanghaa), Karanja and Vitapakaranja, Patola, Kiraatatiktaka and Sushavee (Kaaravella).

   The Aaragwadhaadigana checks Kapha and poison. It is also used in Prameha, Kushta, Jwara, Vami (Vomitting) and Kandu (Itching) and to purify Vrana (Dushta).

3. Varunaadigana: Varuna,Aartagala (Kakubha), Sigru, Madhusigru, Tarkaaree, Mushasringee (Karkatasringee), Pooteeka (Chiribilwa), Naktamaala (Brihat Karanja), Morata (Ankolapushpa), Agnimantha, two kinds of Saireyaka (red and blue flowers), Bimbee, Vasuka (Buka), Vasira (Markatapippalee), Chitraka, Sataavari, Bilwa, Ajasringee (Chagalavishanikaa), Darbha (Kusa), Brihatee and Kantakaari.

   The Varunaadigana checks Kapha and Medas. It is also useful in headache, Gulma and Aabhyantara vidradhi (internal abscess).

4. Veeratarwaadigana: Veerataru (Saaaaaavara), two kinds of Sahachara, Darbha, Vrikshaadanee, Gundraa (Paadaraka bhedah), Nala, Kusa, Kaasa,Asmabhedaka,Agnimandha, Morataa (Ankoila-pushpa), Vasuka, Vasira, Bhallooka, Kurantaka, Indeevara, Kapotavankaa, and Swadamshtra (Gokshura).

   The Veerataraadigana checks vaata diseases.It is useful in Asmaree (Stone), Sarkaraa (Sand in urine etc.), Mootrakrichra (painful passage of urine), Mootraaghaata (retention of urine etc.) and Ruja (a kind of pain connected with passing of urine).

5. Saalasaaraadigana: Saalasaara, Ajakarna (Sarja), Khadira, Kadara (Swetasaara), Kaalaskandha, Kramuka (Pooga), Bhoorja, Meshasringa (Karkatasringi), Tinisa (Syandana), Chandana, Kuchandana, Simsapaa, Sireesha, Asana (Beejaka), Dhava (Sakata), Arjuna (Kakubha) Taala, Saaka, Naktamaala, Pooteeka, Aswakarna, Agaru and Kaaleeyakam (Daaru-haridraa).

   The Saalasaaraadigana is very helpful in Kushta. It also acts as a curative in Prameha and Paandu (anemia) and lessens Kapha and Medas.

6. Rodhraadigana: Rodhra, Saavaralodhra, Palasa, Kutannata, Asoka, Phanjee (Bhaarangee), Katphala, Elavaalukum, Sallakee, Jinginee, Kadamba, Saala and Kadalee.

   The Rodhraadigana lessens medas and Kapha. It is curative of Yoni Dosha (Disorders of Vagina), astringent (Stambhee), develops complextion (Varnya) and mitigates the poisons (Vishvinaasana).

7. Arkaadigana: Arka, Alarka (Sweta Arka), Karanja, Vitapakaranja, Naagadantee, Mayooraka (Apaamaarga), Bhaarngee, Raasnaa, Indrapushpee, Kshudra Swetaa (Sephanda), Mahaaswetaa, Vrischikaalee, Alavanaa (Jyotishmatee) and Taapasa Vriksha (Inguda).

   The Arkaadigana checks kapha, Medas and Visha. It is helful Krimi and Kustha and is a purifier of vranas (Viseshaat Vrana Sodhanah).

8. Surasaadigana: Surasaa (Krishna Tulasee), Sweeta Surasaa (Sweta Tulasee), Phanijjaka (Maruvaka), Arjaka (Swetakutheraka), Bhoosthrina (Angudaaka), Sugandhaka (Dronapushpa), Sumkha (Raajikaa), Kaalamaala (Krishnamallikaa), Kaasamarda, Kshavaka (Chinkini), Kharapushpaa (Kshavakabhedah), Vidanga,Katphala, Surasee (Bilwanasee), Nirgundee, Kulaahala (Mundikaa), Undurukarnika (Mooshikakarnikaa), Phanjee (Bhaarngee), Praacheebala (Matsyaakshaka), Kaakamaachee and Vishnumustika.

   The Suraasadigana checks Kapha and destroys Krimees (parasites). It is also useful in cold, loss of appetite, hardbreathing and cough and it also purify Vrana.

9. Mushkakaadigana: Mushkaka (Kshaaravriksha), Palaasa (Kinsuka), Dhava, Chitraka, Madana, Vrikshaka (Kutaja), Simsapaa,Vajravriksha and Thriphala (Hareetaki, Vibheetaki and Aamalaki).

The Mushkaadigana reduces Medas and useful in Sukradosha (impurity of Semen). It also acts as a curative in Prameha,Arsas, Paanduroga and Sarkara.

10. Pippalyaadigana: Pippalee, Pippaleemoola, Chavya, Chitraka, Sringabera (Sunthee), Maricha, Hastipipalee (Gajapippalee), Harenuka, Elaa, Ajamodaa, Indrayava, Paathaa, Jeeraka, Sarshapa, Mahaanimbaphala, Hingu, Bhaarangee, Madhu-rassa (Moorvaa), Ativisha, Vachaaa, Vidangaa and Katurohinee.

    The Pippalyaadigana reduces Kapha Dosha,Pratisyaaya (cold), Vaata, Anorexia, Gulma, and Soola (pain). It creates appetite (deepana) and digests (resolves) Aama-Dosha (Aama Paachana).

11. Elaadigana: Ela, Tagara, Kushta, Maamsee, Dhyaamaka (Mattrina) Twak, Patra (Patrakam), Naagapushap (Naagakesara), Priyangu, Harenuka, Vyaaghranakha, Sukti (Vyaaghrana-khabheda), Chandana, Sthouneyaka (Thuneraka), Sreeveshtaka (Sareladruma), Chocha, Choraka, Vaaluka, Guggulu, Sarjarasa, Turushka (Sihlaka), Kunduraka (Sallakee Chopa), Agaru,Sprukkaa, Useera, Bhadrudaaru, Kunkumam and Punnaagakesara.

    The Elaadigana checks Vaata and Kapha and Visha (Poison), it creates luster of the skin (Varna Prassaada) and destroys itching (Kantoo), Pindakaa (Diseases with elevated skin on account of an abscess etc.) and Kotha (eruption on skin).

12. Vachaadigana: Vachaa, Mustaa, Ativishaa, Abhayaa, Bhadradaru and Naagakesara.

    The Vachaadigana purifies the breast-milk and acts as amelioratives (Samana) in amaateesaara. This is especiallydigestives of the Doshaas (Viseshaat Doshapaachanou).

13. Haridraadigana: Haridraa, Daaruharidraa, Kalasee (Prisniparnee), Kutajabeej and Yastimadhu.

    The Haridraadigana also purifies the breast-milk and acts as ameliorative (Samana) in Aamaateesaara and is digestive of the Doshaas (Viseshaat Doshapaachanou).

14. Syaamaadigana: Syaama (Sweta Trivrut), Mahaasyaama (Vriddhadaaraka). Trivrut (having red root), Dantee,Sankhinee (Yavatiktabhedah), Tilvaka (Rodhra), Kampilaka (Rochanikaa), Ramyaka (Mahaanimba), Kramuka, Putresrenee, Gavakshee, Raajavriksha, Karaaja, Vitapakaranja, Gudoochi, Saptala (Snuheebheda), Chagalaaantree (Vriddhadaarakabheda), Sudhaa (Sehunda) and Suvarnaksheeri.

    The Syaamaadigana checks Gulma and Visha (poison), Aanaaha (Distension of abdomen), Udara and Udaavartaa. It causes free evacuation of the bowels.

15. Brihatyaadigana: Brihatee, Kantakaarika, Kutajaphala, Paathaa and madhukam (Yasthimadhu).

The Brihatyaadigana is a digestive (Paachaneeya) and it checks Vaata and Pitta. It is alsobeneficial in Arochaka (Anorexia) due to Kapha, Hridroga and Mootrakrichra (Dysuria).

16. Patolaadigana: Pathola, Chandana, Kuchandana (Raktachandana), Moorvaa, Gudoochi, Paathaa and Katurohinee.

    The Patolaadigana cures Pitta, Kapha and Arochaka (Anorexia). It allays jwara and cures vomiting, itching and Visha (Poison) and is also beneficial in Vranas.

17. Kaakolaadigana: Kaakoles, Ksheerakaakolee, Jeevaka, Rishabhaka, Mudgaparnee, Maashaparni, Medaa, Mahaamedaa, Chinnaruhaa (Gudoochi), Karkatasringi, Tugaaksheeree (Vamsaa-lochanaa), Padmaka, Papoundareeka, Riddhi, Vriddhi, Mridweeka (Draakshaa), Jeewantee and Madhukam.

    The Kaakolyaadigana cures Pittasonita (Rakta vitiated by Pitta; or Pitta and Sonita) and Vaayu. It is Jeevana (increases vitality), Brimhana (promotes body growth) and Vrishya (increase semen), increases breast-milk and Kapha.

18. Ooshakaadigana: Ooshaka (Kshaaramrittika), Saindhavalavana, Silaajathu, Kaaseesadwaya (Vaalukaaseesam and Pushpakaaseesam), Hingu and Tuththaka.

    The Ooshakaadigana allays Kapha and diminishes Medas. It acts as curative in Asmaree (stone), Sarkaraa (sand in urine),Mootrakrichchra and Gulma.

19. Saaribaadigana: Saaribaa, Madhuka (Yashtimadhu), Chandana, Kuchandana (Rakhtachandana), Padmaka, Kaasmareephala, Madhookapauspha and Useeram.

    The Saaribaadigana allays thirst and Raktapitta. It is also beneficial in Pittajware and cures daaha (burning sensation).

20. Anjanaadigana: Anjana (Souveeraanjanam), Rasaanjanam, Naagapushpa (Naagakesaram), Priyangu, Neelotpala, Nalada (Maamsee). Nalinakes (Padmakesara) and Madhukam.

    The Anjanaadigana cures Raktapitta and is beneficial in Visha (poisoning). It also allays internal burning sensation.

21. Parooshakaadigana: Parooshakaa, Draakshaa, Katphala, Daadima, Rajaadana (Ksheerika), Katakaphala and Saakaphalas (fruits of Saaka tree).

22. Priyangwaadigana: Priyaangu, Samangaa (Anjali-Kaarikaa of Lajjalu), Dhaatakee, Punnaaga (Tunga), Naagapusphpam, Chandanam, Kuchandanam, Mocharasa (Saalimali chupa), Rasaanjana,Kumbheeka, Srontonjana Padmakesara, Yojanavalli (Manjishta) and Deerghamoolaa (Duraalabhaa).

    The Priyangwaadigana is useful in Pakswaateesara and causes Sandhaana (union of bones etc.). This is also beneficial in Pitta and healing of the ulcers (Vranaanam chaapi Ropanou).

23. Ambashtaadigana:Ambashtaa (Machikaa), Dhaatakikussuma, Samangaa (Lajjaalu), Katvanga (Araluka), Madhuka, Bilwapesikaa (Baala Bhilwagirah), Saavaraprodhra (Lodhra), Palaasa, Nandeevriksha (Kaasmaree) and Padmakesaram.

    The Ambashtaadigana is useful in Pakswaateesara and causes Sandhaana (Union of bones etc.). This is also beneficial in Pitta and healing of the ulcers (Vranaanam chaapi ropanou).

24. Nyagrodhaadigana: Nyagrodha (Vata), Udumbara, Aswattha, Plaksha, Madhuka, Kapeethana (Aamraataka), Kakubha, Aamra, Kosaamra, Chorakapatra (Laakshaavrikshaa), Jamboodways (Raajajamboo and Kaakajamboo), Piyaala (Saaradruma), Madhooka (Gudapushpa), Rohinee (Katphala), Vanjula (Vetasa), Kadamba, Badaree, Tindukee (Tinduka Vriksha), Sallakee, Rodhra, Saavararodhra, Bhallaataka, Palaasa and Nandeevriksha.

    The Nyagrodhaadigana is beneficial in the treatment of Vranas (Vranyah). It is astringent (Sangraahee) and is useful in fractures (Bhagna Sandhaana). Further, It is said to allay Raktapitta (haemorrhage) and Daaha (burning sesation) and reduces Medas (Obesity); it is also useful in Yonidosha (Yonivyaapat).

25. Gudoochyaadigana: Gudoochi,Nimba, Kustumburu (Dhaanyakam), Chandanam and Padmakam.

    The Gudcochyaadigana checks all jwars (fevers) and increases appetite (Deepana). It allays Hrillaasa (nausea accompanied with hawking of mucous etc. from the mouth), Arochaka (Anorexia), Vami (Vomiting), thrist and burning sensation.

26. Utpalaadigana: Utpala (Neelotpala), Raktopala, Kumuda (Swetotpala), Sougandhika (a sweet-scented blue lily different from Neelotpala), Kuvalaya (Utpala having blue-white colour), Pundareeka (White lotus) and Madhukam (Yashtimadhu).

    The Utpalaadigana allays Daaha (burning sensation), Pitta and Rakta, Thirst, Visha (Poisoning), Hridroga, Vami (Vomiting) and Moorchaa (Syncope).

27. Mustraadigana: Mustaa, Haridraa, Daaruharidraa, Hareetakee, Aamalaka, Vibheetakea, Kushta,Haimavatee (Vachaa), Paathaa, Kaaturohinee, Saarngeshtaa (Yavatikaa), Ativishaa, Draavidee (Elaa) Bhallaataka and Chitraka.

    The Mustakadigana reduces Kapha and is digestive (Paachana). It also cures Yoni-dosha and purifies breast-milk.

28. Triphalaa: Hareetakee, Aamalaka and Vibheetaka.

    The Thriphala reduces Kapha and Pittaa. It is useful in Prameha and Kushta and is also beneficial to eyes. It produce the appetite and is also used in Vishamajwara.

29. Trikatukam: Pippalee, Marichaa and Sringabera (Sunthee).

    The Trikatukam is also called Tryooshanam and Vyosha. It reduces Kapha and Medas and is useful in Prameha, Kushta,Twagaamaya (skin diseases) and produce the appetite. It si alsouseful in Gulma, Peenasa and Agnyalpataa (poor digestion).

30. Aamalakyaadigana: Aamalakee, Hareetakee, Pippalee and Chitraka.

    The Aamalakyaadigana is useful in all jwaras (fevers), beneficial to eyes and also used as an apharodisiac. It also cures Kaphaa-rochaka (Anorexia due to Kapha).

31. Trapwaadigana: Trapu (Vangam-Tin), Seesa (Lead), Taamra (Copper), Rajata (Silver), Krishaloha (Steel), Suvarna (Gold) and Lohamala (Mandoor).

    The Trapwaadigana is useful in gara (poisoning usually through worms (Krimi), thirst, Visha, Hridroga, Paandu, and Prameha.

32. Laakshaadigana: Laakshana, Aarevata, (Kiramaalaka), Kutaja, Aswamaaraka (Karaveera), Kathphala, Haridraa, Daaru-haridraa, Nimba, Saptachchada, Maalatee and Traayamaanaa.

    The Laakshaadigana is Kashaaya (astringent), bitter and sweet in taste and it reduces ailments due to Kapha and Pitta. It is also useful in Kushta and Krimi (worms) and purifies Dushta Vranas.

33. Kaneeyahpanchamoola (Hraswapanchamoolam or Laghupanchmool): Trikantaka (Gokshura), Brihatee, Kantakaari, Pithakparnee and Vidaarigandhaa (Saalaparni).

    The Kaneeyahpanchamoola is Kaashaaya, Tikta and Madhura in Rasa (taste) and it reduces Vata and ameliorates Pitta. It is Brimhana (tissue builder) and increases strength (Balavardhanah).

34. Mahaapachmoola (Brihat Panchamoola): Bilwa, Agnimandha, Tuntuka (Syonaaka), Paatala and Kaasmaree.

    The Mahaapanchamoolam is bitter (tikta) and checks Kapha and Vaata. It is Laghu in Paaka and promotes appetite with slightly sweet in Rasa (taste).

35. Dasamoola: Kaneeyahpanchamoola and Mahaapahchamoola collectively are called Dasamoola.

    The Dasamoola gana reduces swaasa (hard breathing) and checks Kapha, Pitta and Vaayu. It digests Aamaa-dosha and is curative of all types of jwaras (fevers).

36. Valleepanchamoola: Vidaaree, Saaribaa, Rajanee (Haridraa), Gudoochee and Ajasringi.

    The Valleepanchamoola is beneficial in Raktapaitta and is used in three kinds of Sopha (Aama, Pachyamaana and Pakwa), Prameha and Sukra.

37. Kantakapanchamoola: Karmarda, TGrikantaka (Gokshura), Saireyaka, Sataavaree and Gridhranakee.

    The Kantakapanchamoola is beneficial in Raktapitta and is used in three kinds of Sopha (Aama, Pachyamaana and Pakwa), Prameha and Sukra.

38. Trinapanchamoola: Kusa, Kaasa, Nala, Darbha and Kaandekshuka.

    The Trinapanchmoola cures disorders of urine (Mootra Dosha) and Raktapitta when it is used especially with milk. Laghupanchamoola and Mahaapanchamoola usually reduce vaata, Trainapanchamoola reduces Pitta; the other two, namely, Valleepanchamoola and Kantaka Pachaa-moola, reduce Kapha.

    It should be understood that the substances in the groups (Ganas) may be altered or individually used and variously combined according to the Doshaas or Dooshyas etc. existing individually or in various combinations in a person.

D. PLANTS AND DRUGS: VAGBHATTA'S CLASSIFICATION

As follows; Vagbhatta has divided the drug plants into 32 ganas(groups)

1. Vamanaoushadhaga: Madana, Madhuka (Yastimadhu), Lambaa (Tumbee), Nimba, Bimbee, Vissalaa (Indravarunee), Trapusa, Kutaja, Moorva, Devadaalee, Krimighna (Vidangam), Vidula (Jalavetasa), Dahana (Chitraka), Chitraa (Mooshikarparnee), Kosavatyou (Ghantalikaa and Dhamargava), Karanja, Kanaa (Pippalli), Lavana (Saindhavalavana), Vacha, Elaa and Sarshapa.
2. Virechanaoushdhagana: Nikumbha (Dantee), Kumbha (Trivruth), Thriphala, Gavakshee (Visalaa), Snuk, Sankhinee (Yavatiktaa), Neelinee, Tilwaka (Rodhra), Samyaka (Aaragwadha), Kampillaka, Hemadugdhaa (Swarnakheeree), Dugdham (Milk) and Mootram (the urine of cow etc.).
3. Niroohana Dravyagana: Madanaphala, Kutaja (Twak), Kushtam, Devadalee, Madhukam, Vachaa, Dashamula,Daaru (Devadaaru), Rasana, Yava, Misi (Satapushpa), Kritavedhanam (Dhaamaargava), Kulmuttha (Madhu), Lavanam (Saindhavam) and Trivruth.
4. Seershavirechaneeyagana: Vella (Vidanga), Apamarga, Vyosha (Trikatutam), Daarvee (Daruharidra), Suraalaa (Sreshta Sarjrasa), Saireeshabeeja, Baarhatabeeja (Brihatibeeja), Saigravam-beejam (Sigru beja), Maadhooka Saara 9Madhupushpasaara), Saindhavam, Taarkshyasailam (Sushkarasanjanam), Trutyou (Sookshma elaa and Sthoola elaa), and Prithveeka (Hingupatree).
5. Vaataharaganas: Bhadradaaru, Natam (Tagaram), Kushtam, Dasamoolaam and Baladwaya (Bala and Atibala).
6. Pittaharaganas (To chek pitta): Doorwaa, Anantaa (Yavass), Nimba, Vassa, Aatma-guptaa (Kapikachhu), Gundraa (Padayerakah), Abheeru (Sataavaree), Seetapaakee (Kaakanantikaabheda), and Priyangu.
7. Kapharaganas: Aragwhaadhigana, Arkaadigana, Mushakaadigana, Surasaadigana, Mustaadigana and Vatsakaadigana.
8. Jeevaneeyaadigana: Jeevantee, Kaakolee, Ksheerakaakolee, Medaa, Mahamedaa, Mudgaparnee, Maashaaparnee, Rishabhaka, Jeevaka and Mudhuka.

9. Vidaayaadigana: Vidaree, Pandhcaangula (Eranda), Vrischikaalee (Meshasringee), Vrischeeva (Kshudravarshaabhoo), Devaahvaya (Devdaaru), Soorpaparnee (Mudga parnee and Masha parnee), Kandookaree (Kapikachoo), Jeevana Panchamoolam (Abheeru, Veeraajeevantee, Jeevaka and Rishsabhaka), Haraswapanchamoolam (Brihatee, Kanatakaaree, Saalaparnee, Prishniparnee and Gokshuraka), Gopasutaa (Sariba) and Tripaadee (Hamsapadee).
10. Saaribadigana: Saaribaa, Useeram, Kaasmarya, Madhooka,a Sisiradwayam (Chandanam and Raktachandam), Yashtee (Yastimadhu) and Parooshakam.
11. Padmaakadigana: Padmaakam, Pundra (Prapoundareekam), Vridhi (Sraavanee), Tunga (Vamsalaochanna), Rodhi (Mahaasraavanee), Sringee (Karkatasringee) and Amritaa (Gududchi).
12. Parooshakadigana: Parooshakam, Varaa (Triphala), Draakshaa, Katphalam, Katakaphalam, Raajaahavam (Aargwadha), Daadimam and Saakam.
13. Anjanaadigana: Anjanam (Srotonjanam), Phalinee (Priyangu), Maamsee (Krishnajataa), Padamam (Lotus), Utpalam (Lily), Rasaanaganam, Elaa, Madhukam (Yastimadhu) and Naagaahavam (Nagakesaram).
14. Patolaaadiagana: Patola,Katurohinee, chandanam, Madhusravaa (Surangee), Gudoochee and Paathaa.
15. Guduochyaadigana: Guduchee, Padmaka, Arista (Nimba), Dhanakaa and Raktachandanam.
16. Aaragwadhaadigana: Aragwadha, Indrayava, Paatali (Vasantha Dootee), Kaakatiktaa (Saarangeshtaa), Nimba, Amritaa (Guduchi), Madhurasaa (Moorva),Sruva Vriksha (Vikamkata), Paathaa (Ambashtaa), Bhoonimba, Sairyaka (Sahachara, Patola,Karanjayugma (Pooteekaranja) and Naktamalaa, Saptachchada (Saptaparna), Agni (Chitraka), Sushavee (Kaaravee), Phala (Madanphala), Baana (Sahachara) and Ghonta (Poogavisesha).
17. Asanaadigana: Asana (Peeta Saala), Tinisha (Syadana), Bhoorja. Swetavaaha (Arjuna), Prakeerya (Pootikaranja), Khadira,Kadara (Swetasaara), Bhandee (Sireesha), Sinsapaa (Mandala-patrika), Meshasringee, Trih (Chandanam, Rakta-chandanam and Daaruharidraa), Tala, Palaa (Kinsuka), Jongaka (Aguru),Saakam (Varadaaru), Saala,Kramuka (Poos), Dhava (Sakata), Kulinga (Sakrayava), Chaagakarna and Aswakarna.
18. Varanaadigana: Varuna,l Sairyakayugma (Kuravaka and Kurantaka), Sataavaree, Dahana (Chitraka), Morata (Moorva), Bilwa, Vishaaniakaa (Ajasringee), Dwibrihatee (Brihatee and Kantakaari), Dwikaranja (Pootikaranja and Naktamala), Jayaadwayam (Tarkaaree and Hareethakee), Bahalapallava (Shobaanjana), Darbha (Kusha), and Ruiaakara (Hitaalu).

19. Ooshakaadigana: Ooshaka(Kallara), Thuththakam (Kitiha), Hingu, Kaaseesadwaya Pushapakaaseesam and Paansukaaseesam), Saindhavam and Silaajatu.
20. Veerataraadigana: Vellaantara (Veeratara of Useera), Aranika (Agnimantha), Booka (Easwara mallika), Vrisha (Vaasa), Asmabheda (Pashanbheda), Gokantaka (Gokshuru), Itkataa, Schaachara, Baana, Kaasa, Vrikshaadanee, Nala, Kusadways (Sthoola Darbha and Sookshma Darbha) Gunthaa (Vrintartrina), Gumdraa (Pada Eraka), Bhallooka (Syonaka), Morata, Kuranta (Stivaaraka), Karambha (Uttamaarain) and Paartha (Suvarchalaa).
21. Rodhraadigana: Rodhra, Saavarka Rodhra, Palaasa (Sathee), Jinginee (Krishnasaalmalee), Sarala (Devadaru), Katphalam, Kutsitaamba (Kadamba), Kadalee,Gatasoka (Asoka), Elavaalu, PAripelavam (Kutannatam) and Mocha.
22. Arkaadiagana: Arka,Alarka, Nagadanti Visalyaa (Langalee), Bhaargee, Rasana,Vrischikalee(Ushtradhoomaka), Prakeeryaa (Karanjaka), Pretyakpushpee (Apamarga), Peetataila (Kakaadanee), Udkeerya (Karanjaka). Swetaa Yugmam (Kinhi and PAllindee) and Taapasavrisksha (Ingudee).
23. Surasaadigana: Sursayugma (Krishna Tulsi and Sweta Tulsi), Pohanijjam (Mareechaka), Kaalamaalaa (Krishnaarjaka), Vidngam, Kharabusa (Maruvaka), Vrishakarnee (Mooshi-kakarnee), Katphalam, Kaasamarda, Kaasamarda, Kshavaka, Sarasee (Tumbarapatrikaa), Bhaargee (Angaaravalee), Kaamukaa (Raktamanjaree), Kaakamachee, Kulahala (Alambusa), Vishamushti (Kuchilaa), Bhoostrina (Atichatra) and Booktakesee (Maamsee Putrachaaraa).
24. Mushkaakaadigana: Mushakaka (Mokshaka),Snuk (Gudaa), Varaa (Triphala), Dweepi (Chitraka), Palaasa,Dhava and Sinsapaa.
25. Vatsakaadigana: Vatsaka (Vanatiktaka), Moorva, Bhaargee, Katukaa (Katurohinee), Maricham, Ghunapriyaa (Ativishaa), Gandeeram (Snuhee), Elaa, Paatha, Ajaajee (Jeerakam), Karvangaphalam (Aralukaphalam), Ajmoda (Deepyaka), Siddhaartha (Goura Sarshapa), Vachaa, Jeeraka, Hingu, Vidangam, Pasugandhaa (Ajagandhaa) and Panchkolakam (Pipalee, Pippaleemoola,Chavya,Chitraka and Sunthee).
26. Vachaadigana: Vachaa, Jalada (Musta), Devaashwa (Kilimam), Naagaram (Sunthee), Ativisha and Abhayaa.
27. Privangawaadigana: Priyangu (Syaama), Pushpaanjanam (Reetipushpam), Anjanayugmam (Srotoanjaman and Souveeranajanam), Padmaa (Padmachaarinee), Padmakesarama, Yojanavalee (Manjishtaa), Anantaa (Yavaasa), Maanadruma (Salmalee), Mochasrasa (Saalmalee Niryasaa), Samangaa (Namaskaaree), Punnaaga (Tunga), Seetham (Chandanam) and Makaneeyahetuh (Dhaatukee).
28. Ambashtaadigana: Ambaashtaa (Mayoorashikaa), Madhukam (Yashti-madhu), Namaskaree (Samangaa), Nandeevriksha (Prarohee), Palaasa,

Kachchuraa (Dhanvayaasa), Rodhram,Dhaatakee, Bilwapesika,Katwanga (Syonaka) and Kamalodhbhavan Rajah (Padmakesaram).

29. Mustaadigana: Mustaa, Vachaa, Agni (Chitrak), Dwinisaa (Haridraa and Daaruharidraa), Dwitikta (Katurohinee and Kaakatikta), Bhallaataka, Pathaa, Triphala, Vishaakhyaa (Suklakandaa), Kushtam, TGrutee (Elaa) and Haimavatee (Swetavachaa).

30. Nyagrodhaadigana: Nyagrodha (Vata), Pipplaa (Aswattha), Sadaaphaala (Udumbara), Rodhrayugman (Rodhra and Saavara Rodhra), Jamboodwaya (Raajajamboo and Kaakajamboo),Arjuna, Kapeetana (Vaaneera), Somavalka (Sitasaarakhadira), Plaksha,Aamra, Vanjul (Vetasa), Piyaala,Palaasa, Nandee (Jayavriksha), Kolee (Bedaree), Kakamba, Viralaa (Tindukee), Madhukam(Yasti-madhu)and Madhookam.

31. Elaadigana: Elaayugmam (Sookshmaa Elaa and Sthoola Elaa), Turushka,Kushtam, Phalinee (Gandha Priyangu),Maamsee (Naladam), Jalam (Hreeberam), Dhyaamakam (Devedaghakam), Sprikkaa (Devee), Chouraka (Gandhiparne), Chocha (Twak), Patram (Gandhapatram), Tagaram (Chakram), Sthouneyam (Tailapeetakam), Jeateerasa (Bola), Sikti (Nakha), Vyaaghranakha, Amaraahawam (Devadaaru), Aguru, Sreevaasakam, Kumkumam, Chandana, Guggulu, Devadhoopa (Sarjarasa), Khapura (Kunduruka), Punnaga (Raktakesara) and Naagaahwayam.

32. Syaamaadigana: Syaamaa (Syaamaa Trivrit), Dante (Chitraa), Dravantee (Indurukarnikaa), Kramuka (Pattikaalodhra), Kutaranee (Suklaa), Sankhinee (Yavatiktaa), Charmasaahwaa (Saatalaa), Swarnaksheeree, Gavaakshee, Sikhari (Apamaarga), Rajanaka, Karanja, Bastaantree (Vrishagandhaa), Vyaadhighaata (Kritamaala), Bahalabahurasa (Ikshu) and Teekshnabrikshaphala.

Appendix II
Medicines of Common Bazar

By the local people or commercially for ayurvedic formulations. The following raw drugs are generally sold by the 'Pansaris' or pharmaceutical concerns and are used as folk medicine. These are listed in alphabetical order as follows:

Drug Name	*Botanical Name*	*Parts Used*
Adrak	*Zingiber officinale*	Tubers
Agar	*Aquillaria agallocha*	Wood/Oil
Agia ghas	*Cymbopoogon citrates*	Leaves
Ajmod	*Apium graveolens*	Roots
Ajwain	*Trachyspermum amni* (*Syn. Carum copticum*)	Fruits
Akarkara	*Anacyclus pyrethrum*	Roots
Akashbel	*Cuscutar reflexa*	Panchang
Amaltas	*Cassia fistula*	Fruits
Amla	*Embelica officinalis*	Fruits
Anardana	*Punica granatum*	Seeds
Ankol	*Alangium salvifolium ssp. salvifolium*	Root bark/seeds
Arjun	*Terminalia arjuna*	Bark
Arand	*Ricinus communis*	Wood, Seed oil
Arni	*Clerodendron phlomoides*	Bark

Drug Name	*Botanical Name*	*Parts Used*
Ashok	*Saraca asoca*	Bark
Atibala	*Abutilon indicum*	Panchang
Atis	*Aconitum heteropohyllum*	Root tubers
BadiElaichi	*Eletttaria cardamomum*	Fruits
Badi Kateli	*Solanum indicum*	Fruits
Bahera	*Terminalia bellirica*	Fruits
Bail	*Aegle marmelos*	Fruits/Seeds
Bakuchi	*Psooralia caryopholia*	Seeds
Bala	*Sida* sp.	Panchang
Bansa	*Justicia adhtoda*	Panchang
Bar	*Ficus benglensts*	Fruits
Ber	*Zizyphus jujube*	Fruits
Bharangi	*Clerodendron serratum*	Panchang
Bhilawa	*Semicarpus anacardium*	Panchang
Bhringraj	*Eclipta prostrate*	Panchang
Bhumi amla	*Phyllanthus fraternus*	Panchang
Brahmi	*Centellaasiatica*	Panchang
Chakramard	*Cassia tora*	Fruits/seeds/ Panchang
Chameli	*Jasminum grandifolium*	Panchang
Champa	*Michelia champaca*	Panchang/Bark/Flowers
Chandan	*Santalum album*	Wood/oil
Chavya	*Piper officinarum*	Fruits
Chir	*Pinus longifolia*	Oil
Chitrak	*Plumbago zeylanica*	Roots
Choti Elaichi	*Amomum subulatum*	Fruits
Choti Kateli	*Solanum surrattense*	Panchang
Dhamasa	*Fagonia indica var. schweinfurthii*	Panchang
Danti	*Croton tiglium/C. oblongifolius/ Baliopspermum montanum*	Fruits
Darbha/Kusha	*Desmostachys bipinnata*	Roots
Dek	*Melia azedarach*	Fruits
Devdaru	*Cedrus deodara*	Wood(Kasth sar)

Drug Name	*Botanical Name*	*Parts Used*
Dhari	*Woodfordia fruticosa*	Bark
Dhaia/Dhatha	*Streblus asper*	Bark
Dhav	*Anogeissus latifolia*	Gum
Gambhari	*Gmelina arborea*	Bark
Ghrit Kumari	*Aloe vera*	Leaves, Oil
Giloe	*Tinospora cordifolia*	Panchang
Gokhru	*Tribulus terrestris*	Panchang
Gorakhmndi	*Sphaericanthus indicus*	Fruits
Guggulu	*Commiphora wightii*	Oleogum resin
Gulab	*Rosa species*	Floower petals
Gular/Udumbar	*Ficus racemosa*	Fruits
Gurhal	*Hibiscus rosa-sinensis*	Flowers
Gurmar	*Gymnema sylvsetre*	Panchang
Haldi	*Curcuma longa*	Roots
Haldu	*Haldinia cordifolia*	Fruits
Harad	*Terminalia chebula*	Fruits
Harsingar	*Nyctanthes arbor-trisits*	Flowers
Hauber	*Juniperus communis*	Fruits
Imli	*Tamarindus indica*	Bark/Fruits
Indrajau/Kutaj	*Holarrhena antidysenterica*	Frutis/Seeds/Bark
Inndrayan	*Citrullus colocynthis*	Fruts/Roots
Jaipal	*Myristica fragrans*	Seed/Oil
Jaiyanti	*Sesbania sesban*	Panchang/Roots/ Seeds
Jaljami	*Cocculus hirsutus*	Roots/Leaves
Jalpippali	*Lippia nudiflora*	Panchang
Jatamansi	*Nardostachys jatamansi*	Roots
Jawasa	*Alhagi pseudalhagi*	Resin
Jeevanti	*Lepotadinia reticulate*	Roots
Jira	*Cuminum cyminum*	Seeds
Kachnar	*Bauhinia varigata*	Bark/Flowers
Kachur	*Curcurma zedoraria*	Tubers
Kadanb	*Anthocephallus cadamba*	Fruits

Drug Name	*Botanical Name*	*Parts Used*
Kaiphal	*Myrica nagi*	Fruits
Kaith	*Limonia elephantum*	Fruits
Kakjungha	*Leea asiatica*	Roots
Kaknasa	*Hygrophilla auriculata*	Panchang/Seeds
Kakrasingi	*Pistacia integerrima*	Galls
Kali Dudhi	*Hemisdus indicus*	Panchang
Kali Jiri	*Centratherum anthelminitcum*	Seed
Kali Mirch	*Piper nigram*	Frutis
Kamal	*Nelumbo nucifera*	Flowers/Seeds
Kampilak/Raini	*Mallotus phillipinensis*	Flower powders
Kaner	*Nerium indicum*	Roots/Root bark
Kantak Chaulai	*Amaranthus spinosus*	Panchang
Kant Karanj	*Derris indica*	Seeds/Bark
Kapoor	*Cinnamomum camphora*	Condensed oil
Kapur Kachri	*Hedychium acuninatum*	Roots
Karanj	*Caesalpoinia bonduc*	Seeds
Kas	*Saccharum spontaneum*	Roots
Kasmard	*Cassia occidentalis*	Seeds
Kevara	*Pandanus fascicularis*	Leaves/Flowers
Khair	*Acacia catchu*	Bark
Khas	*Vetiveria zizanioides*	Roots
Kounch	*Mucuna pruriens*	Seeds
Kuchla	*Strychnos nuxvomica*	Fruits
Kulath	*Doilchos filflorus*	Seeds
Langli	*Gloriosa supreba*	Tubers
Lasora	*Cordia dichotoma*	Fruits
Lawang	*Syzygium aromaticum*	Fruits
Lodhra	*Symplocos racemosa*	Bark
Madanphal	*Catunaregam nutans*	Fruits
Madhvi	*Hiptage benghlensis*	Panchang
Mahua	*Madhuca longifolia*	Dried Flowers
Makoi	*Solanum villosum ssp. villosum*	Fruits
Malkangni	*Celastrus paniculata*	Panchang

Drug Name	*Botanical Name*	*Parts Used*
Malshroni	*Soyamida febriguga*	Bark
Manhishtha	*Rubia codifolia*	Roots
Marorphali	*Helicteres isora*	Fruits
Maruva	*Origanum majorana*	Seeds
Maulshri	*Mimusops elengi*	Fruits/Seeds
Mudgparni	*Puraria lobata*	Seeds
Munnaka	*Vitis vinifera*	Dried Fruits
Musli	*Asparagus adscendens*	Roots
Nagkesar	*Mesua ferrea*	Pollens
Nagarmotha	*Cyperus rotundus*	Roots
Neel	*Indigofera tinctoria*	Seeds
Netrabala	*Pavonia odorata*	Panchang
Nimb	*Azadirachta indica*	Leaves/Wood Pieces/ Fruits
Nirmali	*Strychnos potatorum*	Seeds
Nirtundi/Shambhalu	*Vitex negundo*	Panchang
Nishoth	*Operaculina turpethum*	Panchang
Palash	*Butea monosperma*	Flowers/Gaund
Pan	*Piper betel*	Leaves
Paribhadhara	*Erythrina variegate*	Bark/Leaves
Patharchatta	*Berginia ciliate fa. ligulata*	Rhizomes
Patha	*Cissampelos pareira*	Panchang
Patla	*Sterospermum chelinoides*	Bark
Peelu	*Salvadora persica*	Fruits/Bark/Seed oil
Pitpapra	*Hedyotis corymbosa/ Fumaria indica/Polycarea corymbosa*	Panchang
Prashniparni	*Uraria picta*	Panchang
Priyangu	*Callicarpa macrophylla*	Fruits
Punarnava	*Boerhavia diffusa*	Panchang
Putranjivi	*Drypetes roxburghii*	Bark/Seeds
Rasana	*Pluchea lanceolata/Inula racemosa/ Vanda roxburghii*	Leaves
Ritha	*Sapindus mukorossi*	Fruits

Drug Name	*Botanical Name*	*Parts Used*
Rohitak	*Tecomella undulate/ Aphanamixis polystachya*	Bark
Sahadevi	*Verononia cinerea*	Panchang
Samundraphal	*Barringtonia acutangula ssp. spicata*	Fruits
Sanai	*Cassia senna*	Leaves
Saptparni	*Alstonia scholaris*	Bark
Sarkanda	*Sacchraum munja*	Rootso
Sarpgandha	*Rauvolfia serpentina*	Roots
Shalpoarni	*Desmodium gangeticum*	Panchang
Shankpushpi	*Convolvulus pluricaulis*	Panchang
Sarpankha	*Tepohrosia purpurea*	Panchang
Semul	*Bombax eceiba*	Roots
Shatavar	*Asparagus recemosus*	Roots
Shivlingi	*Bryonopis lacinosa*	Panchang
Shyonakq	*Oroxylum indicum*	Bark
Siris	*Albizzia lebbeck/A. procera*	Seeds
Talsihpatra	*Taxus baccata*	Leaves
Tejbal	*Xanthoxylum alatum*	Panchang
Tejpatra	*Cinnamomum tomala*	Leaves
Til	*Sesamum orientale*	Seeds
Tulsi	*Ocimum sanctum/ Pongostemon bengalense*	Seeds
Ulatkambal	*Abroma agusta*	Panchang
Unnao	*Zizyphus vulgaris*	Bark
Usva	*Smilax ovalifolia*	Roots
Vacha	*Acorus calamus*	Roots
Varahikand	*Dioscorea bulbifera*	Tubers
Varun	*Crateva nurvala*	Bark
Vidarikand	*Ipomoea mauritiana*	Roots

Appendix III
Glossary

Abcess: Collection of pus in a tissue

Abortifacient: Agent that cause abortion

Abrotion: Giving birth to an embryo or foetus prior to stage of viability

Adenitits: Inflammation of a gland

Alexteric: Defensive to infectious diseases

Amenorrhoea: Absence or abnormal cessation of the menses Analgesic Agent that relieves pain

Anaemia: Reduction of erythrocytes below normal number in the blood

Anasarca: Accumulation of fluid is subcutaneoustissues anemia

Angina: A severe constricting pain, commonly used in term of angina pectoris

Anodyne: Agent capable of relieving of pain

Anorexia: Loss of appetite

Anthelmintic: Agent that destroys expulsion of intestinal worms Antidote Counteracts poison

Antiemetic: Counteracting nausea and vomiting

Antiperiodic: Agents tha prevents the regular occurrence of a disease or a symptom

Antiphlogistic: Preventing or relieving inflammation

Antipyretic : Agent that reduces fever

Antiscrobutic : Effective in prevention of scurvy

Antispasmodic: Preventing spasms

Aperient: Mild purgative

Aphrodisiac: Arousing sexual desire

Aphthous: Relating to ulcer on mucous membrane with exudates Ascites Abnormal accumulation of serous flud in peritoneal cavity.

Asthma: A condition marked by recurrent attack of paroxysmal dyspnea with wheezing.

Astringent: Agent that causs contraction, usually after local application

Atonic: Without normal tension

Beriberi: A disease due to deficiency of Vitamin B1

Billiousness: Complex system of nausea, abdominal discomfort, headache and constipation due to excessive bile secretion

Blindness: Loss of ability to see

Bronchitis: Inflammation of bronchi

Bruises: Superficial discolouration of the skin due to haematoma without rupture

Buboes: Inflammatory swelling of lymph-glands in the groin

Carminative: Relieving flatulence

Cataract: Opacity of crystalline lens of the eye

Catarrh: Inflammation of mucous membrane with discharge

Cathartic: Strong purgative

Cephalgia: Headache

Chilblain: A recurrent localized itching, swelling and painful erythema of finger.

Cholagogue: Agent that stimulates gall bladder contraction and promotes the flow of bile into the intestine

Cholera: Acute infectious disease caused by *Vibrio cholerae*

Cirrchosis: Intestinal inflammation fan organ, particularly of liver

Colic: Spasmodic pain in the abdomen

Coma: Unconsciousness from which the patient cannot be aroused even by powerful stimuli

Congestion: Abnormal accumulationof blood in part of body

Conjunctivitis: Inflammation of the mucous membrane covering the anterior surface of the eye ball

Contraceptive: Agent for prevention of conception

Contusion: Bruises without repture of blood vessels

Convulsion: Violent spasm

Coryza: Inflammation of the nasal mucous membrane with discharge

Cystitis: Inflammation of the urinary bladder

Delirium: Mental disturbance of relatively short duration usually reflecting a toxic state marked by illusion and excitement

Demulcent : Soothing

Diarrhoea: Abnormal frequent evacuation of watery stools

Diphoretic: Agent that promotes sweating

Diphteria: An acute disease, caused by *Corynebacterium diphterieae* and marked by formation of grey white pseudomembrane innose and throat or larynx with fever and also respiratory obstruction

Diuretic: Agent that promotes the flow of urine

Dropsy: An abnormal accumulation of fluid in intercellular spaces of the body

Dysentery: Inflammation of intestine with frequent stools with blood

Dysmenorrhoea: Painful menstruation

Dyspepsia: Impairment of power or function of digestion

Dyspnea: Difficult breathing

Dysuria: Discharge of urine with pain and difficulties

Eczema: Superficial inflammation of the skin marked by redness itching and minute papules and later by scaling, lichenification and often de-pigmentation of the skin

Elephantiasis: Inflammationof fibrous tissues caused by filarial worms

Emaciation: Growing pain

Emetic: Agent tha causes vomiting

Emmenagogue: Agent that promotes menustral flow

Emollient: Agent that softens the skin and soothes irrigation of the skin

Epilepsy: Syndromes characterized by paroxysomal transient disturbances of brain function that may be manifested as episodic impairment or loss of consciousness psychic disturbances or perturbation of the autonomic nervous system

Epitaxis: Nose bleeding, usually due to rupture of small vessels over lying anterior part of the cartilaginous nasal septum

Erysipelas: A disease characterized by the inflammation and redness of the skin

Euphoria: Absence of pain or distress

Expectorant: Agent that increases bronchial secretion and facilitates its esxpulsion

Fatigue: A state of diminished capacity to respond effectively to stimulus reducing the efficiency.

Febrifuge: A remedy for fever

Febrile: Feverish

Fistula: Abnormal passage, usually between two organs, especially near anus

Flatulence: Excessive gas formation instomach intestine Galactagogue Agent that promotes the secretion and flow of milk

Glaucoma: Eye disease, characterized by an increase in ocular pressure causing visual defects.

Gonorrhoea: A contagious inflammation of the genital mucous membrane transmitted chiefly by coitus and due to *Neisseraia gonorrhoae*

Haemeplegia: Paralysis of one side of body

Haemoptysis: Spitting of blood from the lungs or bronchial tubes

Haemostasis: Arrest of bleeding

Heliminthiasis: Vomiting of worms

Hemicrania: Unilateral headache

Hemorrhoids: Piles

Herpes: Any inflammatory skin disease marked by formation of small vesicles in clusters.

Hydrocele: A circumscribed collection of fluid, especially in tunica vaginallis of testis or along the spermatic cord

Hydrophobia: Fear of water caused by rabies

Hyperglycaemia: Abnormal high concentration of glucose in blood

Hypertension: Persistent high blood pressure

Hypotension: Low blood pressure

Hyptonic: Decreased tension

Hysteria : A term used widely in psychiatry; sudden onset of dream states, stupor and paralysis

Impetigo: Scabby eruption commonly occurring on the face caused by bacteria

Impotency: Lack of sexual power

Indigestion: Lack of failure of digestion, usually used to dennot vague abdominal discomfort after meads

Indolent : Painless

Inflammation: Injury or destruction to protective tissues, signs are pain, heat, redness, swelling and loss of function etc.

Insomnia: Sleeplessness

Jaundice: Yellowness of skin, mucous membrane and excretion due to hyper biliruinemia and deposition of bile pigment.

Lactagogue: Agent that promotes milk flow

Leprosy: Chronic communicable disease, caused by *Mycobacterium tepreae* characterized by production of mucous membranes and peripheral system.

Leucoderma: Depigmentation of the skin

Lochia: Discharge from the reproductive tract occurring after childbirth

Lumbago: Pain in lumbar region

Mania: A phase of bipolar disorder and charactrerized by expansiveness, elation, agitation, hyperexcitability, hyperactivity and increaasd speed of thought and ideas

Mania: A phase of bipolar disorder and characterized by expansiveness, elation, agitation, hyperexcitability, hyperactivity and increased speed of thought and ideas.

Measles: Contagious viral infection, usually of childhood marked by an eruption of discrete, red papules which become flatten, confluent and brown.

Melancholia: Depressed and unhappy emotional state with abnormal inhibition of mental and bodily activity.

Menorrhagia: Excessive or prolonged menstruation

Nausea: An unpleasant sensation, vaguely referred to epigastriumand abdomen

Neuralgia: Nerve pain

Nyctalopia : Night blindness

Opthalmia: Inflammation of eye

Orchitis: Inflammation of testis

Otalgia: Ear-ache

Otitis: Inflammation of bone

Paralysis: Impairment or loss of motor function in a part due to lesion of the neural or muscular mechanism

Pityriasis: Skin disease marked with formation of fine and braining scales

Pleurisy: Inflammati of pleura, serous membrane investing the lungs

Pneumonia: Inflammation of lungs with exudation and consolidation

Prolapse: Descending of an organ from normal position

Prurigo: Skin disease marked by intense itching

Pruritus: Itching

Psoriasis: Chronic or hereditary recurrent dermatitis marked by discrete vivid red macules, papules or plaques covered with silvery lamellated scales.

Pubescent: Arriving at puberty stage

Puerperal: Post natal period

Purgative : Agent that causs bowel evacuation

Purulent: With pus formation

Rabies: Infectious viral disease of central nervous system results from rabid animal bite *e.g.* dog and Bats

Refrigerant: Agent that causes cooling effect

Rinderpest: Cattle plague

Rubefacient: Reddening of the skin

Scabies: Contagious skin disease, caused by itch mite, *Sarcooptes scabiei*

Sciatica: Neuralgic along the coruse fo the sciatica nerve, most often with pain radiating into buttock and lower limb.

Scorbutic: Relating to scurvy

Scrofula: Tuberculosis with glandular swellings

Scurvy: A disease caused due to vitamin C deficiency

Sedative: Agent that causes excitement

Sialagogue: Agents tha enhances flow of saliva

Spasmolytic: Agent that relieves spasm

Spermatorrhoea: Involuntary discharge of semen

Sprue: Chronic form of malabsorption syndrome

Stomatitis: Inflammation of the mucous membrance of the mouth

Strangury: Slow and painful discharge of urine

Sudorific: Agent that causes sweating

Suppuration: Formation of pus

Syphilis: An acute and chronic infectious disease caused by *Treponema pallidum* (Syn. *Spirocheta pallida*) and transmitted by direct contact, usually through sexual intercourse

Tetanus: An acute often fatal infectious disease caused by *Clostridium tatani* whose spores entered in body through wounds.

Thermoplegia: Heatstroke or sunstroke

Tuberculosis: Infectious disease, caused by *Mycobacterium* and marked by tubercle formations and caseous necrosis in any organ tissue

Tumour: Swollen part of the body, one of the cardinal signs of inflammation

Ulcer: A wound with superficial loss of tissue

Urethritis: Inflammation of urethra

Urothrorrhoea: Abnormal discharge from urethra

Urticaria: An eruption of the skin characterized by transient appearance of slightly elevatedred dish patches associated with severe itching

Vermicidal: Agent that is lethal to intestinal animal parasites

Vertigo: Dizziness

Index

B

M

N

O

P